T0139781

Studies in Computational Intelligence

Volume 854

Series Editor

Janusz Kacprzyk, Polish Academy of Sciences, Warsaw, Poland

The series "Studies in Computational Intelligence" (SCI) publishes new developments and advances in the various areas of computational intelligence—quickly and with a high quality. The intent is to cover the theory, applications, and design methods of computational intelligence, as embedded in the fields of engineering, computer science, physics and life sciences, as well as the methodologies behind them. The series contains monographs, lecture notes and edited volumes in computational intelligence spanning the areas of neural networks, connectionist systems, genetic algorithms, evolutionary computation, artificial intelligence, cellular automata, self-organizing systems, soft computing, fuzzy systems, and hybrid intelligent systems. Of particular value to both the contributors and the readership are the short publication timeframe and the world-wide distribution, which enable both wide and rapid dissemination of research output.

The books of this series are submitted to indexing to Web of Science, EI-Compendex, DBLP, SCOPUS, Google Scholar and Springerlink.

More information about this series at http://www.springer.com/series/7092

Erik Cuevas · Jorge Gálvez ·
Omar Avalos

Recent Metaheuristics Algorithms for Parameter Identification

Erik Cuevas
CUCEI
Universidad de Guadalajara
Guadalajara, Mexico

Jorge Gálvez
CUCEI
Universidad de Guadalajara
Guadalajara, Mexico

Omar Avalos
CUCEI
Universidad de Guadalajara
Guadalajara, Mexico

ISSN 1860-949X ISSN 1860-9503 (electronic)
Studies in Computational Intelligence
ISBN 978-3-030-28919-5 ISBN 978-3-030-28917-1 (eBook)
https://doi.org/10.1007/978-3-030-28917-1

This Springer imprint is published by the registered company Springer Nature Switzerland AG
The registered company address is: Gewerbestrasse 11, 6330 Cham, Switzerland

Preface

Since ancient times, humans have used rules of thumb and strategies extracted from other domains to solve several conflicting problems. Much of these methods have been adopted in different areas such as economy, engineering. These problem-solving techniques are referred under the concept of heuristics. In general, heuristics are specially considered in those situations where there is insufficient or incomplete information about the problem to be solved. As several problems present this characteristic, heuristics-based techniques became quite popular in the last years.

On the other hand, metaheuristics imply a high-level heuristics in which the problem-solving strategy is more general and adaptable to multiple contexts. Metaheuristic methods perform better than simple heuristics, since their mechanisms are guided for information (or knowledge) embedded within the problem-solving process. Metaheuristic methods use as inspiration our scientific understanding of biological, natural, or social systems, which at some level of abstraction can be represented as optimization processes. They intend to serve as general-purpose easy-to-use optimization techniques capable of reaching globally optimal or at least nearly optimal solutions. In their operation, searcher agents emulate a group of biological or social entities which interact with each other based on specialized operators that model a determined biological or social behavior. These operators are applied to a population of candidate solutions (individuals) that are evaluated with respect to an objective function. Thus, in the optimization process individual positions are successively attracted to the optimal solution of the system to be solved.

There exist several features that clearly appear in most of the metaheuristic approaches, such as the use of diversification to force the exploration of regions of the search space, rarely visited until now, and the use of intensification or exploitation, to investigate thoroughly some promising regions. Another interesting feature is the use of memory to store the best solutions encountered. For these reasons, metaheuristics methods quickly became popular among researchers to solve from simple to complex optimization problems in different areas.

Most of the problems in science, engineering, economics, and life can be translated as an optimization or a search problem. According to their characteristics, some problems can be simple that can be solved by traditional optimization methods based on mathematical analysis. However, most of the problems of practical importance such as system identification, parameter estimation, energy systems represent conflicting scenarios so that they are very hard to be solved by using traditional approaches. Under such circumstances, metaheuristic algorithms have emerged as the best alternative to solve this kind of complex formulations. Therefore, metaheuristics techniques have consolidated as a very active research subject in the last ten years. During this time, various new metaheuristic approaches have been introduced. They have been experimentally examined on a set of artificial benchmark problems and in a large number of practical applications. Although metaheuristic methods represent one of the most exploited research paradigms in computational intelligence, there are a large number of open challenges in the area of metaheuristics. They range from premature convergence, inability to maintain population diversity, and the combination of metaheuristics with other algorithmic schemes, toward extending the available techniques to tackle ever more difficult problems.

Among the engineering problems, identification systems of processes, which are nonlinear in nature, represent a challenging formulation. In general, identification systems refer to methods which allow to estimate the parameters that mathematically model a certain process. From an optimization perspective, identification systems are considered extremely complex due to their nonlinearity, discontinuity, and high multimodality. These characteristics make difficult to solve them by using traditional optimization techniques. In the last years, researchers, engineers, and practitioners in identification systems and modeling have faced problems of increasing complexity. These problems can be stated as optimization formulations. Under these circumstances, an objective function is defined to evaluate the quality of each candidate solution composed of the problem parameters. Then, an optimization method is used to find the best solution that minimizes/maximizes the objective function.

Numerous books have been published tacking in account any of the most widely known metaheuristic methods, namely simulated annealing, tabu search, evolutionary algorithms, ant colony algorithms, particle swarm optimization or differential evolution, but attempts to consider the discussion of new alternative approaches are always scarce. Initial metaheuristic schemes maintain in their design several limitations such as premature convergence and inability to maintain population diversity. Recent metaheuristic methods have addressed these difficulties providing in general better results. Many of these novel metaheuristic approaches have also been lately introduced. In general, they propose new models and innovative algorithm combinations for producing an adequate exploration and exploitation of large search spaces considering a significant number of dimensions. Most of the new metaheuristic algorithms present promising results. Nevertheless, they are still in their initial stage. To grow and attain their complete potential, new metaheuristic methods must be applied in a great variety of problems and contexts, so that they do not only perform well in their reported sets of optimization problems, but also in new complex formulations. The only way to accomplish this is making possible the transmission and presentation of these methods in different technical areas as optimization tools. In general, once a researcher, engineer, or practitioner recognizes a problem as a particular instance of a more generic class, he/she can select one of the different metaheuristic algorithms that guarantee an expected optimization performance. Unfortunately, the set of options are concentrated in algorithms whose popularity and high proliferation are better than the new developments.

The excessive publication of developments based on the simple modification of popular metaheuristic methods presents an important disadvantage: They avoid the opportunity to discover new techniques and procedures which can be useful to solve problems formulated by the academic and industrial communities. In the last years, several promising metaheuristic methods that consider very interesting concepts and operators have been introduced. However, they seem to have been completely overlooked in the literature, in favor of the idea of modifying, hybridizing, or restructuring popular metaheuristic approaches.

The first goal of this book is to present advances that discuss new alternative metaheuristic developments which have proved to be effective in their application to several complex problems. The book considers different new metaheuristic methods and their practical applications. This structure is important to us, because we recognize this methodology as the best way to assist researchers, lecturers, engineers, and practitioners in the solution of their own optimization problems.

The second goal of this book is to bridge the gap between recent metaheuristic techniques with interesting identification system methods that profit on the convenient properties of metaheuristic schemes. To do this, at each chapter we endeavor to explain basic ideas of the proposed applications in ways that can be understood by readers who may not possess the necessary backgrounds on either of the fields. Therefore, identification systems and energy practitioners who are not researchers in metaheuristics will appreciate that the techniques discussed are beyond simple theoretical tools since they have been adapted to solve significant problems that commonly arise on such areas. On the other hand, members of the metaheuristic community can learn the way in which system identification and energy problems can be translated into optimization tasks.

This book has been structured so that each chapter can be read independently from the others. Chapter 1 describes the main characteristics and properties of metaheuristic methods. This chapter concentrates on elementary concepts of metaheuristic. Readers that are familiar with metaheuristic algorithms may wish to skip this chapter.

In Chap. 2, an algorithm for the optimal parameter identification of induction motors is presented. In the identification, the parameter estimation process is transformed into a multidimensional optimization problem where the internal parameters of the induction motor are considered as decision variables. Under this approach, the complexity of the optimization problem tends to produce multimodal error surfaces for which their cost functions are significantly difficult to minimize.

To determine the parameters, the presented scheme uses a recent metaheuristic method called the gravitational search algorithm (GSA). Different to the most of existent evolutionary algorithms, GSA presents a better performance in multimodal problems, avoiding critical flaws such as the premature convergence to suboptimal solutions. Numerical simulations have been conducted on several models to show the effectiveness of the proposed scheme.

Chapter 3 presents an improved version of the crow search algorithm (CSA) method to solve complex optimization problems of energy. In the improved algorithm, two features of the original CSA are modified: (I) the awareness probability (AP) and (II) the random perturbation. With the purpose to enhance the exploration–exploitation ratio, the fixed awareness probability (AP) value is replaced (I) by a dynamic awareness probability (DAP), which is adjusted according to the fitness value of each candidate solution. The Lévy flight movement is also incorporated to enhance the search capacities of the original random perturbation (II) of CSA. In order to evaluate its performance, the algorithm has been tested in a set of four optimization problems which involve induction motors and distribution networks. The results demonstrate the high performance of the proposed method when it is compared with other popular approaches.

In Chap. 4, a comparative study between metaheuristic techniques used for solar cells parameter estimation is presented. The comparison evaluates the solar cell models of one diode, two diodes, and three diodes. In the analysis, the solar cell models are evaluated considering different operation conditions. Experimental results obtained during the comparison are also statistically validated.

Chapter 5 considers a nonlinear system identification method based on the Hammerstein model. In the scheme, the system is modeled through the adaptation of an adaptive network-based fuzzy inference system (ANFIS) scheme, taking advantage of the similitude between it and the Hammerstein model. To identify the parameters of the modeled system, the approach uses a recent nature-inspired method called the gravitational search algorithm (GSA). Different to most of existent optimization algorithms, GSA delivers a better performance in complex multimodal problems, avoiding critical flaws such as the premature convergence to suboptimal solutions. To show the effectiveness of the proposed scheme, its modeling accuracy has been compared with other popular evolutionary computing algorithms through numerical simulations on different complex models.

In Chap. 6, a methodology to implement human-knowledge-based optimization strategies is presented. In the scheme, a Takagi-Sugeno Fuzzy inference system is used to reproduce a specific search strategy generated by a human expert. Therefore, the number of rules and its configuration only depends on the expert experience without considering any learning rule process. Under these conditions, each fuzzy rule represents an expert observation that models the conditions under which candidate solutions are modified in order to reach the optimal location. To exhibit the performance and robustness of the method, a comparison to other well-known optimization methods is conducted. The comparison considers several standard benchmark functions which are typically found in the scientific literature. The results suggest a high performance of the proposed methodology.

Chapter 7 presents a recent metaheuristic algorithm called Neighborhood-based Consensus for Continuous Optimization (NCCO). NCCO is based on typical processes present in multi-agent systems, such as local consensus formulations and reactive responses. These operations are conducted by using appropriate operators that are applied in each evolutionary stage. A traditional metaheuristic algorithm considers in its operation the application of every operator without examining its final impact in the searching process. In contrast to other metaheuristic techniques, the proposed method uses additional operators to avoid the undesirable effects produced by the over-exploitation or suboptimal exploration of conventional operations. In order to illustrate the performance and accuracy of the proposed NCCO approach, it is compared to several well-known, state-of-the-art algorithms over a set of benchmark functions, and real-world design applications. The experimental results demonstrate that NCCO's performance is superior to the test algorithms.

Chapter 8 presents a metaheuristic algorithm in which knowledge extracted during its operation is employed to guide its search strategy. In the approach, a self-organizing map (SOM) is used as extracting knowledge technique to identify the promising areas through the reduction of the search space. Therefore, in each generation, the scheme uses a subset of the complete group of generated solutions seen so far to train the SOM. Once trained, the neural unit from the SOM lattice that corresponds to the best solution is identified. Then, by using local information of this neural unit an entire population of candidate solutions is produced. With the use of the extracted knowledge, the new approach improves the convergence to difficult high multimodal optima by using a reduced number of function evaluations. The performance of our approach is compared to several state-of-the-art optimization techniques considering a set of well-known functions and three real-world engineering problems. The results validate that the presented method reaches the best balance regarding accuracy and computational cost over its counterparts.

Guadalajara, Mexico

Erik Cuevas
Jorge Gálvez
Omar Avalos

Contents

1 Introduction to Optimization and Metaheuristic Methods 1
1.1 Description of an Optimization Task 1
1.2 Classical Optimization 2
1.3 Metaheuristic Search Schemes 5
1.3.1 Framework of a Metaheuristic Approach 6
References 7

2 Optimization Techniques in Parameters Setting for Induction Motor 9
2.1 Introduction 9
2.2 Gravitational Search Algorithm (GSA) 10
2.3 Parameter Setting Formulation 12
2.3.1 Approximate Circuit Model 12
2.3.2 Exact Circuit Model 13
2.4 Experimental Results 14
2.4.1 Performance Evaluation Concerning Its Tuning Parameters 15
2.4.2 Induction Motor Inner Parameter Identification 17
2.4.3 Statistical Analysis 18
2.5 Conclusions 24
References 24

3 An Enhanced Crow Search Algorithm Applied to Energy Approaches 27
3.1 Introduction 27
3.2 Crow Search Algorithm (CSA) 28
3.3 The Improved Crow Search Algorithm (ICSA) 30
3.3.1 Dynamic Awareness Probability (DAP) 30
3.3.2 Random Movement—Lévy Flight 30

3.4 Motor Parameter Estimation Formulation . . . 31
3.5 Capacitor Allocation Formulation . . . 32
3.5.1 Load Flow Analysis . . . 32
3.5.2 Mathematical Approach . . . 33
3.5.3 Sensitivity Analysis and Loss Sensitivity Factor . . . 34
3.6 Experimental Results . . . 35
3.6.1 Motor Inner Parameter Estimation . . . 36
3.6.2 Capacitor Allocation Test . . . 39
3.7 Conclusions . . . 47
References . . . 47

4 Comparison of Solar Cells Parameters Estimation Using Several Optimization Algorithms . . . 51
4.1 Introduction . . . 51
4.2 Evolutionary Computation Techniques . . . 53
4.2.1 Artificial Bee Colony (ABC) . . . 53
4.2.2 Differential Evolution (DE) . . . 54
4.2.3 Harmony Search (HS) . . . 54
4.2.4 Gravitational Search Algorithm (GSA) . . . 55
4.2.5 Particle Swarm Optimization (PSO) . . . 56
4.2.6 Cuckoo Search (CS) . . . 57
4.2.7 Differential Search Algorithm (DSA) . . . 58
4.2.8 Crow Search Algorithm (CSA) . . . 58
4.2.9 Covariant Matrix Adaptation with Evolution Strategy (CMA-ES) . . . 59
4.3 Modeling of Solar Cells . . . 60
4.3.1 Single Diode Model (SDM) . . . 60
4.3.2 Double Diode Model (DDM) . . . 61
4.3.3 Three Diode Model (TDM) . . . 62
4.3.4 Solar Cells Parameter Identification as an Optimization Problem . . . 63
4.4 Experimental Results . . . 64
4.5 Conclusions . . . 93
References . . . 93

5 Gravitational Search Algorithm for Non-linear System Identification Using ANFIS-Hammerstein Approach . . . 97
5.1 Introduction . . . 97
5.2 Background . . . 99
5.2.1 Hybrid ANFIS Models . . . 99
5.2.2 Adaptive Neuro-fuzzy Inference System (ANFIS) . . . 100
5.2.3 Gravitational Search Algorithm (GSA) . . . 102

5.3 Hammerstein Model Identification by Applying GSA 103
5.4 Experimental Analysis 107
5.4.1 Test I 109
5.4.2 Test II. 111
5.4.3 Test III 114
5.4.4 Test IV 116
5.4.5 Test V 119
5.4.6 Test VI 124
5.4.7 Test VII 126
5.4.8 Statistical Study 130
5.5 Conclusions 131
References 131

6 Fuzzy Logic Based Optimization Algorithm 135
6.1 Introduction 135
6.2 Reasoning Models 138
6.2.1 Fuzzy Logic Concepts 138
6.2.2 The Takagi-Sugeno (TS) Fuzzy Model 139
6.3 The Proposed Method. 140
6.3.1 Optimization Mechanism 141
6.4 Computational Procedure 147
6.5 Discussion 148
6.5.1 Optimization Algorithm 148
6.5.2 Modeling Characteristics 148
6.6 Experimental Study 149
6.6.1 Performance Evaluation Considering the Tuning Parameters 149
6.6.2 Comparison with Another Optimization Techniques 157
6.6.3 Convergence Analysis 172
6.6.4 Computational Complexity Analysis 175
6.7 Conclusions 178
References 179

7 Neighborhood Based Optimization Algorithm 183
7.1 Introduction 183
7.2 Literature Review 185
7.3 Preliminary Concepts 186
7.3.1 Reactive Models 186
7.4 Neighborhood-Based Consensus for Continuous Optimization. . . . 187
7.4.1 Initialization 189
7.4.2 Reactive Flocking Response 189
7.4.3 Update Mechanism 194
7.5 Computational Procedure 195
7.6 Experimental Study 196

7.6.1 Performance Comparison 197
7.6.2 Computational Time and Convergence Analysis 226
7.6.3 Engineering Design Problems 230
7.7 Conclusions 238
References 241

8 Knowledge-Based Optimization Algorithm 245
8.1 Introduction 245
8.2 Self-organizing Map 247
8.3 The Proposed EA-SOM Method 249
8.3.1 Initialization 251
8.3.2 Training 251
8.3.3 Extracting Information (EK) 251
8.3.4 Solution Production 252
8.3.5 Construction of the New Training Set 253
8.4 Computational Procedure 253
8.5 Experimental Study 254
8.5.1 Performance Comparison 255
8.5.2 Convergence 264
8.5.3 Engineering Design Problems 268
8.6 Conclusions 275
References 275

Appendix A: Systems Data 279

Appendix B: Optimization Problems 283

Chapter 1
Introduction to Optimization and Metaheuristic Methods

This chapter presents an introduction of optimization techniques considering their principal properties. The objective of this chapter is to stimulate the use of metaheuristic approaches for solving complex optimization problems. The analysis is carried out so that it results clear the need of employing metaheuristic schemes for the solution of parameter identification system schemes.

1.1 Description of an Optimization Task

Most of the parameter identification systems use any class of optimization scheme for their operation. Under this perspective, they try to discover a particular solution, which is "best" regarding some objective function. In general, an optimization problem refers to a searching process that expects to find an option from a set of potential alternatives to attain a required benefit reflexed as the minimal cost of an objective function [1].

Assume a public transportation net of a particular town, as an example. Here, it is required to determine the "best" path to a destination location. To evaluate each alternative solution and finally obtain the "best" alternative, a proper criterion must be considered. A practical principle might be the distance among routes. Under such circumstances, it is expected that the optimization approach selects the route with the smallest distance as a final solution. Consider that different evaluation schemes are also probable, which could obtain similar optimal routes, e.g., ticket price, number of transfers or the time that it is required to travel from a location to another.

Formally, optimization is defined as follows: Assuming a function $f : S \to \Re$ which is nominated as the objective function, search the argument that minimizes f:

$$x^* = \arg\min_{x \in S} f(x) \tag{1.1}$$

E. Cuevas et al., *Recent Metaheuristics Algorithms for Parameter Identification*, Studies in Computational Intelligence 854, https://doi.org/10.1007/978-3-030-28917-1_1

S refers to the search space which represents all probable solutions of the optimization problem. In general, the unknown values of x symbolize the decision variables. The function f specifies the optimization evaluation that allows calculating the index which determines the "quality" of a specific value of x.

In the public transportation example, S involves all the bus lines, subway trajectories, etc., collected in the database of the routing system. x corresponds to the route that the system tries to determine. $f(x)$ represents the optimization criterion which evaluates the quality of all probable solutions. Some other constraints can be included in the problem definition such as the distance to the destination or the ticket price (in some cases, it is considered the combination of both indexes, depending on our preferences).

In case of the existence of additional constraints, the optimization problem is defined as a constrained optimization (different to unconstrained optimization where such constraints do not exist). As a summary, an optimization task includes the following elements:

- One or more decision variables $\mathbf{x}$ that represent a candidate solution
- An objective function $f(x)$ that defines the optimization quality of each solution x
- The search space S that specifies the set of all probable solutions
- Constraints that define several feasible areas of the search space S.

From a practical perspective, an optimization method searches for a solution in a moderate period with reasonable precision. Along with the capacities of the used approach, the performance also depends on the nature of the problem to be optimized. In general, considering a numerical solution, an optimization problem is well-defined if the following conditions are fulfilled:

1. There exists a solution.
2. There exist a particular association between the solution and its initial locations so that small perturbations in the original values produce small variations in the objective function $f(x)$.

1.2 Classical Optimization

Once a problem has been converted into an objective function, the subsequent action is to adopt a proper optimizer. Optimization techniques can be classified into two groups: classical methods and metaheuristic schemes [2].

Typically, $f(\mathbf{x})$ maintains a nonlinear relationship regarding the adjustable decision variables $\mathbf{x}$. In classical techniques, an iterative method is used to explore the search space effectively. From all classical methods, the techniques based on the derivative-descent methods are the most popular. In such schemes, the next location x_{k+1} is computed by a step down from the current position x_k in a direction to $\mathbf{d}$:

$$x_{k+1} = x_k + \alpha \mathbf{d}, \tag{1.2}$$

where α corresponds to a learning rate that regulates the magnitude of the search in direction **d**. The searching direction **d** in Eq. (1.2) is calculated considering the gradient (**g**) of the objective function $f(\bullet)$.

The steepest descent scheme is one of the most important methods for solving objective functions due to its simplicity and effectiveness. Many other derivative-based methods use this method as the basis for their construction. Under this approach, Eq. (1.3) represents the well-known gradient formulation:

$$x_{k+1} = x_k - \alpha \mathbf{g}(f(x)), \tag{1.3}$$

Nonetheless, classical derivative-based optimization methods can be applied as long as the objective function maintains two important conditions:

- The objective function is two-time derivable.
- The objective function is unimodal, i.e., it has only one minimum.

The following formulation represents a simple case of a derivable and unimodal objective function.

$$f(x_1, x_2) = 10 - e^{-(x_1^2 + 3 \cdot x_2^2)} \tag{1.4}$$

Figure 1.1 presents the function described in Eq. (1.4).

However, in such conditions, classical approaches are only suitable for a few kinds of optimization problems. One example is combinatorial optimization where there is no differentiation of the objective function.

Moreover, there exist many causes of why an objective function could not be derivable. One example is the "floor" operation presented in Eq. (1.5) which modifies nonlinearly the function exhibited in Eq. (1.4). Under such conditions, Fig. 1.1 is transformed into the stairs shape shown in Fig. 1.2. It is important to remark that at each step the objective function is non-differentiable. Therefore, the formulation is modeled as follows:

$$f(x_1, x_2) = \text{floor}\left(10 - e^{-(x_1^2 + 3 \cdot x_2^2)}\right) \tag{1.5}$$

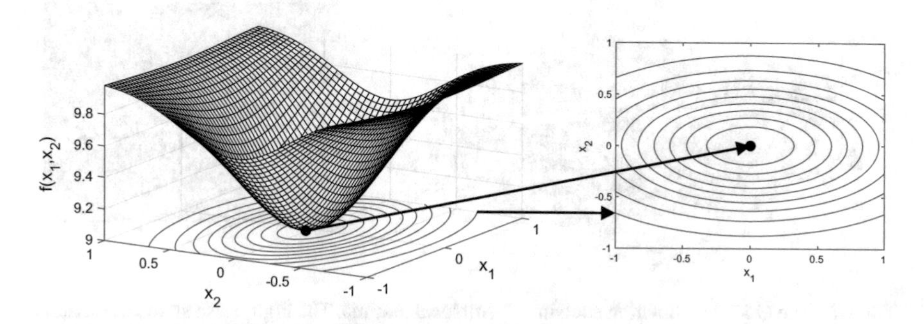

Fig. 1.1 Example of unimodal and derivable objective function

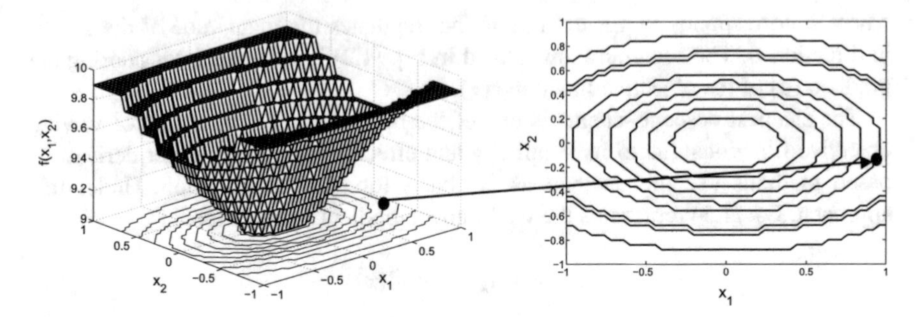

Fig. 1.2 A unimodal function and non-differentiable objective function

Also in derivable objective functions, gradient-based schemes could not properly work. As an example, consider the minimization of the Griewank function which is described as following:

$$\begin{array}{ll} \text{minimize} & f(x_1, x_2) = \frac{x_1^2 + x_2^2}{4000} - \cos(x_1)\cos\left(\frac{x_2}{\sqrt{2}}\right) + 1 \\ \text{subject to} & \begin{array}{l} -30 \leq x_1 \leq 30 \\ -30 \leq x_2 \leq 30 \end{array} \end{array} \tag{1.6}$$

Observing the optimization problem defined in Eq. (1.6), it is clear that the global solution is located in $x_1 = x_2 = 0$. Figure 1.3 exhibits the function described in Eq. (1.6). From Fig. 1.3, it can be seen that the objective function contains several local optimal solutions (multimodal). Under such conditions, gradient methods with a random initial position would converge to one of them with a high probability.

Assuming the restrictions of gradient-based schemes, parameter identification problems present difficulties when they use classical optimization methods for obtaining their suitable solutions. Alternatively, any other methods that do not consider constraints and assumptions can be used in an extensive variety of problems [3].

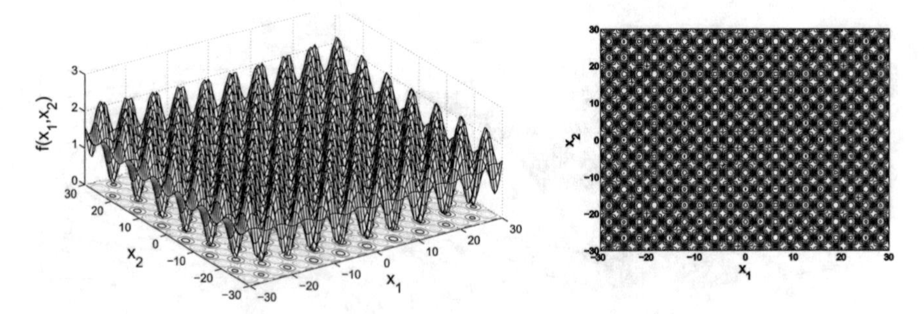

Fig. 1.3 The Griewank function contains several local minima. The Figure also shows the contour map generated by the objective function

1.3 Metaheuristic Search Schemes

Metaheuristic search [4] techniques are derivative-free processes that do not demand some restriction of the objective function. Under such conditions, it is not required neither unimodal nor two-timing differentiable. Hence, metaheuristic search approaches as global optimization methods can solve nonlinear, multimodal and non-convex problems subject to several nonlinear constraints with discrete or continuous design variables.

The metaheuristic area has attracted the attention of the research community in the last years. The advancement of computational science and requirements of technical processes have produced the need to solve optimization problems even though there is not enough information about the optimization task. Therefore, it is not clear if it is possible the application of a classical method. Indeed, most of the parameter identification applications are highly multimodal, with noisy fitness, without an exact deterministic representation so that the objective function could be the effect of a simulation or an experimental process. Under such restrictions, metaheuristic search algorithms have been considered as effective optimization alternatives to solve complex problems.

Metaheuristic search schemes refer to common techniques for solving complex optimization problems. Typically, they consider an objective function without considering any restriction or assumption of its mathematical characteristics. Metaheuristic approaches require neither a characterization of the optimization task nor any previous information of the objective function. They consider an objective function as a "black box" [5] which represents the most notable and important characteristic of such a search strategy.

Metaheuristic schemes get the information about the composition of an optimization problem by using the knowledge acquired from the possible solutions (i.e., candidate solutions) evaluated in terms of the objective function. Then, this knowledge is utilized to build new candidate solutions which maintain a higher probability to have better quality than their previous values.

Lately, many metaheuristic schemes have been introduced with impressive results. Such methods use as motivation our scientific knowledge of natural, biological or social systems, which can be considered, at some level of abstraction, as optimization processes [6]. Some of the most important examples include the collective behavior of bee colonies such as the Artificial Bee Colony (ABC) method [7], the grouping behavior of fish schooling and bird flocking such as the Particle Swarm Optimization (PSO) scheme [8], the improvising process that happens when a composer searches for a better harmony state such as the Harmony Search (HS) [9], the interaction behavior of firefly insects such as the Firefly (FF) technique [10], the simulation of the bat collective behavior such as the Bat Algorithm (BA) algorithm [11], the emulation of the animal interaction of a group such as the Collective Animal Behavior [12], the social-spider collective behavior such as the Social Spider Optimization (SSO) [13] approach, the modeling of immunological elements as the clonal selection algorithm (CSA) [14], the reproduction of the electromagnetism principle as the electromagnetism-Like algorithm [15], the emulation of a colony of locust insects such as the Locust Algorithm (LA) [16] and the simulation of the conventional and differential evolution in some species such as the Genetic Algorithms (GA) [17] and Differential Evolution (DE) [18], respectively.

1.3.1 Framework of a Metaheuristic Approach

From a conventional perspective, a metaheuristic search algorithm is an approach that emulates a biological, natural or social system from at certain abstraction level. In general, a generic metaheuristic method includes:

1. A set of candidate solutions known as population.
2. This population is modified dynamically through the generation of new solutions.
3. An objective function reflects the solution capacity to reproduce and maintain itself.
4. A set of operations are considered as a search strategy to exploit and explore appropriately the search space of possible solutions.

Under the methodology of metaheuristic search methods, a determined number of candidate solutions enhance their quality during the evolution process (i.e., their ability to solve the optimization problem at hand). Therefore, during the optimization process, a population of candidate solutions whose fitness values are properly associated with a particular objective function, improve their fitness values leading the operated population towards the global optimum.

Several optimization schemes have been devised to find the global optimum of a nonlinear optimization problem considering box constraints under the following formulation:

$$\begin{aligned} &\text{maximize } f(\mathbf{x}), \quad \mathbf{x} = (x_1, \ldots, x_d) \in \Re^d \\ &\text{subject to } \mathbf{x} \in \mathbf{X} \end{aligned} \tag{1.7}$$

with $f : \Re^d \rightarrow \Re$ corresponds to a nonlinear objective function while $\mathbf{X} = \left\{\mathbf{x} \in \Re^d | l_i \leq x_i \leq u_i, i = 1, \ldots, d\right\}$ represents a bounded feasible search space, limited by the lower (l_i) and upper (u_i) points.

To solve the optimization problem defined in Eq. (1.7) from a metaheuristic perspective, a population $\mathbf{P}^k\left(\left\{\mathbf{p}_1^k, \mathbf{p}_2^k, \ldots, \mathbf{p}_N^k\right\}\right)$ of N individuals (candidate solutions) is modified from the starting state ($k = 0$) to a total *gen* number of generations ($k = gen$). In the starting point, the process initializes a group of N possible solutions in positions that are randomly and uniformly located within a determined interval defined between a lower (l_i) and upper (u_i) limits. At every generation, a set of stochastic operators are operated over the population $\mathbf{P}^k$ to generate a new solution population $\mathbf{P}^{k+1}$. Each possible solution $\mathbf{p}_i^k$ ($i \in [1, \ldots, N]$) corresponds to a vector of d dimensions $\left\{p_{i,1}^k, p_{i,2}^k, \ldots, p_{i,d}^k\right\}$ where every element corresponds to a design variable of the optimization problem at hand. The quality of each possible solution $\mathbf{p}_i^k$

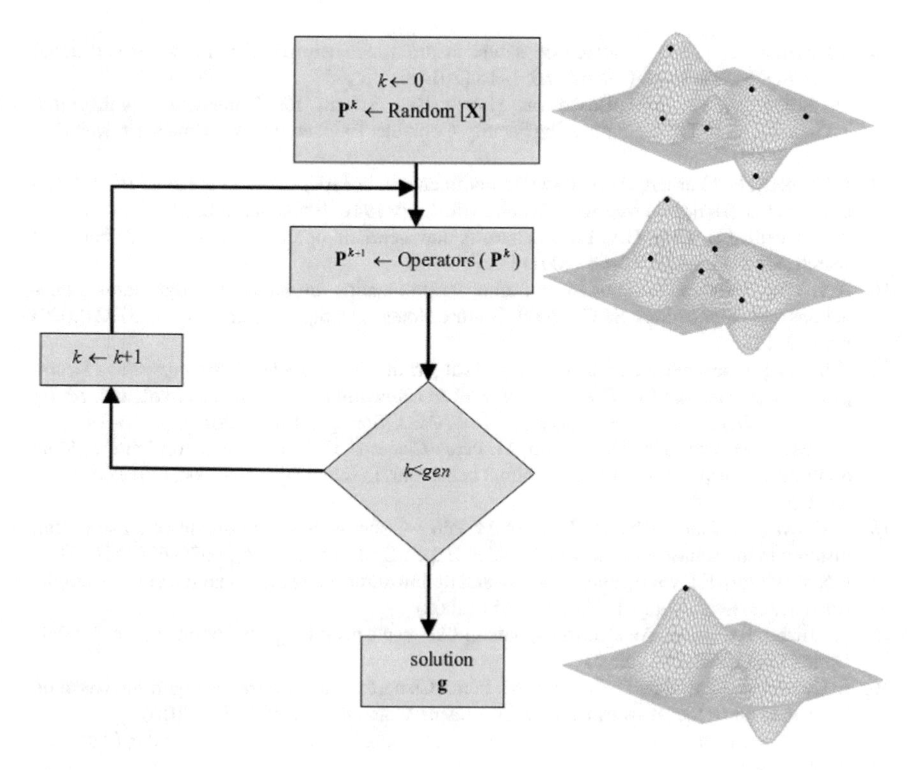

Fig. 1.4 Generic operation of a metaheuristic approach

is continuously assessed in terms of an objective function $f\left(\mathbf{p}_i^k\right)$ whose value symbolizes the fitness magnitude of $\mathbf{p}_i^k$. In the optimization process, the best possible solution $\mathbf{g}$ $(g_1, g_2, \ldots g_d)$ currently seen is stored assuming that it refers to the best available solution [19]. Figure 1.4 exhibits a graphical illustration of the operation of a generic metaheuristic scheme.

References

1. B. Akay, D. Karaboga, A survey on the applications of artificial bee colony in signal, image, and video processing. SIViP **9**(4), 967–990 (2015)
2. X.-S. Yang, in *Engineering Optimization* (Wiley, London, 2010)
3. M.A. Treiber, in *Optimization for Computer Vision An Introduction to Core Concepts and Methods* (Springer, Berlin, 2013)
4. D. Simon, in *Evolutionary Optimization Algorithms* (Wiley, London, 2013)
5. C. Blum, A. Roli, Metaheuristics in combinatorial optimization: overview and conceptual comparison. ACM Comput. Surv. (CSUR) **35**(3), 268–308 (2003). https://doi.org/10.1145/937503.937505

6. S.J. Nanda, G. Panda, A survey on nature inspired metaheuristic algorithms for partitional clustering. Swarm Evol. Comput. **16**, 1–18 (2014)
7. D. Karaboga, An Idea Based on Honey Bee Swarm for Numerical Optimization. TechnicalReport-TR06. Engineering Faculty, Computer Engineering Department, Erciyes University, 2005
8. J. Kennedy, R. Eberhart, Particle swarm optimization, in *Proceedings of the 1995 IEEE International Conference on Neural Networks*, vol. 4, pp. 1942–1948, December 1995
9. Z.W. Geem, J.H. Kim, G.V. Loganathan, A new heuristic optimization algorithm: harmony search. Simulations **76**, 60–68 (2001)
10. X.S. Yang, Firefly algorithms for multimodal optimization, in: Stochastic Algorithms: Foundations and Applications, SAGA 2009, Lecture Notes in Computer Sciences, vol. 5792 (2009) pp. 169–178
11. X.S. Yang, A new metaheuristic bat-inspired algorithm, in *Nature Inspired Cooperative Strategies for Optimization (NISCO 2010), Studies in Computational Intelligence*, vol. 284, ed. by C. Cruz, J. González, G.T.N. Krasnogor, D.A. Pelta (Springer, Berlin, 2010), pp. 65–74
12. E. Cuevas, M. González, D. Zaldivar, M. Pérez-Cisneros, G. García, An algorithm for global optimization inspired by collective animal behaviour, Discrete Dyn. Nat. Soc. (2012). Art. no. 638275
13. E. Cuevas, M. Cienfuegos, D. Zaldívar, M. Pérez-Cisneros, A swarm optimization algorithm inspired in the behavior of the social-spider. Expert Syst. Appl. **40**(16), 6374–6384 (2013)
14. L.N. de Castro, F.J. von Zuben, Learning and optimization using the clonal selection principle. IEEE Trans. Evol. Comput. **6**(3), 239–251 (2002)
15. Ş.I. Birbil, S.C. Fang, An electromagnetism-like mechanism for global optimization. J. Glob. Optim. **25**(1), 263–282 (2003)
16. E. Cuevas, A. González, D. Zaldívar, M. Pérez-Cisneros, An optimisation algorithm based on the behaviour of locust swarms. Int. J. Bio-Inspir. Comput. **7**(6), 402–407 (2015)
17. D.E. Goldberg, in *Genetic Algorithm in Search Optimization and Machine Learning* (Addison-Wesley, 1989)
18. R. Storn, K. Price, Differential Evolution—A Simple and Efficient Adaptive Scheme for Global Optimisation Over Continuous Spaces. TechnicalReportTR-95–012, ICSI, Berkeley, CA, 1995
19. E. Cuevas, Block-matching algorithm based on harmony search optimization for motion estimation. Appl. Intell. **39**(1), 165–183 (2013)

Chapter 2
Optimization Techniques in Parameters Setting for Induction Motor

2.1 Introduction

The environmental consequences for overconsumption of electrical have recently attracted the attention in many disciplines of engineering. Therefore, the increase of efficient machinery and elements that have low electrical energy consumption levels has become an important task nowadays [1].

Induction motors offer diverse benefits such as their ruggedness, moderate price, low-cost maintenance, and easy handling [2]. Nonetheless, more than half of the electrical energy consumed by industrial facilities is due to the use of induction motors. With the extensive use of induction motors, electrical energy consumption has grown exponentially in recent years. This reality has generated the necessity to improve the efficiency of such components, which principally depends on their inner parameters. The parameter calculation of induction motors represents a difficult task due to its non-linearity. For this, several alternatives have been suggested in the literature. Some works include that proposed by Waters and Willoughby [3], where the parameters are computed from the knowledge of specific variables, such as stator resistance and the leakage reactance, the proposed by Ansuj [4], where the estimation is based on a sensitivity analysis, and that proposed by De Kock [5], where the determination is carried through an output error method.

As an alternative to such techniques, the parameter determination of induction motors is also addressed through evolutionary methods. These techniques have demonstrated, under certain conditions, to achieve better results than those based on deterministic methods in terms of efficiency and robustness [6]. Some works reported in literature for the identification of parameters in induction motors include methods, such as genetic algorithms (GAs) [7], particle swarm optimization (PSO) [8, 9], artificial immune system (AIS) [10], the bacterial foraging algorithm (BFA) [11], the locust search (LS) scheme [12], the shuffled frog-leaping algorithm [13], a hybrid of GAs and PSO [14], and multiple-global-best guided artificial bee colony (ABC) [15]. Although these techniques present good results, they have a significant

E. Cuevas et al., *Recent Metaheuristics Algorithms for Parameter Identification*, Studies in Computational Intelligence 854, https://doi.org/10.1007/978-3-030-28917-1_2

deficiency: They frequently obtain sub-optimal solutions as a result of the inadequate balance between exploration and exploitation in their search stage.

Otherwise, the gravitational search algorithm approach (GSA) [16] is a recent evolutionary computation technique which is inspired in the laws of gravity. The GSA different to the reported, considers the gravitational principles to build its operators. Such a difference makes the GSA perform better the problems with high multimodal nature, avoiding critical deficiencies such as the premature convergence to local minima solutions [17, 18]. These advantages have motivated its use to solve a wide diversity of engineering problems such as energy [19], image processing [6], machine learning [20] and so on.

This chapter presents the GSA technique to determine the inner parameters of induction motors. A comparative study with state-of-the-art methods such as ABC [21], differential evolution (DE) [22], and PSO [23] on different induction models has been included to confirm the performance of the suggested approach. The conclusions of the comparative study are verified by statistical tests that properly support the investigation.

2.2 Gravitational Search Algorithm (GSA)

Rashedi proposed the GSA [16] in 2009 motivated by the gravity phenomena to construct its operators. Under this approach the GSA performs better the problems with high multimodal nature, avoiding critical deficiencies such as the premature convergence to local minima solutions. The GSA emulates masses as candidate solutions that attract each other by operators that simulate the gravitational attraction. The mass (quality) of each candidate solution is selected according to its fitness value. The GSA is designed to find the optimal solution of non-linear optimization problems with box constraints in the form:

$$\begin{aligned} &\text{minimize } f(\mathbf{x}), \quad \mathbf{x} = (x^1, \ldots, x^d) \in \mathrm{R}^d, \\ &\text{subject to } \mathbf{x} \in \mathbf{X} \end{aligned} \tag{2.1}$$

where $f : \mathrm{R}^d \rightarrow \mathrm{R}$ is a nonlinear function, whereas $\mathbf{X} = \left\{\mathbf{x} \in \mathrm{R}^d \middle| l_h \leq x^h \leq u_h, h = 1, \ldots, d\right\}$ is a defined available search space, constrained by the lower (l_h) and upper (u_h) boundaries. To solve the problem expressed in Eq. (2.1), the GSA employs a population of N candidate solutions. Each mass (or candidate solution) denotes a d-dimensional vector $\mathbf{x}_i(t) = (x_i^1, \ldots, x_i^d)$ $(i \in 1 \ldots, N)$, where each dimension corresponds to a decision variable of the optimization problem.

In the GSA, the force operating from a mass i to a mass j of the h variable ($h \in 1\ldots, d$) at a time t, is defined as follows:

$$F_{ij}^{h}(t) = G(t)\frac{Mp_i(t) \times Ma_j(t)}{R_{ij}(t) + \varepsilon}\left(x_j^h(t) - x_i^h(t)\right) \tag{2.2}$$

where Ma_j is the actual gravitational mass associated to solution j, Mp_i expresses the latent gravitational mass of solution i, $G(t)$ is the gravitational constant at time t, ε is a small constant, and R_{ij} is the Euclidian distance between the ith and jth candidate. In the GSA, $G(t)$ is a function which is adjusted during the process. The main idea of this adjustment is to regulate the balance between exploration and exploitation through the modification of the gravity forces among candidates.

The total force operating across a candidate solution i is determined as follows:

$$F_i^h(t) = \sum_{j=1,\, j\neq i}^{N} F_{ij}^h(t) \tag{2.3}$$

Then, the acceleration of the candidate solution i at time t is calculated as:

$$a_i^h(t) = \frac{F_i^h(t)}{Mn_i(t)} \tag{2.4}$$

where Mn_i denotes the inertial mass of the solution i. Under such circumstances, the new position of each solution i is determined as follows:

$$\begin{aligned} x_i^h(t+1) &= x_i^h(t) + v_i^h(t+1) \\ v_i^h(t+1) &= \text{rand}() \cdot v_i^h(t) + a_i^h(t) \end{aligned} \tag{2.5}$$

At each iteration, the gravitational and inertial masses of each candidate are estimated in terms of its fitness function. Hence, the gravitational and inertial masses are updated by the following models:

$$Ma_i = Mp_i = M_{ii} = M_i \tag{2.6}$$

$$m_i(t) = \frac{f(\mathbf{x}_i(t)) - worst(t)}{\text{best}(t) - worst(t)} \tag{2.7}$$

$$M_i(t) = \frac{m_i(t)}{\sum_{j=1}^{N} m_j(t)} \tag{2.8}$$

where $f(\cdot)$ denotes the objective function whose last result presents the fitness value. Otherwise, best(t) and worst(t) express the best and worst fitness values respectively determined at time t of the total population. Algorithm 2.1 shows the pseudo code of the GSA technique.

Algorithm 2.1. Gravitational search algorithm (GSA) pseudo code.

Initialization of the population Randomly
Determine the best and worst solutions from the initial population
while (*stop criteria*)
  for $i = 1{:}N$ (for all elements)
    update $G(t)$, $best(t)$, $worst(t)$ and $M_i(t)$ for $i = 1, 2, .., N$
    Compute the mass of individual $M_i(t)$
    Compute the gravitational constant $G(t)$
    Compute acceleration $a_i^h(t)$
    update the velocity and positions of each individual v_i^h, x_i^h
  end (for)
    Find the best Solution
end while
    Display the best solution

2.3 Parameter Setting Formulation

The inner parameters of an induction motor cannot be directly measured. Because of this, some techniques such as identification methods are used to estimate them. Under such strategies, the induction motor operation can be described with equivalent nonlinear circuits. Depending on the accuracy, two different circuit models are used [10]: the approximate circuit model and the exact circuit model. Usually, they provide an adequate relation to the motor parameters for their estimation.

In the estimation process, the determination of the parameters is conducted as multi-dimensional optimization processes where the inner parameters of the induction motor are treated as decision variables. Hence, the main objective is to minimize the error between the determined and the manufacturer data, modifying the parameters of the equivalent circuits. Under this strategy, the complex formulations process tends to produce multimodal error surfaces which are significantly difficult to minimize.

2.3.1 *Approximate Circuit Model*

In the approximate circuit model, the magnetizing reactance and rotor reactance are not considered; therefore, its precision is lower than the exact circuit model. The approximate model employs the manufacturer data maximum torque ($T_{\max}$), full load torque $\left(T_{fl}\right)$ and starting torque (T_{lr}), to determine the rotor resistance (R_2), stator leakage reactance (X_1), stator resistance (R_1), and motor slip (s). Figure 2.1

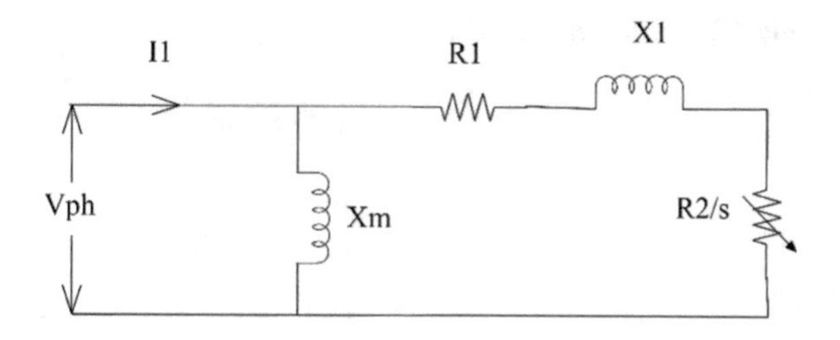

Fig. 2.1 Approximate circuit model

shows the approximate model. Under the approximate model, the identification task is formulated as follows:

$$\begin{aligned}&\text{minimize } F_A(\mathbf{x}),\ \mathbf{x} = (R_1, R_2, X_1, s) \in \mathrm{R}^4,\\&\text{subject to } 0 \le R_1 \le 1, 0 \le R_2 \le 1, 0 \le X_1 \le 10, 0 \le s \le 1\end{aligned} \tag{2.9}$$

where

$$\begin{aligned}F_A(\mathbf{x}) &= (f_1(\mathbf{x}))^2 + (f_2(\mathbf{x}))^2 + (f_3(\mathbf{x}))^2\\ f_1(\mathbf{x}) &= \frac{\frac{K_t R_2}{s\left[(R_1+R_2/s)^2+X_1^2\right]} - T_{fl}}{T_{fl}}\\ f_2(\mathbf{x}) &= \frac{\frac{K_t R_2}{(R_1+R_2)^2+X_1^2} - T_{lr}}{T_{lr}}\\ f_3(\mathbf{x}) &= \frac{\frac{K_t}{2\left[R_1+\sqrt{R_1^2+X_1^2}\right]} - T_{\max}}{T_{\max}}\\ K_t &= \frac{3V_{ph}^2}{\omega_s}\end{aligned} \tag{2.10}$$

2.3.2 *Exact Circuit Model*

In the exact model, the magnetizing reactance and rotor reactance effects are considered in the estimation. In such model, the stator leakage inductance (X_1), rotor leakage reactance (X_2), magnetizing leakage reactance (X_m), stator resistance (R_1), rotor resistance (R_2), and motor slip (s) are calculated to estimate the full load torque (T_{fl}), starting torque (T_{str}), maximum torque ($T_{\max}$), and full load power factor (pf). In Fig. 2.2 the exact model is shown. Considering the exact model, the estimation task is formulated as follows:

$$\begin{aligned}&\text{minimize} \quad F_E(\mathbf{x}), \mathbf{x} = (R_1, R_2, X_1, X_2, X_m, s) \in \mathrm{R}^6,\\&\text{subject to} \quad 0 \le R_1 \le 1, 0 \le R_2 \le 1, 0 \le X_1 \le 1, 0 \le X_2 \le 1,\\&\qquad\qquad\quad 0 \le X_m \le 10, 0 \le s \le 1,\end{aligned} \tag{2.11}$$

Fig. 2.2 Exact circuit model

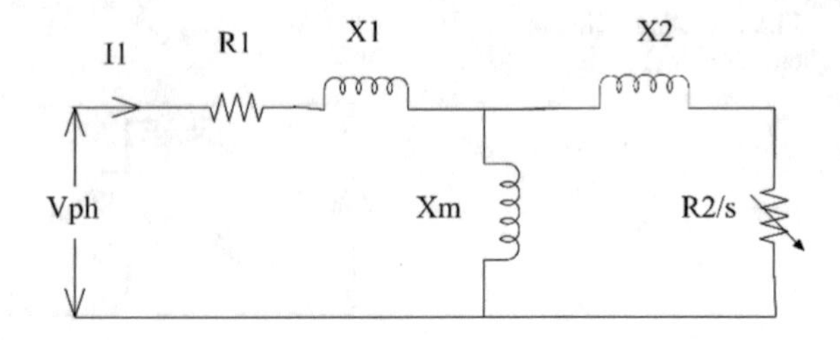

where

$$
\begin{aligned}
F_E(\mathbf{x}) &= (f_1(\mathbf{x}))^2 + (f_2(\mathbf{x}))^2 + (f_3(\mathbf{x}))^2 + (f_4(\mathbf{x}))^2 \\
f_1(\mathbf{x}) &= \frac{\frac{K_t R_2}{s\left[(R_{th}+R_2/s)^2+X^2\right]} - T_{fl}}{T_{fl}}, \\
f_2(\mathbf{x}) &= \frac{\frac{K_t R_2}{(R_{th}+R_2)^2+X^2} - T_{str}}{T_{str}} \\
f_3(\mathbf{x}) &= \frac{\frac{K_t}{2\left[R_{th}+\sqrt{R_{th}^2+X^2}\right]} - T\max}{T\max(mf)}, \\
f_4(\mathbf{x}) &= \frac{\cos\left(\tan^{-1}\left(\frac{X}{R_{th}+R_2/s}\right)\right) - pf}{pf}
\end{aligned}
\tag{2.12}
$$

$$R_{th} = \frac{R_1 X_m}{X_1 + X_m},\ V_{th} = \frac{V_{ph} X_m}{X_1 + X_m},\ X_{th} = \frac{X_1 X_m}{X_1+X_m},\ K_t = \frac{3V_{th}^2}{\omega_s},\ X = X_2 + X_{th}$$

In Eq. (2.11), it is necessary to satisfy a condition, the values of the computed parameters must fulfill the following restriction:

$$\frac{p_{fl} - (I_1^2 R_1 + I_2^2 R_2 + P_{rot})}{p_{fl}} = \eta_{fl} \tag{2.13}$$

where p_{fl} and P_{rot} describes the measured power and rotational losses, respectively. Moreover, η_{fl} expresses the efficiency given by the manufacturer. Considering this restriction, the computed efficiency is forced to be similar to the manufacturer efficiency, maintaining equilibrium between both. Generally, the parameters p_{fl} and P_{rot} are computed by two experimental tests known as No-load-test and Blocked-rotor-test [24, 25]. Nonetheless, to maintain congruity with similar works, the parameters were obtained from Refs. [13, 14, 11].

2.4 Experimental Results

In this chapter, the GSA is used to estimate the optimal inner parameters of two induction motors regarding the approximate model (F_A) and the exact model (F_E). Table 2.1 shows the technical features of the two motors adopted for the experiments. The suggested approach is also evaluated in comparison with other techniques based on evolutionary algorithms. In the experiments, we use the GSA estimator to the inner

Table 2.1 Manufacturer data of the two motors used for the experiments

	Motor 1	Motor 2
Power (HP)	5	40
Voltage (V)	400	400
Current (A)	8	45
Frequency (Hz)	50	50
No. poles	4	4
Full load slip (s)	0.07	0.09
Starting torque (T_{str})	15	260
Max. torque ($T_{\max}$)	42	370
Stator current	22	180
Full load torque (T_{fl})	25	190

parameter identification of the two induction motors, and its results are compared to those generated by ABC [21], DE [22], and PSO [23]. The parameters used for all compared algorithms are taken from their own referenced works. The parameters for each algorithm are described as follows:

1. PSO, parameters $c_1 = 2$, $c_2 = 2$ and weights are set $w_{\max} = 0.9$, and $w_{\min} = 0.4$ [23].
2. ABC, the parameters are provided by [21], *limit* = 100.
3. DE, according to [22] the parameters are set $p_c = 0.5$ and $f = 0.5$.
4. GSA, the parameters are taken according to [16].

The experimental results are divided into three sub-sections. In the first Sect. 2.4.1, the performance of the presented technique is evaluated regarding its operation parameters (sensibility analysis). In Sect. 2.4.2, the overall performance of the presented technique in comparison with similar approaches is presented. Finally, in Sect. 2.4.3, the results are statistically examined and validated by using the Wilcoxon test.

2.4.1 Performance Evaluation Concerning Its Tuning Parameters

The parameters of the GSA affect principally its performance. In this sub-section, the behavior of the GSA over the inner parameter estimation for the induction motors problem is analyzed regarding different setting parameters.

During the analysis, the parameters G_0 and α are set to their reference values as $G_0 = 100$ and $\alpha = 20$. In the study, when one of these parameters is evaluated, the other remains fixed to the reference value. To reduce the stochastic effect of the GSA, all benchmark functions are executed independently 30 times. As a stop criterion, a certain number of iterations is considered, which is set to 3000. For all

Table 2.2 Experimental results achieved by the GSA using different values of G_0

	$G_0 = 80$ $\alpha = 20$	$G_0 = 90$ $\alpha = 20$	$G_0 = 100$ $\alpha = 20$	$G_0 = 110$ $\alpha = 20$	$G_0 = 120$ $\alpha = 20$
Min	0.0044	0.0036	0.0032	0.0036	0.0033
Max	0.0119	0.0103	0.0032	0.0082	0.0088
Std.	0.0016	0.0013	0.0000	0.0012	0.0014
Mean	0.0052	0.0040	0.0032	0.0042	0.0039

executions, the population size N is set to 25 individuals. This process is typical in several metaheuristic approaches [26].

As the first step, the operation of the GSA is analyzed considering different values for G_0. In the study, the values of G_0 are modified from 80 to 120, while the values of α remains fixed at 10 and 30, respectively. In the study, the presented technique is executed independently 30 times for each value of G_0. The results achieved for the parameter combination of G_0 and α are shown in Table 2.2. These values denote the minimum, maximum, standard deviation, and mean values of F_E (exact model), regarding the Motor 1. The best results are marked in boldface. From Table 2.2, we can assume that the presented GSA $G_0 = 100$ maintains the best performance.

As the second step, the performance of the GSA is evaluated regarding different values of α. In the study, the values α are modified from 10 to 30 while the value of G_0 remains set to 100. The statistical results achieved by the GSA using different values α are shown in Table 2.3. The values described are the minimum, maximum, standard deviation, and mean values of F_E (exact model), regarding the Motor 2. The best results are marked in boldface. In Table 2.3, it is visible that the GSA with $\alpha = 20$ outperforms the other parameter configurations.

After the study, the experimental results in Tables 2.2 and 2.3 infer that a proper combination of parameter values can improve the performance of the presented technique and the quality of the solutions. After experimentation, it can be concluded that the optimal GSA parameter is composed of the following values: $G_0 = 100$ and $\alpha = 20$. After determining the optimal parameters experimentally, those are kept for all test functions in the following experiments.

Table 2.3 Experimental results achieved by the GSA using different values of α

	$G_0 = 100$ $\alpha = 10$	$G_0 = 100$ $\alpha = 15$	$G_0 = 100$ $\alpha = 20$	$G_0 = 100$ $\alpha = 25$	$G_0 = 100$ $\alpha = 30$
Min	0.0093	0.0093	0.0071	0.0093	0.0092
Max	0.0730	0.0433	0.0209	0.0435	0.0493
Std.	0.0147	0.0085	0.0043	0.0094	0.0109
Mean	0.0235	0.0164	0.0094	0.0191	0.0215

2.4.2 *Induction Motor Inner Parameter Identification*

In this section, the performance of the GSA method is compared with DE, ABC, and PSO, regarding the parameter estimation of the two equivalent models. In the experiment, all techniques are operated with a population of 25 individuals (N = 25). The total iteration number for all techniques is set to 3000. The stop criterion is selected to maintain similarity to similar works reported in the state of art [21–23]. All the experimental results shown in this section consider the analysis of 35 independent executions of each technique. Thus, the values of (approximate model), deviation standard, and mean obtained by every technique for Motor 1 are reported in Table 2.4, while the results generated by Motor 2 are presented in Table 2.5. On the other hand, the values of (exact model) for Motor 1 and Motor 2 are presented in Tables 2.6 and 2.7, respectively.

According to the results reported in Tables 2.4, 2.5, 2.6 and 2.7, the GSA presents better performance than DE, ABC, and PSO in all tests. This directly related to the better trade-off between exploration and exploitation of the presented method.

Once the induction motor inner parameters of all algorithms were estimated, their estimations were compared with the manufacturer starting torque (T_{str}), maximum

Table 2.4 Results of F_A, regarding the Motor 1

	GSA	DE	ABC	PSO
Min	**3.4768e × 10^{-22}**	1.9687e × 10^{-15}	2.5701e × 10^{-05}	1.07474e × 10^{-04}
Max	**1.6715e × 10^{-20}**	0.0043	0.0126	0.0253
Mean	**5.4439e × 10^{-21}**	1.5408e × 10^{-04}	0.0030	0.0075
Std.	**4.1473e × 10^{-21}**	7.3369e × 10^{-04}	0.0024	0.0075

Table 2.5 Results of F_A, regarding the Motor 2

	GSA	DE	ABC	PSO
Min	**3.7189e × 10^{-20}**	1.1369e ×10^{-13}	3.6127e × 10^{-04}	0.0016
Max	**1.4020e × 10^{-18}**	0.0067	0.0251	0.0829
Mean	**5.3373e × 10^{-19}**	4.5700e ×10^{-04}	0.0078	0.0161
Std.	**3.8914e × 10^{-19}**	0.0013	0.0055	0.0165

Table 2.6 Results of F_E, regarding the Motor 1

	GSA	DE	ABC	PSO
Min	**0.0032**	0.0172	0.0172	0.0174
Max	**0.0032**	0.0288	0.0477	0.0629
Mean	**0.0032**	0.0192	0.0231	0.0330
Std.	**0.0000**	0.0035	0.0103	0.0629

Table 2.7 Results of F_E, regarding the Motor 2

	GSA	DE	ABC	PSO
Min	**0.0071**	0.0091	0.0180	0.0072
Max	**0.0209**	0.0305	0.2720	0.6721
Mean	**0.0094**	0.0190	0.0791	0.0369
Std.	**0.0043**	0.0057	0.0572	0.1108

torque ($T_{\max}$), and full load torque (T_{fl}) values from Table 2.1. The principal objective of this comparison is to estimate the accuracy of each technique concerning the actual motor parameters. Tables 2.8 and 2.9 show the experimental results of F_A for Motors 1 and 2, respectively. On the other hand, Tables 2.10 and 2.11 present the comparative results of F_E for Motors 1 and 2, respectively.

As the convergence rate of evolutionary algorithms is an essential characteristic to estimate their performance for solving optimization problems, the convergence of all techniques facing functions F_A and F_E is compared in Fig. 2.3a, b. The convergence rate of the GSA can be observed in both figures. According to the figures, GSA finds the global optimum faster than other techniques.

Finally, in Fig. 2.4 the relation of the slip versus torque for both models (F_A and F_E) and for both Motors (1 and 2) is shown.

2.4.3 Statistical Analysis

In the statistical analysis, the non-parametric test known as Wilcoxon analysis [27] was carried out. It allows evaluating the differences between two related techniques. In the test, a 5% significance level over the mean fitness values of F_A and F_E, is considered regarding Motors 1 and 2. In Table 2.12 are reported the p-values generated by the Wilcoxon test for the pair-wise comparison between the algorithms. Under these conditions, three groups are formed: GSA versus DE, GSA versus ABC, and GSA versus PSO. In the Wilcoxon analysis, it is considered as the null hypothesis that there is no significant difference between the two algorithms. Otherwise, as an alternative hypothesis, it is considered that there is a significant difference between both techniques. Table 2.12 shows that all p-values are less than 0.05 (5% significance level). This evidence provides solid evidence against the null hypothesis, indicating that the presented technique statistically presents better results than the other approaches.

Table 2.8 Comparison of GSA, DE, ABC, and PSO regarding manufacturer data, F_A, for Motor 1

	True val.	GSA	Error%	DE	Error%	ABC	Error%	PSO	Error%
Tst	**15**	**15.00**	**0**	14.9803	−0.131	14.3800	−4.133	15.4496	2.9973
Tmax	**42**	**42.00**	**0**	42.0568	0.135	40.5726	−3.398	39.6603	−5.570
Tfl	**25**	**25.00**	**0**	24.9608	−0.156	25.0480	0.192	25.7955	3.182

Table 2.9 Comparison of GSA, DE, ABC, and PSO regarding manufacturer data, F_A, for Motor 2

	True val.	GSA	Error%	DE	Error%	ABC	Error%	PSO	Error%
Tst	**260**	**260.00**	**0**	258.470	−0.588	260.636	0.244	288.905	11.117
Tmax	**370**	**370.00**	**0**	372.769	0.748	375.066	1.369	343.538	−7.151
Tfl	**190**	**190.00**	**0**	189.050	−0.49	204.149	7.447	196.117	3.219

Table 2.10 Comparison of GSA, DE, ABC, and PSO regarding manufacturer data, F_E, for Motor 1

	True val.	GSA	Error%	DE	Error%	ABC	Error%	PSO	Error%
Tst	**15**	**14.947**	**−0.353**	15.408	2.726	16.419	9.462	15.646	4.308
Tmax	**42**	**42.00**	**0**	**42.00**	**0**	**42.00**	**0**	**42.00**	**0**
Tfl	**25**	**25.066**	**0.264**	26.082	4.331	25.339	1.358	26.619	6.478

Table 2.11 Comparison of GSA, DE, ABC, and PSO regarding manufacturer data, F_E, for Motor 2

	True val.	GSA	Error%	DE	Error%	ABC	Error%	PSO	Error%
Tst	**260**	**258.158**	**−0.708**	262.056	0.790	246.213	−5.302	281.897	8.422
Tmax	**370**	**370.00**	**0**	**370.00**	**0**	**370.00**	**0**	**370.00**	**0**
Tfl	**190**	**189.884**	**−0.061**	192.291	1.206	207.913	9.428	166.676	−12.27

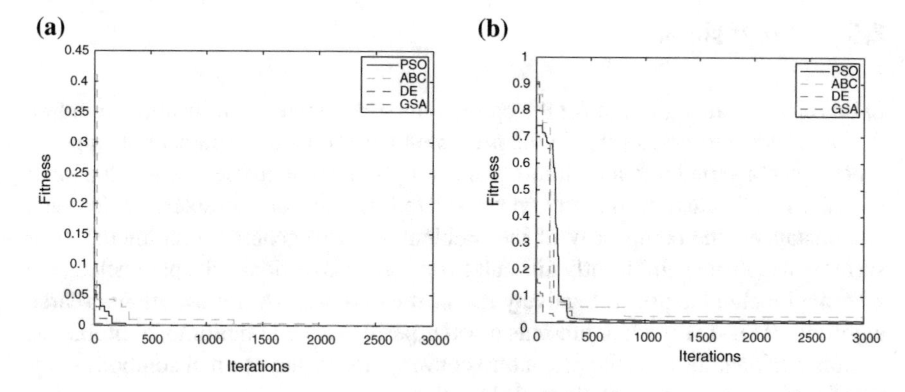

Fig. 2.3 Convergence evolution through iterations: **a** Model 1 (F_A); **b** Model 2 (F_E)

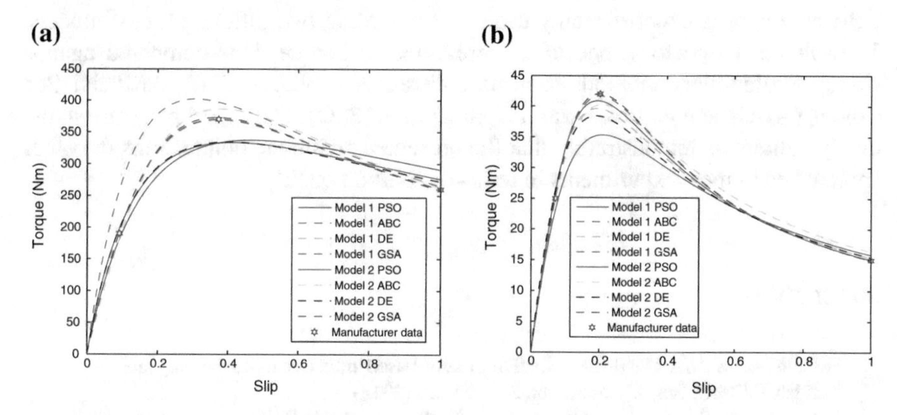

Fig. 2.4 Curve slip versus torque via PSO, ABC, DE, and GSA considering Model 1 (F_A), Model 2 (F_E), **a** Motor 1, and **b** Motor 2

Table 2.12 *p*-values performed by Wilcoxon test comparing GSA versus DE, GSA versus ABC, and GSA versus PSO over the mean fitness values of F_A and F_B regarding Motors 1 and 2 from Tables 2.4, 2.5, 2.6 and 2.7

GSA versus	DE	ABC	PSO
F_A, Motor 1	$6.5855e \times 10^{-13}$	$6.8425e \times 10^{-13}$	$3.5651e \times 10^{-13}$
F_A, Motor 2	0.0091	0.0365	0.0046
F_E, Motor 1	$7.5485e \times 10^{-13}$	$7.3655e \times 10^{-13}$	$5.5436e \times 10^{-13}$
F_E, Motor 2	$1.6127e \times 10^{-09}$	$9.4655e \times 10^{-13}$	$3.4830e \times 10^{-08}$

2.5 Conclusions

In this chapter, an algorithm for the optimal inner parameter identification of induction motors is presented [28]. In the presented technique, the parameter estimation process is converted into a multidimensional optimization problem where the inner parameters of induction motors are considered as decision variables. Under such circumstances, the complexity of the problem tends to generate multimodal error surfaces which are significantly difficult to optimize. To estimate the parameters, the presented technique uses a relatively recent method called the gravitational search algorithm (GSA). The GSA presents a better performance in multimodal problems, avoiding critical flaws as the premature convergence to sub-optimal solutions compared with the existent evolutionary algorithms.

To demonstrate the capabilities and robustness of the presented technique, the GSA estimator is experimentally evaluated regarding two different motor models. To evaluate the performance of the presented technique, it is compared against similar evolutionary methods such as differential evolution (DE), Artificial Bee Colony (ABC), and Particle Swarm Optimization (PSO). The experiments are statistically validated, demonstrating that the presented technique outperforms the other approaches in most experiments in terms of solution quality.

References

1. H. Çaliş, A. Çakir, E. Dandil, Artificial immunity-based induction motor bearing fault diagnosis. Turkish J. Electr. Eng. Comput. Sci. **21**(1), 1–25 (2013)
2. V. Prakash, S. Baskar, S. Sivakumar, K.S. Krishna, A novel efficiency improvement measure in three-phase induction motors, its conservation potential and economic analysis. Energy. Sustain. Dev. **12**(2), 78–87 (2008)
3. S.S. Waters, R.D. Willoughby, Modeling induction motors for system studies. IEEE Trans. Ind. Appl. **IA-19**(5), 875–878 (1983)
4. S. Ansuj, F. Shokooh, R. Schinzinger, Parameter estimation for induction machines based on sensitivity\nanalysis. IEEE Trans. Ind. Appl. **25**(6), 1035–1040 (1989)
5. J. De Kock, F. Van der Merwe, H. Vermeulen, Induction motor parameter estimation through an output error technique. IEEE Trans. Energy Conver. **9**(1), 69–76 (1994)
6. V. Kumar, J.K. Chhabra, D. Kumar, Automatic cluster evolution using gravitational search algorithm and its application on image segmentation. Eng. Appl. Artif. Intell. **29**, 93–103 (2014)
7. R.R. Bishop, G.G. Richards, Identifying induction machine parameters using (1990), pp. 476–479
8. H.M. Emara, W. Elshamy, A. Bahgat, Parameter identification of induction motor using modified particle swarm optimization algorithm, in *2008 IEEE International Symposium on Industrial Electronics*, no. 2 (2008), pp. 841–847
9. V.P. Sakthivel, R. Bhuvaneswari, S. Subramanian, An improved particle swarm optimization for induction motor parameter determination. Int. J. Comput. Appl. **1**(2), 71–76 (2010)
10. V.P. Sakthivel, R. Bhuvaneswari, S. Subramanian, Artificial immune system for parameter estimation of induction motor. Expert Syst. Appl. **37**(8), 6109–6115 (2010)

11. V.P. Sakthivel, R. Bhuvaneswari, S. Subramanian, An accurate and economical approach for induction motor field efficiency estimation using bacterial foraging algorithm. Meas. J. Int. Meas. Confed. **44**(4), 674–684 (2011)
12. E. Cuevas, A. González, D. Zaldívar, M. Pérez-Cisneros, An optimisation algorithm based on the behaviour of locust swarms. Int. J. Bio-Inspired Comput. **7**(6), 402–407 (2015)
13. I. Perez, M. Gomez-Gonzalez, F. Jurado, Estimation of induction motor parameters using shuffled frog-leaping algorithm. Electr. Eng. **95**(3), 267–275 (2013)
14. H.R. Mohammadi, A. Akhavan, Parameter estimation of three-phase induction motor using hybrid of genetic algorithm and particle swarm optimization. **2014** (2014)
15. A.G. Abro, J. Mohamad-Saleh, Multiple-global-best guided artificial bee colony algorithm for induction motor parameter estimation. Turkish J. Electr. Eng. Comput. Sci. **22**, 620–636 (2014)
16. E. Rashedi, H. Nezamabadi-pour, S. Saryazdi, GSA: a gravitational search algorithm. Inf. Sci. (Ny) **179**(13), 2232–2248 (2009)
17. F. Farivar, M.A. Shoorehdeli, Stability analysis of particle dynamics in gravitational search optimization algorithm. Inf. Sci. (Ny) **337–338**, 25–43 (2016)
18. S. Yazdani, H. Nezamabadi-Pour, S. Kamyab, A gravitational search algorithm for multimodal optimization. Swarm Evol. Comput. **14**, 1–14 (2014)
19. A. Yazdani, T. Jayabarathi, V. Ramesh, T. Raghunathan, Combined heat and power economic dispatch problem using firefly algorithm. Front. Energy **7**(2), 133–139 (2013)
20. W. Zhang, P. Niu, G. Li, P. Li, Forecasting of turbine heat rate with online least squares support vector machine based on gravitational search algorithm. Knowl.-Based Syst. **39**, 34–44 (2013)
21. M. Jamadi, F. Merrikh-bayat, New method for accurate parameter estimation of induction motors based on artificial bee colony algorithm
22. R.K. Ursem, P. Vadstrup, Parameter identification of induction motors using differential evolution, in *The 2003 Congress on Evolutionary Computation, 2003. CEC '03*, vol. 2 (2003), pp. 790–796
23. V.P. Sakthivel, S. Subramanian, On-site efficiency evaluation of three-phase induction motor based on particle swarm optimization. Energy **36**(3), 1713–1720 (2011)
24. S. Jurkovic, Induction motor parameters extraction. Michigan State University College Engineering, 2005
25. R. Krishnan, *Electric Motor Drives Modeling, Analysis, and Control* (Prentice Hall, 2001)
26. E. Cuevas, Block-matching algorithm based on harmony search optimization for motion estimation. Appl. Intell. **39**(1), 165–183 (2013)
27. F. Wilcoxon, in *Breakthroughs in Statistics: Methodology and Distribution*, eds. by S. Kotz, N.L. Johnson (Springer, New York, NY, 1992), pp. 196–202
28. O. Avalos, E. Cuevas, J. Gálvez, Induction motor parameter identification using a gravitational search algorithm. Computers **5**(2), 6 (2016). https://doi.org/10.3390/computers5020006

Chapter 3
An Enhanced Crow Search Algorithm Applied to Energy Approaches

3.1 Introduction

The practical use of energy has attracted attention in extensive fields of engineering areas due to its environmental effects. Induction motors and distribution networks are two emblematic problems that have substantial implications by their extensive energy demand.

Induction motors are one of most elements used in industries as electromechanical actuators due to benefits such as ruggedness, cheap maintenance, and easy operation. Nevertheless, studies report that around 2/3 of industrial energy is demanded by the use of induction motors [1, 2]. This increasing consumption has produced the need to improve their performance which depends highly on the configuration of their inner parameters. The estimation of inner parameters for induction motors represents a challenge due to their non-linearity. Under this purpose, the inner parameter identification of induction motors is considered an open research area in the engineering field. As a consequence, several techniques for inner parameter identification in induction motors have been reported in the literature [3, 4].

In contrast, distribution networks are an active study area in electrical systems. The distribution system along with generators and transmission are the three principal elements of a power system. Distribution networks are responsible for the 13% [5, 6] of losses in the produced energy. These losses of energy in distribution networks are essentially caused by a lack of reactive power in the buses. Capacitors bank allocation in distribution networks has shown to decrease the losses of energy generated by the lack of reactive power. The problem of bank allocation can be expressed as a combinatorial optimization problem in which the number of capacitors, their capacities and position can be optimized satisfying the system restrictions. Several techniques are reported to solve this optimization problem, which can be divided into four main methods; analytical [5, 7, 8], numerical [9, 10], heuristic [11–13], and based on artificial intelligence [14, 15]. A detailed study of these methods is described in [15–17].

E. Cuevas et al., *Recent Metaheuristics Algorithms for Parameter Identification*, Studies in Computational Intelligence 854, https://doi.org/10.1007/978-3-030-28917-1_3

From an optimization viewpoint, the problems of inner parameter identification in induction motors and capacitor allocation in distribution networks are considered notably difficult to perform due to their non-linearity, discontinuity, and multimodality. These qualities make difficult to solve them by using classical optimization methods.

Metaheuristic techniques motivated by nature are widely used recently to solve numerous difficult engineering problems obtaining interesting results. These techniques do not need continuity, convexity, differentiability or specific initial conditions, which signifies an advantage over classical techniques. Certainly, inner parameter identification of induction motors and the capacitor allocation express two significant problems that can be represented as optimization tasks. They have been already addressed by using metaheuristic methods. Some examples involve the gravitational search algorithm [18, 19], bacterial foraging [20, 21], crow search algorithm [14], particle swarm optimization [22, 23], genetic algorithm [24, 25], differential evolution [26–28], tabu search [29] and firefly [30].

The Crow Search Algorithm (CSA) [28] is an optimization technique that emulates the intelligent behavior of the crows. Its reported results prove its ability to solve many difficult engineering optimization problems. Some examples of CSA include its application to image processing [29] and water resources [30]. Despite its interesting results, its search strategy presents high difficulties when it addresses high multimodal problems.

In this chapter, an improved variant of the CSA technique is proposed to solve multi-modal problems such as the inner parameter identification of induction motors and capacitor allocation in distribution networks. In the proposed approach, called Improved Crow Search Algorithm (ICSA), two features from the original CSA are restructured: (I) the awareness probability (AP) and (II) the random perturbation. To balance the exploration–exploitation ratio the fixed awareness probability (AP) is substituted (I) by a dynamic awareness probability (DAP), which is modified concerning the fitness value of each element in the population. The Lévy flight is additionally incorporated to improve the search capabilities of the original CSA. With these modifications, the new approach maintains solution diversity and enhances the convergence to complex multi-modal optima. To evaluate the capabilities of the proposed approach several experiments are carried out. In the first set of problems, the algorithm is used to determine the inner parameters of two models of induction motors. For the second set, the proposed technique is tested in the 10-bus, 33-bus, and 64-bus of distribution networks. The results achieved in the experimentation are examined statistically and compared with similar techniques.

3.2 Crow Search Algorithm (CSA)

In this section, a brief description of the original Crow search algorithm is presented. CSA is a metaheuristic technique proposed by Askarzadeh [28], which is inspired on the behavior of crows. In nature, crows show intelligent practices like self-awareness, employing tools, identifying faces, prepare the flock of possibly conflicting ones, advanced communication ways and remembering the food's hidden place after a while. All these behaviors combined with the fact that the brain-body ratio of the crows is insignificantly lower than the human brain have made it acknowledged as one of the most intelligent birds in nature [31, 32].

The CSA evolutionary process mimics the behavior from crows hiding and recovering the excess of food. As a population-based algorithm, the size of the flock is formed by N individuals (crows) that are of n-dimensional where n is the problem dimension. The position $X_{i,k}$ of the crow i in a particular iteration k is expressed in Eq. (3.1) and describes a feasible solution for the problem:

$$X_{i,k} = \left[x_{i,k}^{1}, x_{i,k}^{2}, \ldots, x_{i,k}^{n}\right], \quad i = 1, 2, \ldots, N, \quad k = 1, 2, \ldots, \max Iter; \tag{3.1}$$

where *max Iter* is the maximum number of iterations. Each crow (individual) is considered to have the ability to remember the best-visited location $M_{i,k}$ to hide food from the current iteration:

$$M_{i,k} = \left[m_{i,k}^{1}, m_{i,k}^{2}, \ldots, m_{i,k}^{n}\right] \tag{3.2}$$

The location of each crow is adjusted according to two behaviors: Pursuit and evasion.

Pursuit: A crow j follows crow i to discover its hidden position. The crow i does not notice the presence of the other crow, as result the purpose of crow j is achieved.

Evasion: The crow i knows regarding the presence of crow j and to protect its food, crow i deliberately choose an arbitrary trajectory. This behavior is generated in CSA with a random movement.

The kind of behavior considered by every crow i is determined by an awareness probability (*AP*). Hence, a uniformly distributed random value *ri* between 0 and 1 is sampled. If *ri* is higher or equal than *AP*, behavior 1 is used, otherwise, situation two is taken. This process can be reviewed as follows:

$$X_{i,k+1} = \begin{cases} X_{i,k} + r_i \cdot fl_{i,k} \cdot (M_{j,k} - X_{i,k}) & r_i \geq AP \\ random & \text{otherwise} \end{cases} \tag{3.3}$$

The flight length $fl_{i,k}$ symbolizes the magnitude of movement from crow $X_{i,k}$ towards the best location $M_{j,k}$ of crow j, the r_i is a random number with uniform distribution in the range [0 1].

Once the crows are adjusted, their location is evaluated and the memory vector is updated as:

$$M_{i,k+1} = \begin{cases} F(X_{i,k+1}) & F(X_{i,k+1}) < F(M_{i,k}) \\ M_{i,k} & \text{otherwise} \end{cases} \tag{3.4}$$

where the $F(\cdot)$ describes the objective function to be minimized.

3.3 The Improved Crow Search Algorithm (ICSA)

The CSA has shown its capability to determine optimum solutions in certain search spaces configurations [12, 28, 33]. Nevertheless, its convergence is not guaranteed due to its weak exploration strategy of the search space. Under this situation, its search strategy presents high problems when it faces multi-modal problems. In the original CSA technique, two elements are principally responsible for the search process: The awareness probability (*AP*) and the random movement (evasion). The value of *AP* is responsible for providing the balance between diversification and intensification. Otherwise, the arbitrary movement affects the exploration stage by the re-initialization of candidate solutions. In the proposed ICSA technique, these two elements, the awareness probability (*AP*) and the random movement, are reformulated.

3.3.1 Dynamic Awareness Probability (DAP)

The value of *AP* is determined at the beginning of the original CSA technique and continues fixed during the evolution process. This fact is unfavorable to the diversification–intensification ratio. To improve this relation, the awareness probability (*AP*) is replaced by a dynamic awareness probability (*DAP*), such probability value is adjusted by the fitness quality of each individual. The use of a probability based on fitness values has been strongly used in evolutionary literature [31]. Hence, the dynamic awareness probability (*DAP*) is calculated as follows:

$$DAP_{i,k} = 0.9 \cdot \frac{F(X_{i,k})}{wV} + 0.1 \tag{3.5}$$

where wV describes the worst fitness value. Considering a minimization problem, this value is computed as $wV = \max\,(F(X_{j,k})$. Under this probabilistic method, likely solutions will have a great probability to be exploited. Otherwise, solutions with bad quality will have a high probability to be discarded.

3.3.2 Random Movement—Lévy Flight

The original CSA mimics two behaviors of crows: pursuit and evasion. The evasion is reproduced by an arbitrary movement that is calculated by a random value uniformly distributed.

In real life, the employment of strategies to find food is crucial for survival. A bad search strategy for good sources may be lethal for the animal. Lévy flights, proposed by Paul Lévy in 1937, is a model of a random walk which has been observed in several species as a foraging pattern [34–36]. The Lévy flight step size is regulated by a heavy-tailed probability distribution normally known as Lévy distribution. The Lévy Flights improve the exploration in search space regarding the uniform random distribution [37].

In the proposed ICSA, to produce a better diversification on the search space, Lévy flights are applied instead of uniform arbitrary movements to emulate the evasion behavior. Hence, a new position $X_{i,k+1}$ is produced adding to the actual position $X_{i,j}$ the calculated Lévy flight L.

Using the Mantegna algorithm [34] a symmetric Lévy stable distribution L is generated. Under such method, the first step is to compute the value size Zi as follows:

$$Z_i = \frac{a}{|b|^{1/\beta}} \tag{3.6}$$

where a and b are n-dimensional vectors and $\beta = 3/2$. The components of each vector a and b are sampled from a normal distribution defined as follows:

$$\begin{array}{ll} a \sim N\left(0, \sigma_a^2\right) & b \sim N\left(0, \sigma_b^2\right) \\ \sigma_a = \left\{ \frac{\Gamma(1+\beta)\sin(\pi\beta/2)}{\Gamma[(1+\beta)/2]\beta 2^{(\beta-1)/2}} \right\}^{1/\beta}, & \sigma_b = 1 \end{array} \tag{3.7}$$

where $\Gamma(\cdot)$ denotes the Gamma distribution. After calculating the value of Z_i, the factor L is computed by the following form:

$$L = 0.01 - Z_i \oplus \left(X_{i,k} - X^{best}\right), \tag{3.8}$$

where the product $\oplus$ indicates the element-wise multiplications, X^{best} symbolizes the best solution observed so far in terms of the fitness quality, then, the new location $X_{i,k+1}$ is given by:

$$X_{i,k+1} = X_{i,k} + L \tag{3.9}$$

The proposed ICSA algorithm is shown as a flowchart in Fig. 3.1.

3.4 Motor Parameter Estimation Formulation

The physical properties of an inductor motor make it difficult to obtain the inner parameter values directly. A form to deal with this limitation is estimating them through identification techniques. Two different circuit models allow a suitable configuration to determine the inner motor parameters. The explanation of this problem is described in Sect. 2.3 in detail.

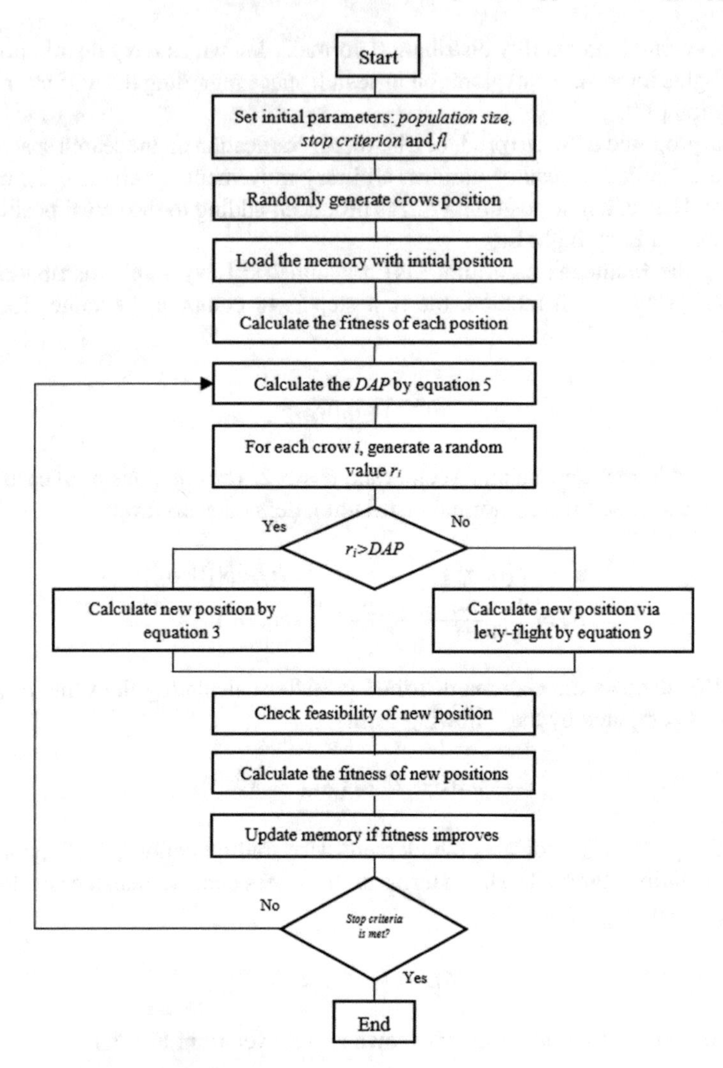

Fig. 3.1 Flowchart of ICSA technique

3.5 Capacitor Allocation Formulation

3.5.1 Load Flow Analysis

In this section, the capacitor allocation problem is explained. Hence, to know the features of voltage profile and power losses in a distribution network a load flow study is carried out. Many methods have been examined to perform the study [38, 39]. For its simplicity, the technique introduced in [40] is adopted to determine the voltages in all buses. Considering the single line diagram of a three balanced

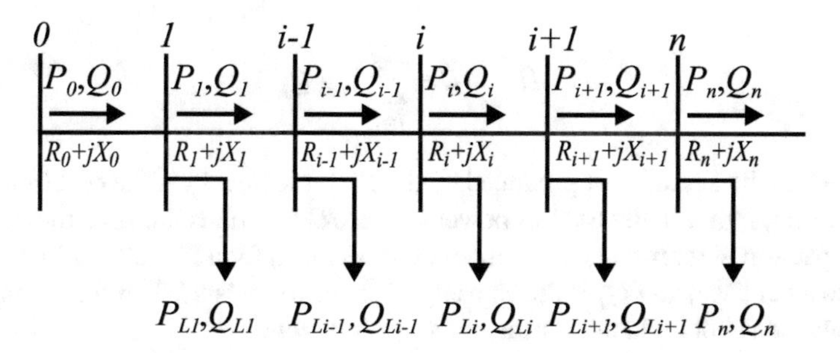

Fig. 3.2 Single radial distribution system

distribution system, as is shown in Fig. 3.2, the values of voltage and power losses are computed as:

$$|V_{i+1}| = -\left[\frac{V_i^2}{2} - R_i \cdot P_{i+1} - X_i \cdot Q_{i+1}\right] + \left[\begin{array}{c}\left(-\frac{V_i^2}{2} + R_i \cdot P_{i+1} + X_i \cdot Q_{i+1}\right) \\ -\left(R_i^2 + X_i^2\right) \cdot \left(P_i^2 + Q_i^2\right)\end{array}\right]^{1/2} \tag{3.10}$$

$$P_{Li} = \frac{R_i \cdot [P_{i+1}^2 + Q_{i+1}^2]}{\left|V_{i+1}^2\right|^2} \tag{3.11}$$

$$Q_{Li} = \frac{X_i \cdot [P_{i+1}^2 + Q_{i+1}^2]}{\left|V_{i+1}^2\right|^2} \tag{3.12}$$

$$P_{LOSS} = \sum_{i=1}^{N} P_{Li} \tag{3.13}$$

where $|V_{i+1}|$ is the voltage magnitude in the i-th + 1 node, X_i and R_i are the reactance and the resistance respectively in the branch i. Q_{i+1} and P_{i+1} are the reactive and real power load flowing through node i-th + 1, Q_{Li} and P_{Li} are the reactive and real power losses at node i, and P_{LOSS} is total loss in the network.

3.5.2 Mathematical Approach

The optimal allocation of capacitors in distribution network can be described by the solution that minimizes the annual cost caused by power losses in the complete system (Eq. 3.14), as well as the cost of capacitor allocation (Eq. 3.15):

$$Min \quad AC = \mathrm{k}_p \cdot P_{LOSS} \tag{3.14}$$

$$Min\ \ IC=\sum_{i=i}^{N} k_i^c \cdot Q_i^c \tag{3.15}$$

where AC is the annual cost produced by the power losses, k_p is the cost of losses in kW per year, P_{LOSS} is the total of power losses, IC describes the investment cost of each capacitor, N represents the number of buses taken for a capacitor allocation, K_i^c is the cost per kVar, and Q_i^c is the capacity of capacitor in bus i. The cost of capacitor maintenance is not included for the objective function.

Hence, the objective function is expressed as:

$$Min\ F=AC + IC \tag{3.16}$$

The optimization process of F is subject to specific voltage constraints given by:

$$V_{\min} \leq |V_i| \leq V_{\max} \tag{3.17}$$

where $V_{\min} = 0.90$ p.u. and $V_{\max} = 1.0$ p.u. are the lower and upper boundaries of voltages, respectively. $|V_i|$ describes the voltage magnitude in bus i. Under this approach, in the optimization process, the optimal selection of capacity, type, number, and location of capacitors must be determined.

3.5.3 Sensitivity Analysis and Loss Sensitivity Factor

The sensitivity study is a procedure used principally to reduce the search space. The main idea is to provide information about the parameters to reduce the optimization process. In the capacitor allocation problem, this study is employed to obtain the parameters with smaller variability [41]. Such information allows identifying the nodes that can be considered as candidates to allocate a capacitor. Besides, the nodes are identified with less variability corresponding to those which will have lower loss with the capacitor installation.

Assuming a single distribution line from Fig. 3.2, as is presented in Fig. 3.3, the equations of active power loss and reactive power loss

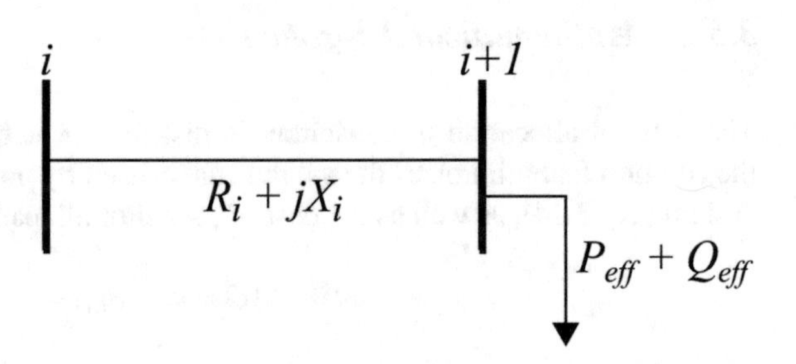

Fig. 3.3 Single distribution line

(Eqs. (3.18)–(3.19)) are rewritten as follows:

$$P_{Lineloss}(i+1) = \frac{\left(P_{eff}^2(i+1) + Q_{eff}^2(i+1)\right) \cdot R(i)}{V^2(i+1)} \tag{3.18}$$

$$Q_{Lineloss}(i+1) = \frac{\left(P_{eff}^2(i+1) + Q_{eff}^2(i+1)\right) \cdot X(i)}{V^2(i+1)} \tag{3.19}$$

where $P_{eff}(i+1)$ corresponds to the effective active power given further the node $i+1$, and $Q_{eff}(i+1)$ is the effective value of reactive power provided further the node $i+1$.

Hence, the loss sensitivity factors are computed from Eqs. (3.20) and (3.21) as follows:

$$\frac{\partial P_{Lineloss}}{\partial Q_{eff}} = \frac{2 \cdot Q_{eff}(i+1) \cdot R(i)}{V^2(i+1)} \tag{3.20}$$

$$\frac{\partial Q_{Lineloss}}{\partial Q_{eff}} = \frac{2 \cdot Q_{eff}(i+1) \cdot X(i)}{V^2(i+1)} \tag{3.21}$$

Now, the process to determine the possible candidate nodes is summarized in the following steps:

Step 1. Calculate the Loss Sensitivity Factors for all nodes.

Step 2. Sort in descending order the Loss Sensitivity Factors with its corresponding node index.

Step 3. Compute the normalized voltage magnitudes for all nodes applying:

$$norm(\mathrm{i}) = \frac{V(i)}{0.95} \tag{3.22}$$

Step 4. Establish a node as possible solution those nodes whose norm value (computed in the previous step) is lower than 1.01.

3.6 Experimental Results

To evaluate the proposed technique, a set of different experiments is carried out regarding two energy problems. The first problem is the estimation of the inner parameter of two induction motor models. The second problem is the optimal capacitor allocation over three distribution networks to reduce the power losses and improving the voltage profile. All experiments are executed on a Pentium dual-core computer with a 2.53 GHz CPU and 8-GB RAM.

3.6.1 Motor Inner Parameter Estimation

The performance of the ICSA is analyzed over two induction motors to estimate their optimal inner parameters. In the experimental process, the approximate (F_A) and exact (F_E) models are used. The manufacturer features of the motors are shown in Table 3.1.

In the study, the results of the ICSA are compared to those presented by the traditional algorithms such as the DE, ABC, and GSA. The parameters setting of the techniques has been kept as in their references [20, 27] in order to maintain compatibility and are shown in Table 3.2.

The parameter FL in the ICSA is chosen as a result of a sensitivity study that experimentally evidences the best algorithm response. Table 3.3 shows the result of the sensitivity analysis in the two energy problems.

In the comparative study, all algorithms use a population size of 25 individuals, and a maximum number of iteration is set in 3000. The parameters setting for all algorithms have been maintained concording the literature [16, 24, 38]. Moreover, the results are validated statistically using the Wilcoxon test.

The results for the approximate model (F_A) to the motor 1 and motor 2 are shown in Tables 3.4 and 3.5, respectively. For the exact model (F_E) the standard deviation and mean from all techniques for the motor 1 are resented in Table 3.6, and the results for motor 2 in Table 3.7. The reported results are based on statistical analysis after 35 independent executions for each algorithm. The results show that the proposed ICSA technique outperforms its competitors in terms of accuracy (Mean) and robustness (Std.).

Table 3.1 Manufacturer's motor data

Characteristics	Motor 1	Motor 2
Power, HP	5	40
Voltage, V	400	400
Current, A	8	45
Frequency, Hz	50	50
No. Poles	4	4
Full load split	0.07	0.09
Starting torque (T_{str})	15	260
Max. Torque (T_{max})	42	370
Stator current	22	180
Full load torque (T_{fl})	25	190

Table 3.2 Algorithms parameters

DE	ABC	GSA	ICSA
$CR = 0.5$ $F = 0.2$	$\varphi_{ni} = 0.5$ $SN = 120$	$Go = 100$ $Alpha = 20$	FL = 2.0

Table 3.3 Sensitivity analysis of FL parameter

Parameter	Analysis	Motor 2 Exact Model	Motor 2 Approximate Model	10-Bus	33-Bus	69-Bus
$FL = 1.0$	Min	7.1142×10^{-3}	1.4884×10^{-13}	696.61	139.92	145.87
	Max	1.1447×10^{-2}	2.6554×10^{-5}	699.56	146.06	164.12
	Mean	7.2753×10^{-3}	1.6165×10^{-5}	697.22	140.52	147.89
	Std	7.4248×10^{-4}	5.6961×10^{-5}	0.6749	**1.4291**	3.2862
FL = 1.5	Min	7.1142×10^{-3}	0.0000	696.61	139.42	146.08
	Max	7.2485×10^{-3}	2.6172×10^{-3}	701.03	144.55	161.03
	Mean	7.1237×10^{-3}	7.47886×10^{-5}	697.65	140.70	147.54
	Std	2.5728×10^{-5}	4.4239×10^{-4}	0.8045	1.9417	2.8354
FL = 2.0	Min	7.1142×10^{-3}	**0.0000**	**696.61**	**139.21**	**145.77**
	Max	$\mathbf{7.1142 \times 10^{-3}}$	$\mathbf{1.1675 \times 10^{-25}}$	**698.07**	**140.31**	**156.36**
	Mean	$\mathbf{7.1142 \times 10^{-3}}$	$\mathbf{3.4339 \times 10^{-27}}$	**696.76**	**139.49**	**146.20**
	Std	$\mathbf{2.2919 \times 10^{-11}}$	$\mathbf{1.9726 \times 10^{-26}}$	**0.3808**	1.9374	**2.1202**
FL = 2.5	Min	7.1142×10^{-3}	4.3700×10^{-27}	696.82	140.40	145.91
	Max	7.1170×10^{-3}	2.3971×10^{-16}	698.76	148.80	158.72
	Mean	7.1142×10^{-3}	1.7077×10^{-17}	697.89	141.21	147.49
	Std	5.3411×10^{-7}	5.0152×10^{-17}	0.5282	2.069	2.5549
FL = 3.0	Min	7.1142×10^{-3}	6.8124×10^{-21}	696.78	140.45	145.96
	Max	7.1121×10^{-3}	2.9549×10^{-13}	703.36	158.66	164.12
	Mean	7.1142×10^{-3}	1.3936×10^{-14}	697.25	141.40	147.32
	Std	1.5398×10^{-7}	5.1283×10^{-14}	1.6921	3.309	3.2862

Table 3.4 Results of approximate model F_A, for motor 1

Analysis	DE	ABC	GSA	ICSA
Mean	1.5408×10^{-4}	0.0030	5.4439×10^{-21}	1.9404×10^{-30}
Std.	7.3369×10^{-4}	0.0024	4.1473×10^{-21}	1.0674×10^{-29}
Min	1.9687×10^{-15}	2.5701×10^{-5}	3.4768×10^{-22}	1.4024×10^{-32}
Max	0.0043	0.0126	1.6715×10^{-20}	6.3192×10^{-29}

Table 3.5 Results of approximate circuit model F_A, for motor 2

Analysis	DE	ABC	GSA	ICSA
Mean	4.5700×10^{-4}	0.0078	5.3373×10^{-19}	3.4339×10^{-27}
Std.	0.0013	0.0055	3.8914×10^{-19}	1.9726×10^{-26}
Min	1.1369×10^{-13}	3.6127×10^{-4}	3.7189×10^{-20}	0.0000
Max	0.0067	0.0251	1.4020×10^{-18}	1.1675×10^{-25}

Table 3.6 Results of exact model F_E for motor 1

Analysis	DE	ABC	GSA	ICSA
Mean	0.0192	0.0231	0.0032	0.0019
Std.	0.0035	0.0103	0.0000	4.0313×10^{-16}
Min	0.0172	0.0172	0.0032	0.0019
Max	0.0288	0.0477	0.0032	0.0019

Table 3.7 Results of exact model F_E for motor 2

Analysis	DE	ABC	GSA	ICSA
Mean	0.0190	0.0791	0.0094	0.0071
Std.	0.0057	0.0572	0.0043	2.2919×10^{-11}
Min	0.0091	0.0180	0.0071	0.0071
Max	0.0305	0.2720	0.0209	0.0071

In the comparative study, the final fitness values of different algorithms are not enough to validate the new approach. A graphical representation of the convergence analysis is presented, which shows the evolution of the fitness function during the optimization process. Hence, it indicates which technique reach faster the optimal solutions. Figure 3.4 shows the convergence comparison between the techniques in a logarithmic scale for a better appreciation, being the ICSA method which offers a faster response to the global optimal.

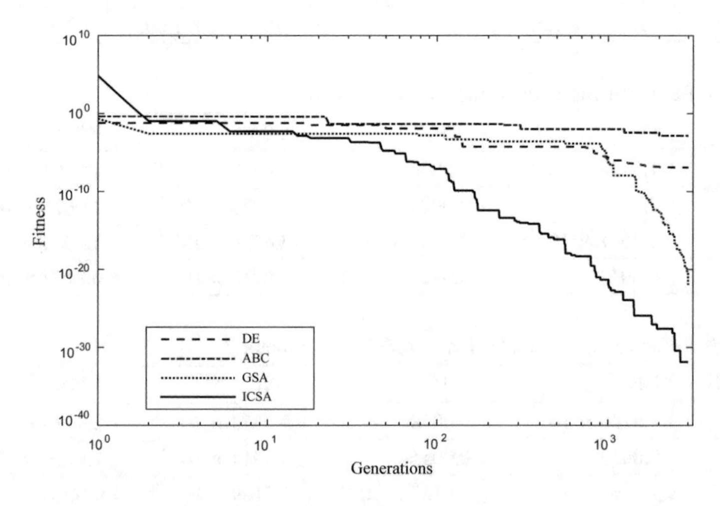

Fig. 3.4 Graphical convergence comparison for approximate model (F_A) for motor 1

3.6.2 Capacitor Allocation Test

In order to determine the performance of the ICSA method a set of three distribution networks, the 10-bus [39], 33-bus [42], and 69-bus [43] is employed in this experiment.

In the experimentation, the algorithms DE, ABC, GSA are used for comparative study. The parameters setting used are shown in Table 3.2. The number of individuals and the maximum number of iterations for all techniques is 25 and 100 respectively.

3.6.2.1 10-Bus System

For the first distribution network with a 10-bus system and nine lines. This bus, shown in Fig. 3.5, is considered as a sub-station bus. Table A1 in the Appendix A describes the system specifications of resistance and reactance for each line, as well as the real and reactive loads to each bus. The system presents a total active and reactive power load of 12,368 kW and 4186 kVAr, while the voltage supplied by the sub-station is 23 kV.

The network in uncompensated mode i.e., before the allocation of any capacitor, has a power loss of 783.77 kW, the minimum voltage is 0.8404 p.u. located on 10th bus and the maximum is 0.9930 p.u. located on 2nd bus. The price per kW lost for this case and the remainder is \$168 with a loss of 783.77 kW, while the annual cost is \$131,674. At the beginning of the methodology, the sensitivity analysis is used to identify the candidate nodes with a high probability to allocate a capacitor. In the case of the 10-bus system, the buses 6, 5, 9, 10, 8 and 7 are considered as possible candidates. Based on a set of 27 standard capacitor capacities and their corresponding annual cost per KVAr, Table 3.8 describes the capacitor investment in each candidate node obtained by ICSA. After the identification process, the corresponding capacitor capacities are 1200, 3900, 150, 600, 450 kVAr allocated in the optimal buses 6, 5, 9, 10, 8 respectively. The total power loss is 696.76 kW with an annual cost of \$117,055.68. The obtained results from the comparative study are presented in Table 3.9. Figure 3.6 shows the convergence evolution of all techniques.

CSA versus ICSA

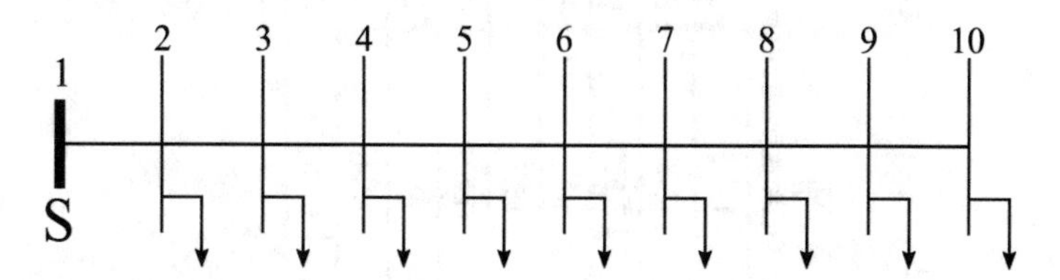

Fig. 3.5 10-bus distribution system

Table 3.8 Candidate capacitor capacity and cost ($/kVAr)

j	1	2	3	4	5	6	7	8	9
Q	150	350	450	600	750	900	1050	1200	1350
$/kVAr	0.500	0.350	0.253	0.220	0.276	0.183	0.228	0.170	0.207
j	10	11	12	13	14	15	16	17	18
Q	1500	1650	1800	1950	2100	2250	2400	2550	2700
$/kVAr	0.201	0.193	0.187	0.211	0.176	0.197	0.170	0.189	0.187
j	19	20	21	22	23	24	25	26	27
Q	2850	3000	3150	3300	3450	3600	3750	3900	4050
$/kVAr	0.183	0.180	0.195	0.174	0.188	0.170	0.183	0.182	0.179

Table 3.9 Experimental results of 10-bus system

Items algorithms	Base case	Compensated			
		DE	ABC	GSA	ICSA
Total power losses (P_{Loss}), kW	783.77	700.24	697.18	699.67	696.76
Total power losses cost, $	131,673.36	117,640.32	117,126.24	117,544.56	117,055.68
Optimal buses	–	6, 5, 9, 10	6, 5, 10, 8	6, 5, 9, 10, 7	6, 5, 9, 10, 8
Optimal capacitor size	–	900, 4050, 600, 600	1200, 4050, 600, 600	1650, 3150, 600, 450, 150	1200, 3900, 150, 600, 450
Total kVAr	–	6150	6450	6000	6300
Capacitor cost, $	–	1153.65	1192.95	1253.55	1189.8
Total annual cost	131,673.36	118,793.97	118,329.19	118,798.11	118,245.48
Net saving, $	–	12,879.38	13,344.17	12,875.24	13,427.88
% saving		9.78	10.10	9.77	10.19
Minimum voltage, p.u.	0.8375 (bus 10)	0.9005 (bus 10)	0.9001 (bus 10)	0.9002 (bus 10)	0.9000 (bus 10)
Maximum voltage, p.u.	0.9929 (bus 2)	0.9995 (bus 3)	1.0001 (bus 3)	0.9992 (bus 3)	0.9997 (bus 3)

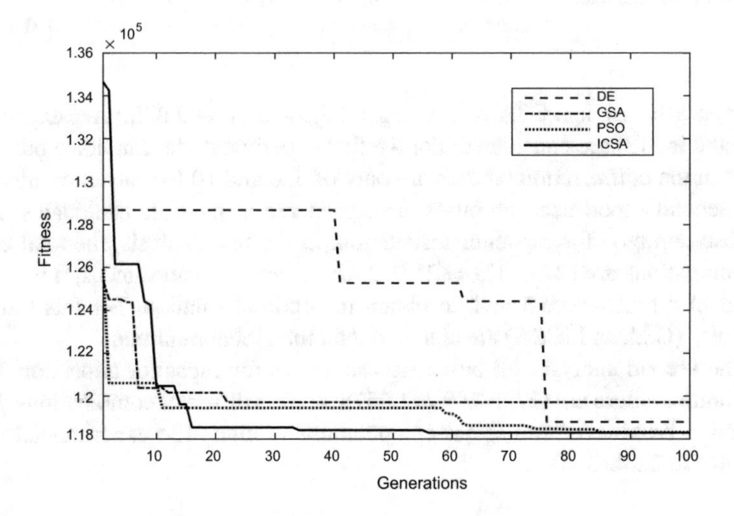

Fig. 3.6 Graphical convergence evolution 10-bus

Table 3.10 Experiments results of ICSA and CSA in 10-bus system

Items algorithms	Base case	Experiment 1		Experiment 2	
		CSA	ICSA	CSA	ICSA
Total power losses (PLoss), kW	783.77	698.14	698.14	676.02	675.78
Total power losses cost, $	131,673.36	117,287.52	117,287.52	113,571.36	113,531.04
Optimal buses	–	5, 6, 10	5, 6, 10	3, 4, 5, 6, 7, 10	3, 4, 5, 6, 8, 10
Optimal capacitor size	–	4050, 1650, 750	4050, 1650, 750	4050, 2100, 1950, 900, 450, 600	4050, 2400, 1650, 1200, 450, 450
Total kVAr	–	6150	6450	10,050	10,200
Capacitor cost, $	–	6450	6450	10,050	10,200
Total annual cost	–	118,537.92	118537.92	115,487.91	115,414.14
Net saving, $	131,673.36	13,135.43	13,135.43	16,185.44	16,259.21
% saving	–	9.9	9.9	12.29	12.35
Minimum voltage, p.u.	0.8375 (bus 10)	0.9000 (bus 10)	0.9000 (bus 10)	0.9003 (bus 10)	0.9000 (bus 10)
Maximum voltage, p.u.	0.9929 (bus 2)	1.0001 (bus 3)	1.0001 (bus 3)	1.0070 (bus 3)	1.0070 (bus 3)

To compare the original CSA version regarding the proposed ICSA, the experiments conducted in [12] are considered. For the first experiment, the candidate buses considered in the optimization process are only of 5, 6 and 10 for capacitor allocation. In the second experiment, all buses are considered as possible candidates (except the substation bus) for capacitor installation. In the first analysis, the total capacitor combinations are $(27 + 1)3 = 21{,}952$. Under such circumstances, it is possible to conduct a brute-force search to obtain the optimal solution. For this test, both techniques (CSA and ICSA) are able to obtain the global minimum.

In the second analysis, all buses are candidates for capacitor allocation. Under this condition, there are $(27 + 1)\,9 = 1.0578 \times 1013$ different combinations. In this situation, a brute-force strategy is computationally high. The experimental results are shown in Table 3.10.

3.6.2.2 33-Bus System

In this analysis, a system with 33 buses and 32 lines is considered. In the system, the first bus is assumed as a sub-station bus with a voltage of 12.66 kV. The network configuration is represented in Fig. 3.7.

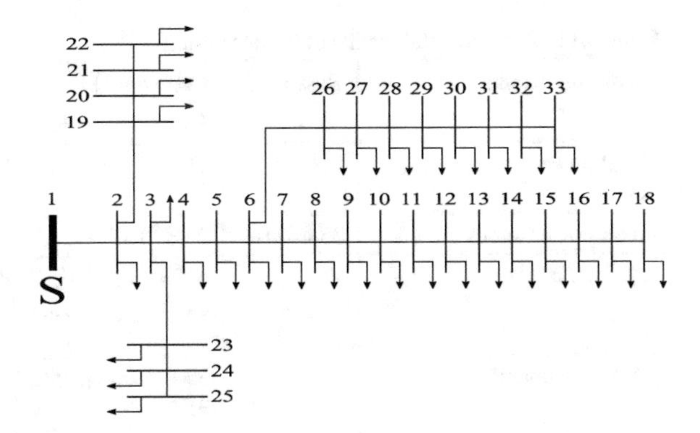

Fig. 3.7 33-Bus distribution test system

The knowledge about line resistance and reactance, as well as the corresponding load profile is presented in the Appendix A in Table A2. The 33-bus distribution network before the capacitor allocation shows a total power loss of 210.97 kW with an annual cost of $35,442.96 and a total active power of 3715 kW.

After the optimization process is carried out, the buses 6, 30, 13 are selected as optimal locations with the corresponding capacities of 450, 900, 350 kVAr respectively. The total power loss after the capacitor allocation decrease from 210.91 to 138.74 kW, saving 32.59% regarding the original cost. The results of the experiment and the comparison between the techniques are shown in Table 3.11. Figure 3.8 presents the graphical convergence evolution of all algorithms.

CSA versus ICSA

This section presents a comparison between the original crow search algorithm (CSA) and the proposed method (ICSA). The 33-bus network is analyzed as is presented in [12] where the buses 11, 24, 30 and 33 are considered as candidates solutions and the capacitor capacities and their kVar values are shown in Table 3.12.

The results achieved from both techniques are compared in Table 3.13. The table shows that the ICSA is able to obtain better results than the original version CSA.

3.6.2.3 69-Bus System

In the third experiment, a network of 68 buses and 69 lines is examined. Before to allocate any capacitor, the network shows a total active power loss of 225 kW, a total active and reactive power load of 3801.89 kW and 2693.60 kVAr. The annual cost for the 225 kW power loss is $37,800.00. The system offers a minimum voltage of 0.9091 p.u. at the 64th bus and a maximum 0.9999 p.u. at 2nd bus. As in the 10 and 33 bus test, the possible capacitor capacities and the cost per kVAr are given in Table 3.8. The network diagram is presented in Fig. 3.9 and the line and load data are shown in Table A3 in the Appendix.

Applying the ICSA method, the optimal buses selected are the 57, 61 and 18 with the capacitor capacities of 150, 1200, 350 kVAr respectively. With this reactance

Table 3.11 Experimental results of 33-bus system

Items algorithms	Base case	Compensated			
		DE	ABC	GSA	ICSA
Total power losses (P_{Loss}), kW	210.97	152.92	141.13	140.27	139.49
Total power losses cost, $	35,442.96	25,690.56	23,740.08	23,565.60	23,434.32
Optimal buses	–	6, 29, 30, 14	6, 29, 8, 13, 27, 31, 14	30, 26, 15	30, 7, 12, 15
Optimal capacitor size	–	350, 750, 350, 750	150, 150, 150, 150, 600, 450, 150	900, 450, 350	900, 600, 150, 150
Total kVAr	–	2200	1800	1700	1800
Capacitor cost, $	–	659	620.85	401.05	446.70
Total annual cost	35,442.96	26,349.56	25,540.00	23,966.65	23,881.02
Net saving, $	–	9093.39	9902.95	11,476.31	11,561.94
% saving	–	25.65	27.94	32.37	32.62
Minimum voltage, p.u.	0.9037 (bus 18)	0.9518 (bus 18)	0.9339 (bus 18)	0.9348 (bus 18)	0.9339 (bus 18)
Maximum voltage, p.u.	0.9970 (bus 2)	0.9977 (bus 2)	0.9976 (bus 2)	0.9975 (bus 2)	0.9976 (bus 2)

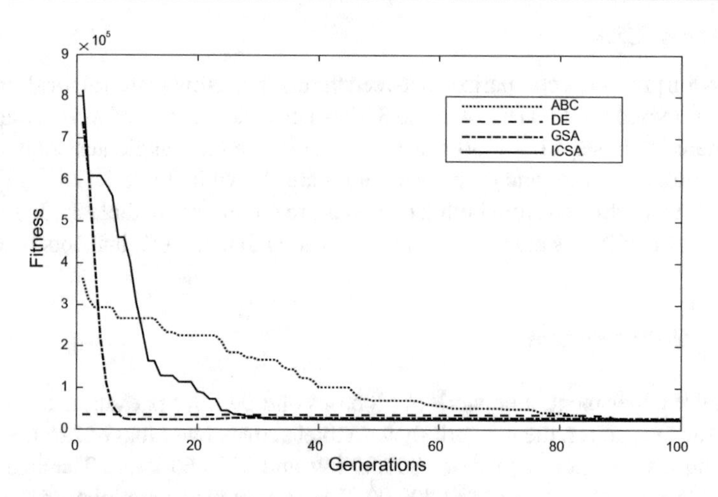

Fig. 3.8 Graphical convergence evolution 33-bus

Table 3.12 Possible capacitor capacities and cost ($/kVAr)

j	1	2	3	4	5	6
Q	150	300	450	600	750	900
$/kVAr	0.500	0.350	0.253	0.220	0.276	0.183

Table 3.13 Experiment result of ICSA over CSA in 33-bus system

Items algorithms	Base case	Compensated	
		CSA	ICSA
Total power losses (P_{Loss}), kW	210.97	139.30	138.14
Total power losses cost, $	35,442.96	23,402.40	23,207.52
Optimal buses	–	11, 24, 30, 33	11, 24, 30, 33
Optimal capacitor size	–	600, 450, 600, 300	450, 450, 900, 150
Total kVAr	–	1950	1950
Capacitor cost, $	–	482.85	467.40
Total annual cost	35,442.96	23,885.25	23,674.92
Net saving, $	–	11,557.71	11,768.04
% saving	–	32.60	33.20
Minimum voltage, p.u.	0.9037 (bus 18)	0.9336 (bus 18)	0.9302 (bus 18)
Maximum voltage, p.u.	0.9970 (bus 2)	0.9976 (bus 2)	0.9976 (bus 2)

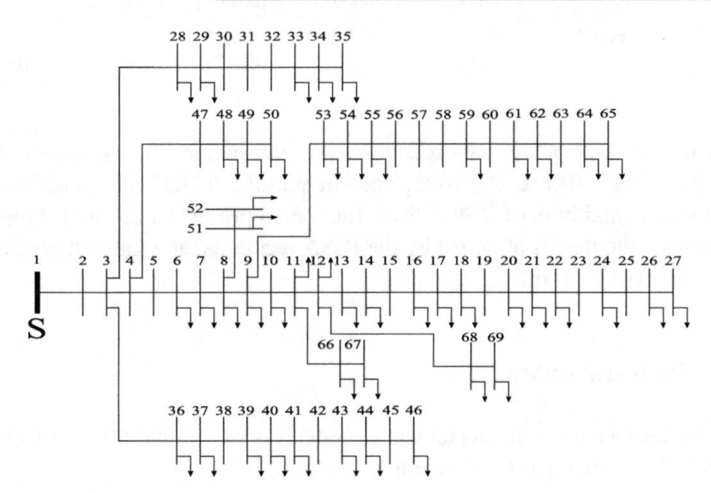

Fig. 3.9 69-Bus distribution test system

Table 3.14 Experiment result of 69-bus system

Items algorithms	Base case	Compensated			
		DE	ABC	GSA	ICSA
Total power losses (P_{Loss}), kW	225	210.02	149.36	147.1017	146.20
Total power losses cost, $	37,800.00	35,283.36	25,092.48	24,712.80	24,561.60
Optimal buses	–	57, 58, 64, 21, 63, 20, 62, 26	58, 59, 62, 24	61, 27, 22	57, 61, 18
Optimal capacitor size	–	750, 600, 900, 350, 150, 150, 350, 150	150, 150, 900, 150	1200, 150, 150	150, 1200, 350
Total kVAr	–	3400	1350	1500	1700
Capacitor cost, $	–	973.70	389.70	354.00	401.50
Total annual cost	37,800.00	36257.06	25,482.18	25,066.8	24,961.10
Net saving, $	–	1542.94	12,317.82	12,733.20	12,836.90
% saving	–	4.08	32.58	33.69	33.96
Minimum voltage, p.u.	0.9091 (bus 65)	0.9504 (bus 61)	0.9287 (bus 65)	0.9298 (bus 65)	0.9313 (bus 65)
Maximum voltage, p.u.	0.9999 (bus 2)	0.9999 (bus 2)	0.9999 (bus 2)	0.9999 (bus 2)	0.9999 (bus 2)

adjustment, the total power loss reduces from 225 to 146.20 kW, saving 33.96% regarding the original cost. The voltage profile presents a minimum of 0.9313 p.u. at bus 65th and a maximum of 0.9999 2nd. Table 3.14 presents a detailed comparative study between the results obtained by the ICSA technique and the results of the DE, ABC and GSA algorithms.

3.6.2.4 Statistical Analysis

In order to verify the results, a statistical analysis between the different techniques is conducted and the results are shown in Table 3.15.

Table 3.15 Wilcoxon statistical analysis

Wilcoxon	DE-ICSA	ABC-ICSA	GSA-ICSA
10-Bus	2.5576×10^{-34}	2.4788×10^{-34}	2.5566×10^{-34}
33-Bus	6.1019×10^{-34}	3.4570×10^{-32}	7.6490×10^{-24}
69-Bus	1.0853×10^{-29}	8.6857×10^{-28}	3.6802×10^{-20}

3.7 Conclusions

In this chapter, an improved variant of the Crow Search Algorithm (CSA) is introduced to solve difficult multi-modal optimization problems of energy: Inner parameters identification of induction motors and capacitor allocation in distribution networks. In the proposed approach, two features from the original CSA are modified: (I) the awareness probability (AP) and (II) the arbitrary perturbation. With the purpose to improve the exploration–exploitation ratio the fixed awareness probability (AP) value is replaced (I) by a dynamic awareness probability (DAP), which is modified according to the fitness value of each individual. The Lévy flight movement is also included to improve the search capabilities regarding the original arbitrary perturbation (II) of CSA. With such modifications, the new method maintains solution diversity and enhances the convergence to complex multi-modal optima.

To evaluate its performance, the introduced method is compared with other popular search techniques such as the DE, ABC, and GSA. The results show the high performance of the introduced ICSA in terms of accuracy and robustness.

References

1. V. Prakash, S. Baskar, S. Sivakumar, K.S. Krishna, A novel efficiency improvement measure in three-phase induction motors, its conservation potential and economic analysis. Energy. Sustain. Dev. **12**(2), 78–87 (2008)
2. I. Perez, M. Gomez-Gonzalez, F. Jurado, Estimation of induction motor parameters using shuffled frog-leaping algorithm. Electr. Eng. **95**(3), 267–275 (2013)
3. V.P. Sakthivel, R. Bhuvaneswari, S. Subramanian, Artificial immune system for parameter estimation of induction motor. Expert Syst. Appl. **37**(8), 6109–6115 (2010)
4. S. Lee, J. Grainger, Optimum placement of fixed and switched capacitors on primary distribution feeders. IEEE Trans. Power Appar. Syst. **PAS-100**(1), 345–352 (1981)
5. J. Grainger, S. Lee, Optimum size and location of shunt capacitors for reduction of losses on distribution feeders. IEEE Trans. Power Appar. Syst. **PAS-100**(3), 1105–1118 (1981)
6. N.M. Neagle, D.R. Samson, Loss reduction from capacitors installed on primary feeders. Trans. Am. Inst. Electr. Eng. Part III Power Appar. Syst. **75**(3) (1956)
7. H. Dura, Optimum number, location, and size of shunt capacitors in radial distribution feeders a dynamic programming approach. IEEE Trans. Power Appar. Syst. **PAS-87**(9), 1769–1774 (1968)
8. P.M. Hogan, J.D. Rettkowski, J.L Bala, Optimal capacitor placement using branch and bound. in *Proceedings of the* 37*th Annual North American Power Symposium*, pp. 84–89 (2005)
9. S.F. Mekhamer, M.E. El-Hawary, S.A. Soliman, M.A. Moustafa, M.M. Mansour, New heuristic strategies for reactive power compensation of radial distribution feeders. IEEE Trans. Power Deliv. **17**(4), 1128–1135 (2002)
10. I.C. da Silva, S. Carneiro, E.J. de Oliveira, J. de Souza Costa, J.L.R. Pereira, P.A.N. Garcia, A heuristic constructive algorithm for capacitor placement on distribution systems. IEEE Trans. Power Syst. **23**(4), 1619–1626 (2008)
11. M. Chis, M.M.A. Salama, S. Jayaram, Capacitor placement in distribution systems using heuristic search strategies. IEE Proc. Gener. Transm. Distrib. **144**(3), 225 (1997)
12. A. Askarzadeh, Capacitor placement in distribution systems for power loss reduction and voltage improvement: a new methodology. IET Gener. Transm. Distrib. **10**(14), 3631–3638 (2016)

13. H.N. Ng, M.M.A. Salama, A.Y. Chikhani, Classification of capacitor allocation techniques. IEEE Trans. Power Deliv. **15**(1), 387–392 (2000)
14. A.R. Jordehi, Optimisation of electric distribution systems: a review. Renew. Sustain. Energy Rev. **51**, 1088–1100 (2015)
15. M.M. Aman, G.B. Jasmon, A.H.A. Bakar, H. Mokhlis, M. Karimi, Optimum shunt capacitor placement in distribution system—A review and comparative study. Renew. Sustain. Energy Rev. **30**, 429–439 (2014)
16. O. Avalos, E. Cuevas, J. Gálvez, Induction motor parameter identification using a gravitational search algorithm. Computers **5**(2), 6 (2016)
17. Y.M. Shuaib, M.S. Kalavathi, C.C.A. Rajan, Optimal capacitor placement in radial distribution system using gravitational search algorithm. Int. J. Electr. Power Energy Syst. **64**, 384–397 (2015)
18. V.P. Sakthivel, R. Bhuvaneswari, S. Subramanian, An accurate and economical approach for induction motor field efficiency estimation using bacterial foraging algorithm. Meas. J. Int. Meas. Confed. **44**(4), 674–684 (2011)
19. K.R. Devabalaji, K. Ravi, D.P. Kothari, Optimal location and sizing of capacitor placement in radial distribution system using bacterial foraging optimization algorithm. Int. J. Electr. Power Energy Syst. **71**, 383–390 (2015)
20. C. Picardi, N. Rogano, Parameter identification of induction motor based on particle swarm optimization. in *International Symposium on Power Electronics, Electrical Drives, Automation and Motion, SPEEDAM 2006*, pp. 968–973 (2006)
21. S.P. Singh, A.R. Rao, Optimal allocation of capacitors in distribution systems using particle swarm optimization. Int. J. Electr. Power Energy Syst. **43**(1), 1267–1275 (2012)
22. K. Prakash, M. Sydulu, Particle swarm optimization based capacitor placement on radial distribution systems. in *2007 IEEE Power Engineering Society General Meeting* (2007), pp. 1–5
23. K. Swarup, Genetic algorithm for optimal capacitor allocation in radial distribution systems. Proc. Sixth WSEAS Int. Conf. Evol. Comput. (2005)
24. R. Ursem, P. Vadstrup, Parameter identification of induction motors using differential evolution. Comput. 2003. CEC'03. 2003 … (2003)
25. C. Su, C. Lee, Modified differential evolution method for capacitor placement of distribution systems. Conf. Exhib. 2002 Asia Pacific. … (2002)
26. J. Chiou, C. Chang, C. Su, Capacitor placement in large-scale distribution systems using variable scaling hybrid differential evolution. Int. J. Electr. Power. (2006)
27. Y.-C. Huang, H.-T. Yang, C.-L. Huang, Solving the capacitor placement problem in a radial distribution system using Tabu search approach. IEEE Trans. Power Syst. **11**(4), 1868–1873 (1996)
28. A. Askarzadeh, A novel metaheuristic method for solving constrained engineering optimization problems: crow search algorithm. Comput. Struct. **169**, 1–12 (2016)
29. D. Oliva, S. Hinojosa, E. Cuevas, G. Pajares, O. Avalos, J. Gálvez, Cross entropy based thresholding for magnetic resonance brain images using crow search algorithm. Expert Syst. Appl. **79**, 164–180 (2017)
30. D. Liu, C. Liu, Q. Fu, T. Li, K. Imran, S. Cui, F. Abrar, ELM evaluation model of regional groundwater quality based on the crow search algorithm. Ecol. Indic. (2017)
31. D. Karaboga, B. Basturk, A powerful and efficient algorithm for numerical function optimization: artificial bee colony (ABC) algorithm. J. Glob. Optim. **39**(3), 459–471 (2007)
32. A. Baronchelli, F. Radicchi, Lévy flights in human behavior and cognition. Chaos Solitons Fractals (2013)
33. S. Rajput, M. Parashar, H.M. Dubey, M. Pandit, Optimization of benchmark functions and practical problems using crow search algorithm. in *2016 Fifth International Conference on Eco-friendly Computing and Communication Systems (ICECCS)*, (2016), pp. 73–78
34. X.-S. Yang, *Nature-Inspired Metaheuristic Algorithms*. Luniver Press (2010)
35. B. Venkatesh, R. Ranjan, Data structure for radial distribution system load flow analysis. IEE Proc. Gener. Transm. (2003)

36. T. Chen, M. Chen, K. Hwang, Distribution system power flow analysis-a rigid approach. IEEE Trans. (1991)
37. X.-S. Yang, T.O. Ting, M. Karamanoglu, *Random Walks, Lévy Flights, Markov Chains and Metaheuristic Optimization* (Springer, Netherlands, 2013), pp. 1055–1064
38. M. Jamadi, F. Merrikh-bayat, New method for accurate parameter estimation of induction motors based on artificial bee colony algorithm
39. J.J. Grainger, S.H. Lee, Capacity release by Shunt capacitor placement on distribution feeders: a new voltage-dependent model. IEEE Trans. Power Appar. Syst. **PAS-101**(5), 1236–1244 (1982)
40. S. Ghosh, D. Das, Method for load-flow solution of radial distribution networks. Gener, Transm. Distrib. (1999)
41. R.S. Rao, S.V.L. Narasimham, M. Ramalingaraju, Optimal capacitor placement in a radial distribution system using plant growth simulation algorithm. Int. J. Electr. Power Energy Syst. **33**(5), 1133–1139 (2011)
42. M.E. Baran, F.F. Wu, Network reconfiguration in distribution systems for loss reduction and load balancing. IEEE Trans. Power Deliv. **4**(2), 1401–1407 (1989)
43. M. Baran, F.F. Wu, Optimal sizing of capacitors placed on a radial distribution system. IEEE Trans. Power Deliv. **4**(1), 735–743 (1989)

Chapter 4
Comparison of Solar Cells Parameters Estimation Using Several Optimization Algorithms

The increasing use of renewable energy due to the excessive use of fossil fuels causing high quantities pollution has caused the growth of research fields in this area. One of the most important is the use of solar cells because of their unlimited source of power. The performance of a solar cell directly depends on its design parameters, so that, the solar cells parameter estimation is a complex task due to its non-linearity and high multimodality. Optimization techniques are widely used to solve complex problems efficiently. On the other hand, such techniques are developed under certain conditions for some specific applications and their performance is weakly tested in real life problems. Some optimization techniques have been used to determine the parameters of solar cells. However, the inappropriate balance between exploration and exploitation is a limitation that generates sub-optimal solutions. In this chapter a comparison between several optimization techniques is carried out considering different operating conditions in tree models; one diode, two diode, and three diode models, all results are statistically validated.

4.1 Introduction

The growing demand and the shortage of fossil fuels [1], air pollution, global warming and several environmental effects have forced that scientists search for alternative energy sources. Solar energy is one of the most promising renewable energy due its unlimited source of power. Nowadays, the solar photovoltaic energy has attacked attention around the world in many areas of engineering, increasing their use through the years [2], principally for the free-emission electrical power generation [3], and easy maintenance, also being a good options in isolated areas.

The solar cells modeling is a complex task due the non-linearity of the current versus voltage curve $(I - V)$ and the dimensionality of the problem, moreover, the solar cells are susceptible to different operation conditions such as the temperature of the cell, the partial shaded conditions [4], just for mention a few. The most efficient method for their modeling is with an electrical equivalent model, in the literature,

E. Cuevas et al., *Recent Metaheuristics Algorithms for Parameter Identification*, Studies in Computational Intelligence 854, https://doi.org/10.1007/978-3-030-28917-1_4

there are two models reported: the single and the double diode model. The single diode model is the most used having just five design parameters [5], while the double diode model has seven design parameters [6], hindering its modeling. A third model is considered in this study, the three diode model which was recently proposed [7], this model has ten different design parameters.

Some methods have been reported in literature for the modeling of solar cells, such as the solar cell parameters extraction based on Newton's method [8], a method for the parameters estimation analytically using the Lambert function [9], and some other methods that use analytical approaches to determinate the parameters for the solar cell modeling [10, 11]. These methods can model the solar cells with a good precision, but their implementation become tedious and inefficient, for this, alternative techniques have been proposed to solve this problem competitively. Traditional optimization methods for the solar cells modeling such as gradient-based [12], have been used for parameters extraction in some applications, although, these techniques are susceptible to find sub-optimal solutions due their limitations in its search strategy and the error surface generated by the complex applications.

Evolutionary computation techniques are optimizations approaches that solve difficult engineering problems due their operators that generate a more efficient search strategy, which normally avoid the local minima, finding optimal solutions. Several evolutionary computation techniques and their variations have been proposed for the solar cell parameters extraction as Genetic Algorithms (GA) [13–15], where basically the parameter identification is focused on the single diode equivalent circuit, and the results are reported individually, the Particle Swarm Optimization (PSO) [16–18], Artificial Bee Colony (ABC) [19, 20], Differential Evolution (DE) [21–23], Harmony Search (HS) [24, 25], Cuckoo Search (CS) [26, 27], just for mention a few, are used to determinate de solar cell parameters such as single as double diode model. The parameter estimation of single and double diode model have been widely explored recently [28–30]. In the other hand the three diode model has been poorly reported [7]. The three models have been design using several techniques reported in the literature, but the comparison between them has not validated properly.

This chapter presents a comparative study of popular and recently evolutionary computation techniques for parameter estimation of single, double and three diode models solar cells such as; Artificial Bee Colony (ABC) [31], Differential Evolution (DE) [32], Harmony Search (HS) [33], Gravitational Search Algorithm (GSA) [34], Cuckoo Search (CS) [35], Differential Search Algorithm (DSA) [36], the Crow Search Algorithm (CSA) [37], and Covariant Matrix Adaptation with Evolution Strategy (CMA-ES) [38, 39], which have shown good performance in benchmark of well-known test functions, and some of them have solved competitively real world applications. Since the error surface generated by the solar cells parameter estimation tends to be highly multimodal, furthermore, the non-linearity, the dimensionality and the complexity of the application, a non-parametric statistical study is included to corroborate the accuracy of the results obtained by each algorithm.

4.2 Evolutionary Computation Techniques

The evolutionary computation techniques are useful tools that help to solve complex problems with a high accuracy. These algorithms are elaborated to solve a peculiar task with specific characteristics, for this, a single method is not able to solve all problems efficiently, to overcome with this problematic different evolutionary techniques have been developed which are tested in well-known benchmark of functions with exact solution without considering the real problems that normally tend to multimodal error surfaces and high dimensionality. The main problem when a new evolutionary technique is developed is the correct valance between the exploration and exploitation in the search strategy, for this, a study for the best technique for a specific application have to be developed, and the results obtained by these techniques must be statistical validated to corroborate the accuracy and consistency of the techniques.

4.2.1 Artificial Bee Colony (ABC)

The Artificial Bee Colony was proposed by Karaboga [31] inspired in the behavior of honeybee swarm. The ABC employs a population $\mathbf{S}^k$ $\left(\left\{\mathbf{s}_1^k, \mathbf{s}_2^k, \ldots \mathbf{s}_N^k\right\}\right)$ of N food sources randomly distributed from an initial point $k = 0$ to a total number of iterations $k = iterations$. Where each food source $\mathbf{s}_i^k$ $(i \in [1, \ldots, N])$ is represented as a m-dimensional vector $\left\{s_{i,1}^k, s_{i,2}^k, \ldots, s_{i,m}^k\right\}$ where m is the number of decision variables of the optimization problem. After initialization, the food source quality is evaluated for a fitness function to determinate if is a feasible solution or not, if the solution s_i^k is candidate, the operators of ABC evolved this candidate to generate a new food source v_i that is defined as follows:

$$v_i = \mathbf{s}_i^k + \phi\left(\mathbf{s}_i^k - \mathbf{s}_j^k\right), \quad i, j \in (1, 2, \ldots, N), \tag{4.1}$$

where s_j^k is a random food source that satisfy $i \neq j$ and ϕ is a random scale factor between $[-1, 1]$. The fitness function for a minimization problem assigned to a candidate solution can be defined as follow:

$$fit\left(s_i^k\right) = \begin{cases} \frac{1}{1+f\left(s_i^k\right)}, & \text{if } f\left(s_i^k\right) \geq 0, \\ 1 + \text{abs}\left(f\left(s_i^k\right)\right), & \text{if } f\left(s_i^k\right) < 0, \end{cases} \tag{4.2}$$

where $f(\cdot)$ is the fitness function to be minimized. When a new food source is computed, a greedy selection handle, if the new food source v_i is better than the actual s_i^k, the actual food source s_i^k is replaced for the new one v_i, otherwise the actual food source s_i^k reminds.

4.2.2 Differential Evolution (DE)

Storn and Price developed the Differential evolution [32], which is an stochastic vector-based evolutionary technique, that utilizes m-dimensional vectors defined as follow:

$$\mathbf{x}_i^k = \left(x_{i,1}^k, x_{i,2}^k, \ldots, x_{i,m}^k\right), \quad i = 1, 2, \ldots, N, \tag{4.3}$$

where $\mathbf{x}_i^k$ is the *i-th* vector at iteration k. The DE uses the weighted difference between two vectors to a third vector for generate a new one; this process is known as mutation and is described below:

$$\mathbf{v}_i^{k+1} = \mathbf{x}_p^k + F \cdot (\mathbf{x}_q^k - \mathbf{x}_r^k), \tag{4.4}$$

where $F(\cdot)$ is a constant that controls the amplification of differential variation between [0, 2]. In the other hand to increment the diversity in the mutated vector, a crossover is incorporated, and is represented as follows:

$$\mathbf{u}_{j,i}^{k+1} = \begin{cases} \mathbf{v}_{j,i} \text{ if } r_i \leq C_r \\ \mathbf{x}_{j,i}^k \text{ otherwise} \end{cases}, \; j = 1, 2, \ldots, d, \; i = 1, 2, \ldots, n, \tag{4.5}$$

then, the vector selection is carry out comparing the fitness values between the candidate vector against the original as follows:

$$\mathbf{x}_i^{t+1} = \begin{cases} \mathbf{u}_i^{t+1} \text{ if } f\left(\mathbf{u}_i^{t+1}\right) \leq f\left(\mathbf{x}_i^t\right), \\ \mathbf{x}_i^t \quad \text{otherwise}, \end{cases} \tag{4.6}$$

4.2.3 Harmony Search (HS)

The harmony search algorithm was submitted by Geem [33]. This evolutionary algorithm is particularity based in the harmonies of improvisation in jazz music, where the harmony memory or the population of N individuals is represented as $\mathbf{HM}^k\left(\left\{\mathbf{H}_1^k, \mathbf{H}_2^k, \ldots, \mathbf{H}_N^k\right\}\right)$. Each harmony $\mathbf{H}_i^k$ represents a m-dimensional vector of the decision variables to be treated. The HS operates by generating an initial harmony-memory which is evaluated by the fitness function, and then new a new solution is generated considering a pitch adjustment or with a random re-initialization. When a new harmony is generated, and an uniform random number between [0, 1] is generated r_1, if this number is less than the harmony-memory consideration rate (HMCR), the new harmony is generated with memory considerations, otherwise, the new harmony is generated with random values between bound limits. The generation of a new harmony is described below:

$$H_{new} = \begin{cases} H_j \in \{x_{1,j}, x_{2,j}, \ldots, x_{HMS,j}\} & \text{with probability HMCR}, \\ lower + (uper - lower) \cdot rand & \text{with probability } (1 - \text{HMCR}), \end{cases} \tag{4.7}$$

to determinate which element should be pitch rated further examinations must be considered. For this rated, a pitch-adjusting rated (PAR) is defined as the frequency of adjustment a bandwidth factor (BW) that controls the local search of selected elements, and can be described as follows:

$$H_{new} = \begin{cases} H_{new} \pm \text{rand}(0, 1) \cdot BW & \text{with probability PAR}, \\ H_{new} & \text{with probability(1 - PAR)}, \end{cases} \tag{4.8}$$

The pitch-adjust is which generate new potential feasible solutions by modifying the original harmonies, this process can be interpreted as mutation in typical evolutionary techniques.

4.2.4 Gravitational Search Algorithm (GSA)

Rashedi proposed the Gravitational Search Algorithm [34] based in the laws of gravity. This technique emulate the candidate solutions as masses, which are attacked one each other by the gravitational forces. Under this approach, the quality (mass) of an individual is determined by its fitness value. The GSA uses a population of N individuals that represent an m-dimensional vector $\mathbf{x}_i^k(\{\mathbf{x}_1, \mathbf{x}_2, \ldots \mathbf{x}_N\})$, where the dimension is the number of decision variables. A force from a mass i to a mass j in a defined time t is determined as follows:

$$F_{ij}^h(t) = G(t)\frac{Mp_i(t) \times Ma_j(t)}{R_{ij}(t) + \varepsilon}(x_j^h(t) - x_i^h(t)), \tag{4.9}$$

where Ma_j is the active gravitational mass related to individual j, Mp_i is the passive gravitational mass of individual i, $G(t)$ is the gravitational constant, ε is a constant, and R_{ij} is the Euclidian distant between i and j individuals. The valance between exploration and exploitation in this approach is made by modifying $G(t)$ through the iterations. The sum of all forces acting on individual i is expressed bellow:

$$F_i^h(t) = \sum_{j=1, j \neq i}^{N} F_{ij}^h(t), \tag{4.10}$$

the acceleration of each individual at time t is given bellow:

$$a_i^h(t) = \frac{F_i^h(t)}{Mn_i(t)}, \tag{4.11}$$

where, Mn_i is the inertia mass of individual *i,* with this, the velocity and position for each individual are computed as follows:

$$x_i^h(t+1) = x_i^h(t) + v_i^h(t+1),$$
$$v_i^h(t+1) = \text{rand}() \cdot v_i^h(t) + a_i^h(t), \tag{4.12}$$

After evaluate the fitness of each individual, their inertia and gravitational masses are updated, where a heavier individual means a better solution, fort this, GSA uses the follows equations:

$$Ma_i = Mp_i = M_{ii} = M_i, \tag{4.13}$$

$$m_i(t) = \frac{f(\mathbf{x}_i(t)) - \text{worst}(t)}{\text{best}(t) - \text{worst}(t)}, \tag{4.14}$$

$$M_i(t) = \frac{m_i(t)}{\sum_{j=1}^{N} m_j(t)}, \tag{4.15}$$

where $f(\cdot)$ is the fitness function and best(t) and worst (t) represent the best and worst solution of the complete population at time *t.*

4.2.5 Particle Swarm Optimization (PSO)

The Particle Swarm Optimization was introduced by Kennedy [40] based in the behavior of birds flocking. The PSO uses a swarm of N particles $\mathbf{P}^k(\{\mathbf{p}_1^k, \mathbf{p}_2^k, \ldots, \mathbf{p}_N^k\})$ which after being evaluated by a cost function, the best positions are storage $\mathbf{p}_{i,j}^*$. To calculate the velocity of the next generation, the velocity operator of PSO is used as follows:

$$v_{i,j}^{k+1} = \omega \cdot v_{i,j}^k + c_1 \cdot r_1 \cdot (p_{i,j}^* - p_{i,j}^k) + c_2 \cdot r_2 \cdot (g_{i,j}^* - p_{i,j}^k), \tag{4.16}$$

where ω is the inertia weighs attendant to control of the updated velocity ($i = 1, 2, .., N$), ($j = 1, 2, \ldots, p$), c_1 and c_2 are the acceleration coefficients for the movement of each individual in the position g and p^* respectively, r_1 and r_2 are random numbers between [0, 1]. To calculate the new position of the individuals is used the follow equation:

$$p_{i,j}^{k+1} = p_{i,j}^k + v_{i,j}^{k+1}, \tag{4.17}$$

when the new position is computed, it is evaluated by a cost function, if the new solution is better that the last one, then it is replace otherwise remains.

4.2.6 Cuckoo Search (CS)

The cuckoo search was proposed in 2009 by Deb and Yang [35], this technique is based in the parasite behavior of cuckoo birds using the Lévy flights [41]. The CS uses a population of $\mathbf{E}^k(\mathbf{e}_1^k, \mathbf{e}_2^k, \ldots, \mathbf{e}_N^k)$ individuals (eggs) in a determined number of generations ($N = gen$), where each individual $\mathbf{e}_i^k (i = 1, 2, \ldots, N)$ represent a m-dimensional vector $\left(\left\{\mathbf{e}_{i,1}^k, \mathbf{e}_{i,2}^k, \ldots, \mathbf{e}_{i,m}^k\right\}\right)$, and all candidate solutions are evaluated by a fitness function to determinate their quality. In order to improve the exploration in the solution space, the cuckoo search includes the Lévy flights which disturb each individual with a position c_i using a random step s_i generated with a symmetric distribution computed as follows:

$$s_i = \frac{\mathbf{u}}{|\mathbf{u}|^{1/\beta}}, \tag{4.18}$$

where $\mathbf{u}(\{u_1, u_2, \ldots, u_n\})$ and $\mathbf{v}(\{v_1, v_2, \ldots, v_n\})$ are n-dimensional vectors and $\beta = 3/2$. The elements of $\mathbf{u}$ and $\mathbf{v}$ are calculated by the follow normal distribution:

$$u \sim N\left(0, \sigma_u^2\right), \quad v \sim N\left(0, \sigma_u^2\right), \tag{4.19}$$

$$\sigma_u = \left(\frac{\Gamma(1+\beta) \cdot \sin(\pi \cdot \beta/2)}{\Gamma(1+\beta/2) \cdot \beta \cdot 2^{(\beta-1)/2}}\right), \quad \sigma_v = 1, \tag{4.20}$$

where $\Gamma(\cdot)$ is the Gamma distribution. When s_i is calculated, c_i is computed as follows:

$$c_i = 0.01 \cdot s_i \oplus \left(e_i^k - e^{best}\right), \tag{4.21}$$

where $\oplus$ is the entrywise multiplications. Ones c_i is obtained, the new candidate solution is estimated as bellow:

$$e_i^{k+1} = e_i^k + c_i, \tag{4.22}$$

The new candidate solution is evaluated by a fitness function, if this solution improves the last one, this is replaced, otherwise remains.

4.2.7 *Differential Search Algorithm (DSA)*

Differential search algorithm was introduced by Civicioglu [36], which imitate the Brownian-like random-walk movement used by an organism to migrate. In the DSA a population $\mathbf{X}^k(\mathbf{x}_1^k, \mathbf{x}_2^k, \ldots, \mathbf{x}_N^k)$ individuals (artificial organisms) are initialized randomly through the search space. After the initialization, a *stopover* vector is generated described by the Brownian-like random-walks for each element of the population, this stopover is described below:

$$\mathbf{s}_{i,N} = \mathbf{X}_{i,N} + A \cdot (\mathbf{X}_{ri,N} - \mathbf{X}_{i,N}), \tag{4.23}$$

where $ri \in [1, NP]$ is a random integer within the population range and $ri \neq i$. A is a scale factor that regulate the position changes of the individuals. For the search process, the stopover position is determined as below:

$$\mathbf{s}'_{i,j} = \begin{cases} \mathbf{s}_{i,j} & \text{if } r_{i,j} = 0, \\ \mathbf{X}_{i,j} & \text{if } r_{i,j} = 1, \end{cases} \tag{4.24}$$

where $j = [1, .., d]$ and $r_{i,j}$ can take the value of 0 or 1. After the selection of the candidate solution, each individual is evaluated by a cost function $f(\cdot)$ to determinate their quality, then a criterion of selection is used which is described as follows:

$$\mathbf{X}_i^{k+1} = \begin{cases} \mathbf{s}_i^k & \text{if } f(\mathbf{s}'_i) \leq f(\mathbf{X}_i^k), \\ \mathbf{X}_i^k & \text{if } f(\mathbf{s}'_i) > f(\mathbf{X}_i^k), \end{cases} \tag{4.25}$$

4.2.8 *Crow Search Algorithm (CSA)*

The crow search algorithm was proposed by Askarzadeh [37], based in the intelligent behavior of crows. The crow search algorithm uses a population of $\mathbf{C}^k(\{\mathbf{c}_1^k, \mathbf{c}_2^k, .., \mathbf{c}_N^k\})$ N individuals (crows), where each individual represents a m-dimensional vector $(\{\mathbf{c}_{i,1}^k, \mathbf{c}_{i,2}^k, .., \mathbf{c}_{i,m}^k\})$. The search strategy of CSA can be summarized in two steps: in the first one is when a crow is aware that is being following for another crow and the second is when is not aware. The states of each crow are determined by a probability factor AP_i^k. The new candidate solution is computed as follows:

$$\mathbf{c}_i^{k+1} = \begin{cases} \mathbf{c}_i^k + r_i \times fl \times \left(\mathbf{m}_j^k - \mathbf{c}_i^k\right) & r_j \geq AP_i^k, \\ \text{random position} & \text{otherwise}, \end{cases} \tag{4.26}$$

where r_i and r_j are random numbers between [0, 1], fl is a parameter that controls the flight length. $\mathbf{m}_j^k$ is the memory of the crow j where is stored the best solution at iteration k.

4.2.9 Covariant Matrix Adaptation with Evolution Strategy (CMA-ES)

The CMA-ES [38, 39] is an evolutionary algorithm proposed by Hansen based in the covariant matrix of the problem data. The CMA-ES uses a population of $\mathbf{X}^k(\{\mathbf{x}_1^k, \mathbf{x}_2^k, \ldots, \mathbf{x}_N^k\})$ N individuals, which are randomly initialized. In the next generation, λ individuals are selected to be updated using the following equation:

$$\mathbf{x}_N^{k+1} \sim N\left(\mathbf{x}_w^k, \sigma^{k^2} C^k\right), \tag{4.27}$$

where $N(\mu, C)$ is a normally distributed random vector with a mean μ and a covariance matrix C. The next weighted mean $\mathbf{x}_w^k$ selected as the best interval is computed as follows:

$$\mathbf{x}_w^k = \sum_{i=1}^{\mu} w_i \mathbf{x}_i^k, \tag{4.28}$$

where $\sum_{i=1}^{\mu} w_i = 1$. To carry out the modification in the parameters, the CMA-ES uses two different adaptations, on the covariance matrix C^k and on the global step size σ^k. For the covariance matrix adaptation case, an evolution path P_c^{k+1} is used, which depends from the parents separation with $\mathbf{x}_w^k$ and the recombination points $\mathbf{x}_w^{k+1}$ as is shown below:

$$p_c^{k+1} = (1 - c_c)P_c^k + H_c^{k+1}\sqrt{c_c(2 - c_c)}\frac{\sqrt{\mu_{eff}}}{\sigma^k}\left(\mathbf{x}_w^{k+1} - \mathbf{x}_w^k\right), \tag{4.29}$$

$$C_c^{k+1} = (1 - c_{\text{cov}})C^k c_{\text{cov}}\frac{1}{\mu_{\text{cov}}} p_c^{k+1}\left(p_c^{k+1}\right)^T + c_{\text{cov}}\left(1 - \frac{1}{\mu_{\text{cov}}}\right)$$
$$\cdot \sum_{i=1}^{\mu} \frac{w_i}{\sigma(k)}\left(\mathbf{x}_i^{k+1} - \mathbf{x}_i^k\right)\left(\mathbf{x}_i^{k+1} - \mathbf{x}_i^k\right), \tag{4.30}$$

$$H_\sigma^{k+1}\begin{cases} 1 \ \frac{p_\sigma^{k+1}}{1-(1-c_\sigma)^{2(k+1)}} < \left(1.5 + \frac{1}{n-0.5}\right)E(\|N(0, I)\|), \\ 0 \text{ otherwise}, \end{cases} \tag{4.31}$$

where $\mu_{eff} = \left(\sum_{i=1}^{\mu} w_i\right)^2 / \sum_{i=1}^{\mu} w_i$ is the effective variance selection and $c_{\text{cov}} \approx \min\left(1, 2\mu_{eff}/n^2\right)$ is the learning rate. For the global step size adaptation a parallel path is use to modify σ^k, this process is described below:

$$p_\sigma^{k+1} = (1 - c_\sigma)p_\sigma^k + \sqrt{c_\sigma(2 - c_\sigma)}B^k D^{k^{-1}} B^{k^T} \times \frac{\mu_{eff}}{\sigma^k} \times \left(\mathbf{x}_w^{k+1} - \mathbf{x}_w^k\right), \quad (4.32)$$

where B^k is an orthogonal matrix and D^k is a diagonal matrix. The adaptation of global step size for the next generation is given by the follow equation:

$$\sigma^{k+1} = \sigma^k \exp\left(\frac{c_\sigma}{d_\sigma}\left(\frac{\| p_\sigma^{k+1} \|}{E(\|N(0, I)\|)} - 1\right)\right), \quad (4.33)$$

here, $E(\|N(0, I)\|) = \sqrt{2}\Gamma(n + 1/2)\Big/\Gamma(n/2) \approx \sqrt{n}(1 - 1/4n + 1/21n^2)$ is the length of p_σ.

4.3 Modeling of Solar Cells

Solar cells are one of the most important and increasing clean energy sources, for this, their correct modeling has become in and important task nowadays. Some alternatives for the solar cells modeling have been proposed in literature, as the use of neural networks [42–44]. However, the most common models are the equivalent circuits models [5–7], which are descripted below.

4.3.1 *Single Diode Model (SDM)*

The single diode equivalent model is the basic and most used model for the representation of solar cells behavior. This model uses a diode connected in parallel with the source of current, which is represented in Fig. 4.1. Following the circuit theory, the total current of single diode model is computed as follows:

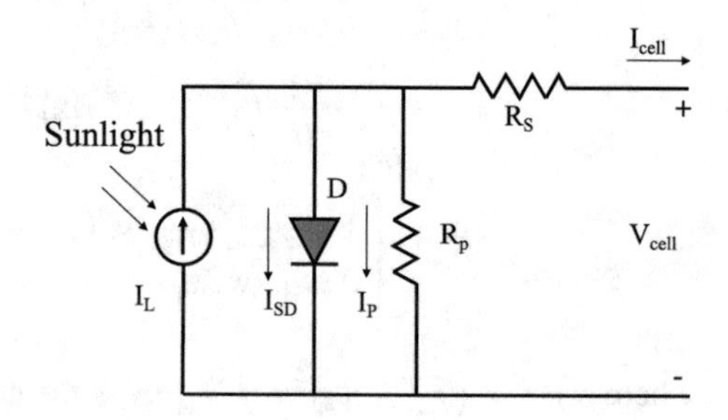

Fig. 4.1 Single diode model

$$I_{cell} = I_L - I_{SD}\left\{\exp\left[\frac{q(V_{cell} + I_{cell}R_S)}{nkT}\right] - 1\right\} - \frac{V_{cell} + I_{cell}R_S}{R_p}, \tag{4.34}$$

where k is the Boltzmann constant, q is the electron charge, I_{SD} is the diffusion current, V_{cell} is the terminal voltage, R_p and R_S are the parallel and serial resistances. For the single diode model the parameters that determinate the accuracy of the solar cell is given by five parameters; R_S, R_p, I_L, I_{SD} and n.

4.3.2 Double Diode Model (DDM)

The double diode model is another alternative to represent the solar cell characterization, where instead one diode uses two in a parallel array as is shown in Fig. 4.2. The total current of this model is computed as:

$$I_{cell} = I_L - I_{D1} - I_{D2} - I_p, \tag{4.35}$$

where the diodes and leakage currents are calculated as follows:

$$I_{D1} = I_{SD1}\left[\exp\left(\frac{q(V_{cell} + I_{cell}R_s)}{n_1kT}\right) - 1\right], \tag{4.36}$$

$$I_{D2} = I_{SD2}\left[\exp\left(\frac{q(V_{cell} + I_{cell}R_s)}{n_2kT}\right) - 1\right], \tag{4.37}$$

$$I_p = \frac{V_{cell} + I_{cell}R_s}{R_p}, \tag{4.38}$$

In the double diode model the elements that must be carefully tuned are R_S, R_P, I_L, I_{SD1}, I_{SD2}, n_1 and n_2 which determinate the performance of the system.

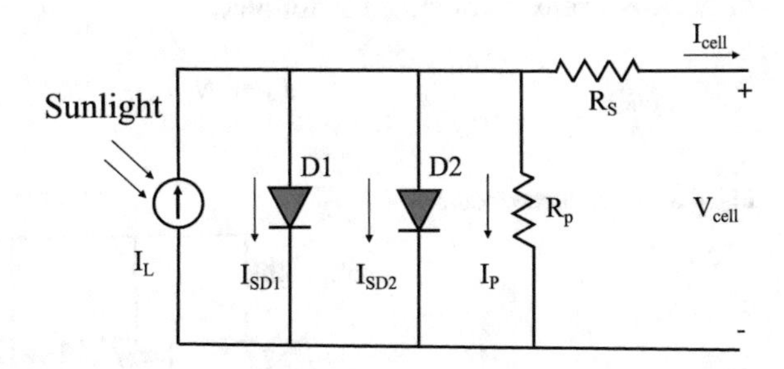

Fig. 4.2 Double diode model

4.3.3 Three Diode Model (TDM)

The three diode model is a representation of the solar cell models which include a third diode in parallel added to the two diodes considering the contribution of the current I_{D3}, where a parameter n_3 must be considered in addition to the elements of the two diodes as can be shown in Fig. 4.3. Similarly to the two diode model, the total current is calculated as:

$$I_{cell} = I_L - I_{D1} - I_{D2} - I_{D3} - I_p, \tag{4.39}$$

where

$$I_{D1} = I_{SD1}\left[\exp\left(\frac{q(V_{cell} + I_{cell}R_{so}(1 + KI))}{n_1 kT}\right) - 1\right], \tag{4.40}$$

$$I_{D2} = I_{SD2}\left[\exp\left(\frac{q(V_{cell} + I_{cell}R_{so}(1 + KI))}{n_2 kT}\right) - 1\right], \tag{4.41}$$

$$I_{D3} = I_{SD3}\left[\exp\left(\frac{q(V_{cell} + Icell R_{so}(1 + KI))}{n_3 kT}\right) - 1\right], \tag{4.42}$$

$$I_p = \frac{V_{cell} + Icell R_{so}(1 + KI)}{R_p}, \tag{4.43}$$

In the case of three diode model, the parameter R_s was replaced by $R_{so}(1 + KI)$ to find de variation of R_s with $Icell$. Where $Icell$ is the load current and K is a parameter that must be determined as the other parameters, for this, the parameters to be tuned are R_{so}, R_p, I_L, I_{D1}, I_{D2}, I_{D3}, n_1, n_2, n_3 and K.

The solar cells can be configured as modules [45, 46], which are an array of individual solar cells connected in serial and parallel. When the cells are connected in serial the voltages increases N_s times, in the case of cells connected in parallel only the current components increases in N_p times. So that, the output of a module of $N_s \times N_p$ cells is computed as follows:

$$I_m = N_p I_{cell}, \tag{4.44}$$

Fig. 4.3 Three diode model

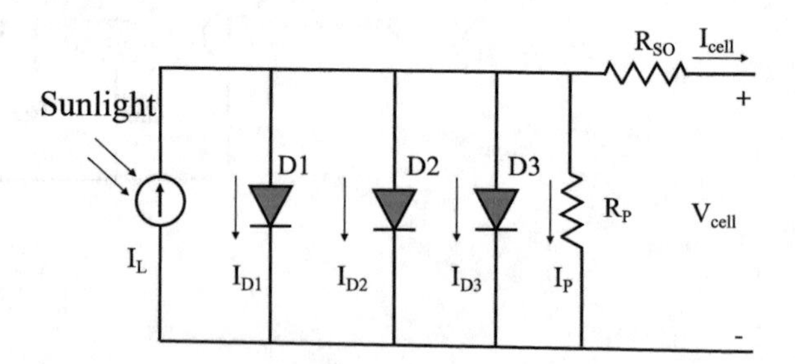

$$V_m = N_s V_{cell}, \tag{4.45}$$

$$R_{sm} = \frac{N_s}{N_p} R_s, \quad R_{pm} = \frac{N_s}{N_p} R_p, \tag{4.46}$$

4.3.4 Solar Cells Parameter Identification as an Optimization Problem

The identification of solar cells can be treated as an optimization problem, where the aim objective is the correct approximation of the $I - V$ output between the true model and the equivalent circuit model, for this, the results obtained by the optimization technique is evaluated by a cost function to determinate the quality of the solutions. To carry out the solar cell parameter identification, Eqs. (4.23), (4.24) and (4.28) are rewritten in order to reflex the difference of the experimental data as follows:

$$f_{SDM}(V_{cell}, I_{cell}, \mathbf{x}) = I_{cell} - I_L + I_D + I_p, \tag{4.47}$$

$$f_{DDM}(V_{cell}, I_{cell}, \mathbf{x}) = I_{cell} - I_L + I_{D1} + I_{D2} + I_p, \tag{4.48}$$

$$f_{TDM}(V_{cell}, I_{cell}, \mathbf{x}) = I_{cell} - I_L + I_{D1} + I_{D2} + I_{D3} + I_p, \tag{4.49}$$

For the three models $\mathbf{x}$ represents the parameters to be optimized, for the single diode model (SDM) $\mathbf{x} = [R_s, R_p, I_L, I_{SD}, n]$, in the double diode model (DDM) $\mathbf{x} = [R_s, R_p, I_L, I_{SD1}, I_{SD2}, n_1, n_2]$ and for the three diode model (TDM) $\mathbf{x} = [R_{so}, R_p, I_L, I_{SD1}, I_{SD2}, I_{SD3}, n_1, n_2, n_3, K]$. The criterion used for the quantification of the solutions is the root mean square error (RMSE) which is defined as follows:

$$RMSE = \sqrt{\frac{1}{N} \sum_{i=1}^{N} (f_i(V_{cell}, I_{cell}, \mathbf{x}))^2}, \tag{4.50}$$

During the optimization process, the parameters are adjusted to minimize the cost function until a stop criteria is reached. Since the data acquisition is in variant environmental conditions, the objective function presents noisy characteristics and multimodal error surface [47], the minimization becomes a complex task in which optimization techniques have complications to find the optimal solution [48].

4.4 Experimental Results

In order to carry out the experimental results, the C60 mono-crystalline solar cell (SUNPOWER) was used for the single and double diode models, in the case of single, double and three diode models the D6P Multi-crystalline Photovoltaic Cell (DelSolar) was adopted, due the three diode model cells limitations [49], a KC200GT Photovoltaic Solar Module (Shell Solar) was included just for the Single and Double Diode Models. The lower and upper solar cells parameters ranges for each model are shown in Table 4.1. For the Single Diode Model the number of parameters to be determined is five, in the Double Diode Model is 7 and finally for the Three Diode Model is 10, in the case of Three Diode Model 3 parameters are fixed [50]. I_{SC} is the short circuit current and R_s was replace with $R_{so}(1 + KI)$ to find the variation of R_s respect to the I. The data collection was made considering the following operation conditions: For the C60 Mono-crystalline two different sun irradiations are considered; $(1000\ \text{W/m}^2)$, $(800\ \text{W/m}^2)$, $(500\ \text{W/m}^2)$ and $(300\ \text{W/m}^2)$ at $T = 25\,^\circ\text{C}$, the RMSE average and standard deviation after 40 individual executions for the parameters obtained from each technique for the SDM and DDM are reported from Tables 4.2, 4.3, 4.4 and 4.5 respectively, where is shown the higher accuracy of the CMA-ES in all approaches such as in the minimum value of the RSME as in the Average. For the all case the number of $Search\ Agents = 50$ and the number of $Iterations = 2000$ in each execution.

The parameters setting for the experimental analysis are defined according to the references which have demonstrated through experimentation the best combination of them for solar cell parameter identification, these parameters are summarized below:

1. ABC: the parameter setting is limit = 100 using [20].
2. DE: the parameters are set to $F = 0.4$ and $C_r = 0.4$ [23].

Table 4.1 Solar cells parameters ranges for single, double and three diode model

Single diode model			Double diode model			Three diode model		
Parameter	Lower	Upper	Parameter	Lower	Upper	Parameter	Lower	Upper
$R_s(\Omega)$	0	0.5	$R_s(\Omega)$	0	0.5	$R_{so}(\Omega)$	0	1
$R_p(\Omega)$	0	200	$R_p(\Omega)$	0	100	$R_p(\Omega)$	0	100
$I_L(A)$	I_{SC}		$I_L(A)$	I_{SC}		$I_L(A)$	I_{SC}	
$I_{SD}(\mu A)$	0	1	$I_{SD1}(\mu A)$	0	1	$I_{SD1}(\mu A)$	0	1
n_1	1	2	$I_{SD1}(\mu A)$	0	1	$I_{SD2}(\mu A)$	0	1
–	–	–	n_1	0	2/3	$I_{SD3}(\mu A)$	0	1
–	–	–	n_2	0	2/3	n_1	1	
–	–	–	–	–	–	n_2	2	
–	–	–	–	–	–	n_3	0	3
–	–	–	–	–	–	K	0	1

Table 4.2 Solar cells (mono-crystalline C60) parameters estimation, mean and standard deviation for the single diode model (SDM) and double diode model (DDM) at 1000 W/m^2

SDM									
Parameters	ABC	CSA	CS	DE	DSA	GSA	HS	PSO	CMA-ES
RS (Ω)	0.005	0.006	0.001	0.004	0.0051	0.0054	0.009	0.003	0.006
Rp (Ω)	183.46	114.25	199.3	199.4	152.35	79.83	164.62	39.481	200
IL (A)	6.240	6.214	6.54	6.231	6.229	3.537	6.415	6.234	6.21
ID (A)	2.0E-06	2.7E-07	6.2E-08	3.3E-06	2.1E-06	2.4E-06	5.3E-06	3.5E-06	2.4E-07
n	1.697	1.496	1.345	1.750	1.703	1.878	1.821	1.752	1.483
Min RMSE	0.010	0.009	0.267	0.010	0.009	0.033	0.013	0.011	**0.009**
Max RMSE	0.014	0.011	0.268	0.012	0.011	0.204	0.079	0.036	**0.010**
Average RMSE	0.011	0.010	0.268	0.011	0.010	0.105	0.028	0.016	**0.009**
Std	0.0006	0.0004	0.0001	0.0003	0.0004	0.044	0.013	0.004	**0.0001**
DDM									
Parameters	ABC	CSA	CS	DE	DSA	GSA	HS	PSO	CMA-ES
RS (Ω)	0.005	0.006	0.002	0.006	0.006	4.6E-05	0.005	0.005	6.6E-03
Rp (Ω)	104.5	191.2	188.4	135.4	161.6	89.74	23.23	70.06	2.0E+02
IL (A)	6.226	6.214	6.546	6.216	6.216	3.293	6.097	6.225	6.214
ID1 (A)	2.5E-06	2.9E-09	6.4E-07	1.2E-06	2.5E-07	3.5E-06	8.3E-07	8.3E-07	2.4E-07
ID2 (A)	1.1E-06	2.8E-07	2.7E-08	1.9E-07	1.9E-07	9.9E-07	9.2E-07	5.3E-07	4.1E-16
n1	2.996	1.548	1.936	2.132	1.499	1.967	1.708	1.805	1.483
n2	1.619	1.499	1.290	1.466	1.677	1.901	1.667	1.563	1.483
Min RMSE	0.010	0.009	0.267	0.009	0.009	0.017	0.012	0.010	**0.009**

(continued)

Table 4.2 (continued)

DDM									
Parameters	ABC	CSA	CS	DE	DSA	GSA	HS	PSO	CMA-ES
Max RMSE	0.011	0.021	0.268	0.031	0.030	0.053	0.043	0.013	**0.011**
Average RMSE	0.0108	0.018	0.268	0.014	0.012	0.032	0.018	0.011	**0.009**
Std	0.0005	0.0002	**0.0001**	0.0003	0.0002	0.0083	0.006	0.0008	0.001

Table 4.3 Solar cells (mono-crystalline C60) parameters estimation, mean and standard deviation for the single diode model (SDM) and double diode model (DDM) at 800 W/m^2

SDM									
Parameters	ABC	CSA	CS	DE	DSA	GSA	HS	PSO	CMA-ES
RS (Ω)	0.0004	0.0008	2E-05	0.0006	0.0009	0.005	0.0003	0.0206	0.001
Rp (Ω)	120.7	98.41	192.0	199.9	131.5	105.7	77.0	66.39	24.893
IL (A)	4.882	4.881	5.13	4.878	4.877	3.766	4.700	4.847	4.882
ID (A)	2.1E-06	1.8E-06	2.9E-09	2.1E-06	1.4E-06	2.9E-06	3.8E-06	6.6E-06	6.8E-07
n	1.728	1.709	1.170	1.727	1.684	1.854	1.8	1.871	1.604
Min RMSE	0.004	0.004	0.490	0.004	0.004	0.017	0.008	0.005	**0.004**
Max RMSE	0.006	0.006	0.490	0.005	0.005	0.168	0.082	0.032	**0.005**
Average RMSE	0.005	0.005	0.490	0.005	0.004	0.086	0.027	0.016	**0.004**
Std	0.0002	0.0003	7.4E-05	0.0001	7.7E-05	0.033	0.016	0.007	**3.0E-05**
DDM									
Parameters	ABC	CSA	CS	DE	DSA	GSA	HS	PSO	CMA-ES
RS (Ω)	0.001	0.0017	4.0E-05	0.001	0.001	0.0001	0.0007	0.0005	0.001
Rp (Ω)	119.4	74.604	192.19	31.65	92.17	94.78	38.13	54.42	24.89
IL (A)	4.87	4.876	5.130	4.880	4.876	2.947	4.876	4.882	4.882
ID1 (A)	1.3E-06	1.7E-06	6.3E-10	1.2E-06	7.4E-07	1.1E-06	5.4E-06	1.9E-06	4.1E-20
ID2 (A)	0.004	2.6E-07	1.9E-08	6.0E-07	4.0E-07	9.9E-07	2.4E-07	3.1E-07	6.8E-07
n1	1.674	1.920	1.090	2.045	1.667	1.939	1.896	1.729	1.969
n2	51.21	1.533	1.999	1.598	1.646	1.810	1.651	1.812	1.604
Min RMSE	0.004	0.00487	0.489	0.004	0.004	0.0102	0.0066	0.004	**0.0048**

(continued)

Table 4.3 (continued)

DDM									
Parameters	ABC	CSA	CS	DE	DSA	GSA	HS	PSO	CMA-ES
Max RMSE	0.005	0.0059	0.490	0.005	0.005	0.0387	0.0229	0.007	**0.0050**
Average RMSE	0.005	0.0049	0.490	0.004	0.005	0.0235	0.0115	0.005	**0.0049**
Std	0.0001	0.0001	0.0001	3.7E-05	0.0001	0.0073	0.0038	0.000	**3.6E-05**

Table 4.4 Solar cells (mono-crystalline C60) parameters estimation, mean and standard deviation for the single diode model (SDM) and double diode model (DDM) at 500 W/m^2

SDM									
Parameters	ABC	CSA	CS	DE	DSA	GSA	HS	PSO	CMA-ES
RS (Ω)	0.001	0.003	1.1E-05	0.003	0.003	0.008	0.006	0.418	0.004
Rp (Ω)	163.4	199.9	198.2	200	130	115.9	109.51	168.01	48.9
IL (A)	3.038	3.037	3.364	3.039	3.038	1.597	3.092	3.036	3.040
ID (A)	2.3E-06	7.6E-07	4.8E-11	1.1E-06	7.4E-07	8.4E-07	7.4E-06	1.7E-06	4.9E-07
n	1.769	1.640	1.00	1.685	1.637	1.945	1.937	1.738	1.594
Min RMSE	0.004	0.003	0.775	0.004	0.003	0.006	0.005	0.004	**0.003**
Max RMSE	0.004	0.004	0.775	0.004	0.004	0.055	0.031	0.007	**0.004**
Average RMSE	0.004	0.004	0.775	0.004	0.004	0.032	0.011	0.005	**0.003**
Std	9.5E-05	0.0001	3.7E-05	8.6E-05	9.0E-05	0.011	0.005	0.0009	**1.2E-05**
DDM									
Parameters	ABC	CSA	CS	DE	DSA	GSA	HS	PSO	CMA-ES
RS (Ω)	0.004	0.0041	7.2E-05	0.003	0.0041	0.0005	6.1E-05	0.0009	0.004
Rp (Ω)	77.26	53.46	199.3	190.1	76.43	83.60	105.24	193.65	48.923
IL (A)	3.045	3.039	3.366	3.038	3.039	1.848	3.045	3.041	3.040
ID1 (A)	6.9E-07	3.5E-06	6.2E-11	8.1E-07	7.9E-10	1.8E-06	5.2E-06	3.4E-06	4.9E-07
ID2 (A)	0.0003	6.3E-08	1.0E-10	5.5E-07	5.9E-07	9.9E-07	6.8E-07	9.4E-07	2.0E-20
n1	1.630	1.995	1.639	1.950	1.743	1.985	1.889	1.915	1.594
n2	30.69	1.439	1.036	1.617	1.613	1.911	1.994	1.742	1.997
Min RMSE	0.003	0.003	0.775	0.003	0.0039	0.005	0.004	0.004	**0.003**

(continued)

Table 4.4 (continued)

DDM									
Parameters	ABC	CSA	CS	DE	DSA	GSA	HS	PSO	CMA-ES
Max RMSE	0.004	0.004	0.776	0.004	0.004	0.018	0.012	0.004	**0.004**
Average RMSE	0.004	0.004	0.775	0.004	0.004	0.011	0.006	0.004	**0.004**
Std	0.0001	9.3E-05	9.8E-05	**2.2E-05**	5.4E-05	0.003	0.001	0.0001	0.0001

Table 4.5 Solar cells (mono-crystalline C60) parameters estimation, mean and standard deviation for the single diode model (SDM) and double diode model (DDM) at 300 W/m^2

SDM									
Parameters	ABC	CSA	CS	DE	DSA	GSA	HS	PSO	CMA- ES
RS (Ω)	0.022	0.024	1.2E-07	0.021	0.019	0.005	0.010	0.015	0.037
Rp (Ω)	89.60	96.997	199.9	200	196.57	95.138	117.8	198.3	20.366
IL (A)	1.815	1.799	2.164	1.804	1.804	0.647	1.810	1.807	1.806
ID (A)	6.6E-07	1.6E-07	1.7E-11	6.0E-07	1.3E-06	2.2E-08	4.8E-06	4.2E-06	4.8E-11
n	1.680	1.537	1.00	1.669	1.764	1.978	1.935	1.919	1.025
Min RMSE	0.005	0.004	0.972	0.004	0.004	0.006	0.006	0.005	**0.003**
Max RMSE	0.006	0.005	0.972	0.005	0.005	0.026	0.014	0.006	**0.004**
Average RMSE	0.005	0.004	0.972	0.005	0.004	0.014	0.008	0.005	**0.003**
Std	0.0001	0.0002	**7.8E-08**	0.0001	0.0003	0.005	0.001	0.0003	0.0006
DDM									
Parameters	ABC	CSA	CS	DE	DSA	GSA	HS	PSO	CMA- ES
RS (Ω)	0.019	0.009	1.1E-06	0.019	0.009	0.002	0.005	0.004	0.034
Rp (Ω)	108.4	190.9	198.73	199.9	199.4	89.632	29.672	60.11	1.2E+02
IL (A)	1.808	1.801	2.164	1.806	1.802	1.641	1.796	1.796	1.8E+00
ID1 (A)	1.4E-06	5.3E-06	1.3E-09	1.1E-06	6.4E-06	2.6E-06	2.3E-06	5.2E-06	2.8E-08
ID2 (A)	9.5E-07	9.9E-07	1.77E-11	6.0E-07	9.0E-07	9.8E-07	7.7E-07	5.8E-07	1.1E-09
n1	1.767	1.999	1.941	1.744	1.999	1.975	1.849	1.968	1.850
n2	4.789	1.892	1.0004	2.260	1.999	1.813	1.966	1.929	1.1E+00
Min RMSE	0.004	0.006	0.972	0.004	0.006	0.007	0.007	0.006	**0.003**

(continued)

Table 4.5 (continued)

DDM									
Parameters	ABC	CSA	CS	DE	DSA	GSA	HS	PSO	CMA- ES
Max RMSE	0.005	0.007	0.972	0.005	0.006	0.015	0.012	0.007	**0.005**
Average RMSE	0.005	0.006	0.972	0.005	0.006	0.010	0.008	0.007	**0.004**
std	0.0002	8.9E-05	2.3E-06	0.0002	**1.2E-06**	0.001	0.0009	0.0001	0.0005

3. HS: following the setting parameters in [25] $HMRC = 0.95$ and $PAR = 0.3$.
4. GSA: the parameters used for this technique are $G_0 = 100$ and $\alpha = 20$ [34].
5. PSO: according to [51] the parameters are set to $c_1 = 0.5$, $c_2 = 2.5$, the weight factor decreases from 0.9 to 0.4.
6. CS: the parameter setting is $p_a = 0.25$ in concordance with [27].
7. DSA: the parameters setting according [36] are $p_1 = 0.3$ and $p_2 = 0.3$.
8. CSA: the parameter are set to $AP = 0.1$ and $fl = 2$.
9. CMA-ES: The technique used is the reported in [38, 39].

For the case of the Multi-crystalline solar cell D6P two different solar irradiations were considered; $\left(1000\ \mathrm{W/m^2}\right)$ and $\left(500\ \mathrm{W/m^2}\right)$ at $T = 25\,°\mathrm{C}$ for the Single Diode Module (SDM), double Diode Model (DDM) and the Three Diode Model (TDM), where the evolutionary computational techniques determined the best solar cell parameters which are shown in Tables 4.6 and 4.7 respectively with the mean and standard deviation to corroborate their accuracy. The CMA-ES improved the results obtained by the others techniques for the SDM, DDM and TDM.

In order to determine the capabilities of the techniques used, we opt to use the Multi-crystalline module KC200GT for the Single Diode Model (SDM) and the Double Diode Model (DDM) at $\left(1000\ \mathrm{W/m^2}\right)$, $\left(800\ \mathrm{W/m^2}\right)$, $\left(600\ \mathrm{W/m^2}\right)$, $\left(400\ \mathrm{W/m^2}\right)$ and $\left(200\ \mathrm{W/m^2}\right)$, where the mean, average and standard deviation are reported from Tables 4.8, 4.9, 4.10, 4.11 and 4.12 respectively. In the Multi-crystalline module the CMA-ES shows competitive results finding the minimum average in the most of the cases, but the CS and DE are able to find good solution with considerable precision.

For the validation of the results obtained, we proceed to statistically analyze the data acquired for the evolutionary computational techniques. For this, we used a non-parametric test known as Wilcoxon analysis [52, 53], which measure the difference between two related methods. This test considers the 5% of significance level of the Average RSME between the techniques, where the p-value set for the pair-wise comparison between algorithms is 0.05. After taking certain considerations among the techniques used for the comparison, we considered the follows groups for the test: CMA-ES versus ABC, CMA-ES versus CSA, CMA-ES versus CS, CMA-ES versus DE, CMA-ES versus DSA, CMA-ES versus HS and CMA-ES versus PSO, this, due the performance showed in the experimental results, where the CMA-ES carried out with a better consistency the data determined for the solar cells models. In the Wilcoxon test the null hypothesis is considered as there is not difference enough between approaches, and as an alternative hypothesis if exists significance difference between both approaches. In the other hand, if the number of elements that intervene in the test is high, the chance to error type 1 increases, for this, the significance value must be reconsidered using the Bonferroni correction [54, 55]. Once we have the new n-value, the result obtained by Wilcoxon is compared with the n-value, if is lower the null hypothesis is rejected, avoiding the error type 1. With the intention to simplify the analysis, in Table 4.13, the symbols ▲, ▶ and ▼ are used, where ▲ indicates that the technique considered as the better performs better than the rest of the algorithms, ▼ means that the technique carry out worse than the compared

Table 4.6 Solar cells (mono-crystalline D6P) parameters estimation, mean and standard deviation for the single diode model (SDM), double diode model (DDM) and three diode model (TDM) at 1000 W/m^2

SDM									
Parameters	ABC	CSA	CS	DE	DSA	GSA	HS	PSO	CMA-ES
RS (Ω)	4.9E-03	0.005	0.006	0.005	0.005	0.005	0.001	0.470	0.470
Rp (Ω)	7.5E+01	100	99.45	100	99.99	66.077	80.98	57.76	57.76
IL (A)	8.3E+00	8.302	8.299	8.300	8.294	6.897	8.233	8.214	8.214
ID (A)	1.5E-06	5.3E-07	6.5E-08	3.5E-07	1.6E-07	1E-05	7.3E-06	8.2E-06	8.2E-06
n	1.669	1.503	1.335	1.466	1.404	1.903	1.776	1.796	1.796
Min RMSE	0.023	0.023	0.015	0.019	0.019	0.034	0.034	0.026	**0.015**
Max RMSE	0.025	0.025	0.023	0.023	0.024	0.216	0.216	0.044	**0.020**
Average RMSE	0.024	0.024	0.019	0.020	0.021	0.108	0.108	0.031	**0.018**
Std	**0.0006**	0.0006	0.002	0.0007	0.001	0.047	0.047	0.003	0.0008
DDM									
Parameters	ABC	CSA	CS	DE	DSA	GSA	HS	PSO	CMA-ES
RS (Ω)	5.7E-03	0.005	0.006	0.006	0.006	0.002	0.003	0.004	0.004
Rp (Ω)	6.7E+01	100	89.09	100	99.96	35.31	55.90	79.05	79.05
IL (A)	8.2E+00	8.301	8.281	8.296	8.288	7.172	8.287	8.319	8.319
ID1 (A)	3.8E-06	3.9E-07	3.5E-08	1.5E-07	1.7E-10	3.9E-06	4.9E-06	4.8E-06	4.8E-06
ID2 (A)	2.4E-07	4.3E-07	2.2E-11	1.8E-07	4.6E-08	1E-06	9.8E-07	1.1E-08	1.1E-08
n1	5.7E+00	1.889	1.293	23.35	1.176	1.867	1.770	1.733	1.733
n2	1.438	1.486	1.389	1.412	1.313	1.621	1.712	1.848	1.848
Min RMSE	0.021	0.021	0.017	0.019	0.017	0.029	0.029	0.021	**0.017**

(continued)

Table 4.6 (continued)

DDM									
Parameters	ABC	CSA	CS	DE	DSA	GSA	HS	PSO	CMA-ES
Max RMSE	0.025	0.025	0.027	0.023	0.023	0.060	0.060	0.028	**0.021**
Average RMSE	0.023	0.023	0.024	0.021	0.020	0.039	0.039	0.025	**0.019**
Std	0.001	0.001	0.002	0.001	0.001	0.007	0.007	0.001	**0.0009**

TDM									
Parameters	ABC	CSA	CS	DE	DSA	GSA	HS	PSO	CMA-ES
ID1 (A)	1.2E-10	1.1E-10	1.2E-10	8.5E-26	1.2E-10	1.2E-10	1.2E-10	1.2E-08	1.2E-08
ID2 (A)	8.2E-11	2.1E-07	2.1E-13	3.3E-05	3.8E-21	9.2E-13	7.9E-07	2.3E-07	2.3E-07
ID3 (A)	4.9E-04	9.9E-06	1.2E-15	2.5E-19	3.1E-14	2.3E-10	8.6E-06	7.0E-06	7.0E-06
n3	8.8E+01	2.414	0.863	2.993	98.52	2.996	2.974	2.650	2.650
Rso (Ω)	8.0E-03	0.007	0.008	0.002	0.008	0.008	0.005	0.006	0.006
K	6.7E-04	0.014	9.3E-09	1.7E-16	1.9E-15	0.004	0.016	0.047	0.047
Rsh (Ω)	1.8E+00	2.128	1.819	4.077	1.818	99.99	36.85	69.68	69.68
Min RMSE	0.020	0.020	0.020	0.020	0.019	0.023	0.062	0.026	**0.019**
Max RMSE	0.020	0.020	0.020	0.020	0.037	0.033	39.09	0.055	**0.020**
Average RMSE	0.020	0.020	0.020	0.020	0.025	0.053	9.684	0.040	**0.020**
Std	1.6E-05	1.6E-05	9.3E-07	**5.6E-09**	0.006	0.0004	8.682	0.005	0.0002

(continued)

Table 4.6 (continued)

TDM									
Parameters	ABC	CSA	CS	DE	DSA	GSA	HS	PSO	CMA-ES
ID1 (A)	1.2E-10	1.1E-10	1.2E-10	8.5E-26	1.2E-10	1.2E-10	1.2E-10	1.2E-08	1.2E-08
ID2 (A)	8.2E-11	2.1E-07	2.1E-13	3.3E-05	3.8E-21	9.2E-13	7.9E-07	2.3E-07	2.3E-07
ID3 (A)	4.9E-04	9.9E-06	1.2E-15	2.5E-19	3.1E-14	2.3E-10	8.6E-06	7.0E-06	7.0E-06
n3	8.8E+01	2.414	0.863	2.993	98.52	2.996	2.974	2.650	2.650
Rso (Ω)	8.0E-03	0.007	0.008	0.002	0.008	0.008	0.005	0.006	0.006
K	6.7E-04	0.014	9.3E-09	1.7E-16	1.9E-15	0.004	0.016	0.047	0.047
Rsh (Ω)	1.8E+00	2.128	1.819	4.077	1.818	99.99	36.85	69.68	69.68
Min RMSE	0.020	0.020	0.020	0.020	0.019	0.023	0.062	0.026	**0.019**
Max RMSE	0.020	0.020	0.020	0.020	0.037	0.033	39.09	0.055	**0.020**
Average RMSE	0.020	0.020	0.020	0.020	0.025	0.053	9.684	0.040	**0.020**
Std	1.6E-05	1.6E-05	9.3E-07	**5.6E-09**	0.006	0.0004	8.682	0.005	0.0002

Table 4.7 Solar cells (mono-crystalline D6P) parameters estimation, mean and standard deviation for the single diode model (SDM), double diode model (DDM) and three diode model (TDM) at 500 W/m^2

SDM									
Parameters	ABC	CSA	CS	DE	DSA	GSA	HS	PSO	CMA-ES
RS (Ω)	0.004	0.011	2E-09	0.005	0.006	0.001	0.0008	0.002	0.007
Rp (Ω)	69.17	100	99.99	100	99.99	35.77	94.14	43.577	100
IL (A)	4.156	4.131	4.647	4.142	4.141	2.647	4.202	4.154	4.140
ID (A)	1.4E-06	2.2E-10	7.2E-11	3.1E-07	1.5E-07	7.0E-06	9.5E-06	3.2E-06	7.2E-08
n	1.586	1	1.00	1.435	1.376	1.933	1.807	1.673	1.319
Min RMSE	0.017	0.017	0.567	0.016	0.015	0.018	0.018	0.017	**0.014**
Max RMSE	0.018	0.018	0.567	**0.017**	0.017	0.076	0.076	0.018	0.017
Average RMSE	0.017	0.017	0.567	0.016	0.016	0.043	0.043	0.018	**0.016**
Std	0.0001	0.0001	**8.5E-08**	0.0002	0.0003	0.013	0.013	0.0002	0.001
DDM									
Parameters	ABC	CSA	CS	DE	DSA	GSA	HS	PSO	CMA-ES
RS (Ω)	6.5E-03	0.007	7.2E-05	0.005	0.006	0.002	0.004	0.005	0.005
Rp (Ω)	7.9E+01	100	99.91	100	61.38	49.47	82.65	82.816	82.816
IL (A)	4.1E+00	4.142	4.646	4.144	4.144	4.158	4.146	4.146	4.146
ID1 (A)	8.0E-07	1.1E-07	8.6E-11	1.5E-06	2.0E-07	4.3E-06	9.2E-07	7.1E-06	7.1E-06
ID2 (A)	1.8E-07	9.5E-08	9.0E-10	1.3E-07	9.9E-07	9.9E-07	8.0E-07	3.5E-07	3.5E-07
n1	3.6E+01	1.359	1.007	1.80	1.404	1.814	1.980	1.579	1.579
n2	1.392	1.772	1.950	1.380	1.99	1.603	1.528	1.449	1.449
Min RMSE	0.0158	0.015	0.567	0.016	0.016	0.017	0.017	0.016	**0.0151**

(continued)

Table 4.7 (continued)

DDM									
Parameters	ABC	CSA	CS	DE	DSA	GSA	HS	PSO	CMA-ES
Max RMSE	0.017	0.017	0.567	**0.017**	0.018	0.019	0.019	0.0174	0.017
Average RMSE	0.017	0.017	0.567	0.016	0.017	0.018	0.018	0.017	**0.0162**
Std	0.0004	0.0004	**2.7E-05**	0.0002	0.0003	0.0005	0.0005	0.0002	0.0004
TDM									
Parameters	ABC	CSA	CS	DE	DSA	GSA	HS	PSO	CMA-ES
ID1 (A)	2.1E-10	2.1E-10	1.5E-13	1.5E-10	6.7E-11	1.2E-10	1.6E-10	4.1E-08	4.1E-08
ID2 (A)	3.2E-15	2.1E-07	2.9E-10	6.8E-21	6.0E-19	5.4E-07	6.3E-07	4.8E-07	4.8E-07
ID3 (A)	4.8E-10	3.4E-11	7.6E-14	1.0E-15	3.1E-13	1.1E-14	9.2E-06	4.8E-06	4.8E-06
n3	9.3E+01	1.785	0.784	0.684	0.792	0.990	1.945	2.112	2.112
Rso (Ω)	1.0E-02	0.0099	0.0016	0.011	0.01	0.007	0.002	0.007	0.007
K	4.3E-11	2.4E-06	0.002	4.1E-15	0.039	0.003	0.098	0.054	0.054
Rsh (Ω)	7.1E-01	0.704	99.97	0.707	0.692	67.37	0.632	70.81	70.81
Min RMSE	0.042	0.042	0.567	0.042	0.041	0.043	0.089	0.043	**0.040**
Max RMSE	0.042	0.042	0.567	0.042	0.042	1.624	5.652	0.080	**0.041**
Average RMSE	0.042	0.042	0.567	0.042	0.041	0.289	1.757	0.059	**0.041**
Std	1.0E-07	1.0E-07	3.1E-05	**9.8E-18**	0.0001	0.0004	1.565	0.010	0.0003

Table 4.8 Solar cells module (multi-crystalline KC200GT) parameters estimation, Mean and standard deviation for the single diode model (SDM) and double diode model (DDM) at 1000 W/m^2

SDM									
Parameters	ABC	CSA	CS	DE	DSA	GSA	HS	PSO	CMA-ES
RS (Ω)	0.008	0.008	0.009	0.008	0.008	0.005	0.006	0.008	0.008
Rp (Ω)	14.27	287.7	278.3	300	274.2	162.5	145.7	70.23	500
IL (A)	8.452	8.401	8.395	8.405	8.408	5.513	8.166	8.451	8.402
ID (A)	2.1E-06	8.8E-07	1.0E-07	1.2E-06	1.2E-06	4.8E-06	9.1E-06	5.2E-06	1.2E-06
n	1.633	1.542	1.361	1.575	1.574	1.836	1.817	1.746	1.575
Min RMSE	0.006	0.005	**0.004**	0.006	0.005	0.016	0.007	0.007	0.006
Max RMSE	0.008	0.006	**0.005**	0.006	0.006	0.084	0.038	0.018	0.006
Average RMSE	0.006	0.006	**0.005**	0.006	0.005	0.051	0.016	0.011	0.006
Std	0.0005	0.0003	0.0002	0.0001	0.0003	0.016	0.007	0.002	**0.0001**
DDM									
Parameters	ABC	CSA	CS	DE	DSA	GSA	HS	PSO	CMA-ES
RS (Ω)	0.010	0.010	0.010	0.010	0.010	0.005	0.012	0.475	0.010
Rp (Ω)	136.8	133.9	264.3	299.9	299.9	169.4	245.0	141.2	197.3
IL (A)	8.374	8.371	8.371	8.371	8.370	4.126	8.015	8.074	8.375
ID1 (A)	6.4E-08	2.3E-15	7.8E-09	1.3E-08	3.2E-13	9.9E-09	4.8E-09	2.5E-09	1.1E-10
ID2 (A)	2.2E-08	8.4E-09	2.5E-09	2.6E-08	8.6E-09	9.9E-09	5.1E-09	1.7E-09	1.2E-08
n1	8.731	1.351	1.194	1.227	1.126	1.693	1.175	1.513	1.240
n2	1.257	1.199	1.875	1.747	1.200	1.890	1.514	1.110	1.222
Min RMSE	0.004	0.004	**0.004**	0.004	0.004	0.013	0.005	0.006	0.004

(continued)

Table 4.8 (continued)

DDM									
Parameters	ABC	CSA	CS	DE	DSA	GSA	HS	PSO	CMA-ES
Max RMSE	0.005	0.004	**0.004**	0.004	0.004	0.097	0.067	0.019	0.004
Average RMSE	0.004	0.004	**0.004**	0.004	0.004	0.052	0.027	0.011	0.004
Std	0.0001	2.8E-05	8.6E-07	8.9E-06	6.5E-06	0.016	0.013	0.004	**7.1E-06**

Table 4.9 Solar cells module (multi-crystalline KC200GT) parameters estimation, mean and standard deviation for the single diode model (SDM) and double diode model (DDM) at 800 W/m^2

SDM									
Parameters	ABC	CSA	CS	DE	DSA	GSA	HS	PSO	CMA-ES
RS (Ω)	0.005	7.8E-05	2.8E-05	5.7E-03	1.0E-03	0.004	1.7E-03	0.531	0.003
Rp (Ω)	156.4	499.9	472.2	500	499.9	2.8E+02	115.8	98.14	500
IL (A)	6.748	5.1E+00	5.130	6.745	5.131	1.353	5.130	2.931	5.131
ID (A)	3.5E-06	5.1E-09	3.6E-07	2.5E-06	1.8E-08	7.5E-08	4.5E-06	3.4E-06	2.6E-10
n	1.648	1.187	1.441	1.610	1.215	1.8E+00	1.715	1.713	1.00
Min RMSE	0.006	0.153	0.153	**0.005**	0.153	0.158	0.154	0.154	0.006
Max RMSE	0.007	0.154	0.153	0.006	0.153	0.213	0.161	0.160	**0.006**
Average RMSE	0.006	0.153	0.153	**0.006**	0.153	0.169	0.156	0.157	0.006
Std	0.0003	0.0004	**8.8E-06**	0.0002	3.4E-05	0.010	0.001	0.001	0.0001
DDM									
Parameters	ABC	CSA	CS	DE	DSA	GSA	HS	PSO	CMA-ES
RS (Ω)	0.006	0.002	0.004	0.007	0.003	0.003	8.3E-04	2.6E-01	0.004
Rp (Ω)	290.3	471.3	499.4	5.564	5.0E+02	2.7E+02	2.5E+02	365.1	500
IL (A)	6.729	5.130	5.131	6.775	5.130	2.8E+00	5.1E+00	3.8E+00	5.131
ID1 (A)	7.7E-07	1.4E-10	4.8E-10	4.0E-08	2.5E-10	1.9E-10	9.6E-09	6.7E-09	2.5E-10
ID2 (A)	7.3E-07	1.7E-10	2.5E-10	6.3E-08	3.0E-11	9.9E-09	4.4E-09	5.2E-09	8.8E-21
n1	1.492	1.014	1.787	1.347	1.00	1.730	1.178	1.808	1
n2	50.40	1.764	1.0E+00	1.317	1.989	1.618	1.668	1.161	1.99
Min RMSE	0.004	0.153	0.152	0.004	0.152	0.155	0.153	0.153	**0.004**

(continued)

Table 4.9 (continued)

DDM									
Parameters	ABC	CSA	CS	DE	DSA	GSA	HS	PSO	CMA-ES
Max RMSE	0.005	0.153	0.153	0.004	0.153	0.168	0.163	0.160	**0.004**
Average RMSE	0.004	0.153	0.153	0.004	0.153	0.161	0.155	0.155	**0.004**
Std	0.0001	6.0E-05	9.6E-06	1.5E-05	1.2E-05	0.002	0.002	0.001	**7.1E-06**

Table 4.10 Solar cells module (multi-crystalline KC200GT) parameters estimation, mean and standard deviation for the single diode model (SDM) and double diode model (DDM) at 600 W/m^2

SDM									
Parameters	ABC	CSA	CS	DE	DSA	GSA	HS	PSO	CMA-ES
RS (Ω)	0.005	4.6E-06	0.0003	0.004	1.9E-09	0.023	3.2E-04	0.436	9.2E-19
Rp (Ω)	103.3	494.4	4.1E+02	11.75	499.9	2.7E+02	297.4	2.4E+02	500
IL (A)	5.032	3.364	3.364	5.064	3.364	1.522	3.3E+00	2.9E+00	3.364
ID (A)	4.3E-07	5.7E-11	6.0E-08	1.2E-06	4.8E-10	1.7E-09	2.1E-06	2.7E-06	2.6E-10
n	1.438	1.001	1.320	1.535	1.029	1.926	1.682	1.7E+00	1
Min RMSE	0.006	0.240	0.240	**0.005**	0.240	0.241	0.241	0.241	0.006
Max RMSE	0.009	0.241	0.240	0.007	0.240	0.266	0.244	0.244	**0.006**
Average RMSE	0.007	0.241	0.240	0.006	0.240	0.249	0.242	0.242	**0.006**
Std	0.0006	0.0003	**8.2E-06**	0.0002	5.6E-05	0.005	0.0007	0.0006	0.0001
DDM									
Parameters	ABC	CSA	CS	DE	DSA	GSA	HS	PSO	CMA-ES
RS (Ω)	0.006	1.8E-06	7.4E-06	7.6E-03	5.3E-12	3.9E-03	1.6E-03	0.0156	6.6E-18
Rp (Ω)	6.475	493.0	4.8E+02	3.829	499.9	2.5E+02	3.6E+02	2.4E+02	500
IL (A)	5.067	3.364	3.364	5.1E+00	3.364	2.089	3.364	8.6E-01	3.364
ID1 (A)	1.1E-07	5.2E-11	3.0E-10	2.8E-10	2.6E-10	4.0E-10	6.2E-09	5.5E-09	2.6E-10
ID2 (A)	1.2E-08	5.6E-11	9.7E-09	1.5E-08	5.0E-15	4.1E-10	4.4E-09	4.1E-09	1.6E-20
n1	1.326	1.813	1.007	1.287	1	1.672	1.163	1.437	1
n2	16.91	1.00	1.772	1.195	1.998	1.583	1.612	1.138	1.986
Min RMSE	0.004	0.240	0.240	0.004	0.240	0.241	0.240	0.241	**0.004**

(continued)

Table 4.10 (continued)

DDM									
Parameters	ABC	CSA	CS	DE	DSA	GSA	HS	PSO	CMA-ES
Max RMSE	0.005	0.241	0.240	**0.004**	0.240	0.248	0.249	0.242	0.004
Average RMSE	0.004	0.240	0.240	0.004	0.240	0.243	0.241	0.241	**0.004**
Std	0.0001	0.0004	9.5E-07	6.2E-06	**2.0E-09**	0.001	0.001	0.0002	7.1E-06

Table 4.11 Solar cells module (multi-crystalline KC200GT) parameters estimation, mean and standard deviation for the single diode model (SDM) and double diode model (DDM) at 400 W/m^2

SDM									
Parameters	ABC	CSA	CS	DE	DSA	GSA	HS	PSO	CMA-ES
RS (Ω)	0.008	8.7E-09	0.0001	0.008	1.0E-09	8.8E-03	0.002	0.407	2.4E-16
Rp (Ω)	244.7	499.9	5.0E+02	15.25	499.9	226.4	364.4	341.5	5.0E+02
IL (A)	3.371	2.164	2.164	3.388	2.1E+00	1.479	2.1E+00	1.109	2.1E+00
ID (A)	2.2E-06	1.6E-11	1.0E-08	2.0E-06	4.6E-10	8.9E-09	9.9E-07	1.7E-06	2.4E-10
n	1.616	1.000	1.2E+00	1.606	1.029	1.698	1.636	1.649	1
Min RMSE	0.006	0.300	0.300	**0.005**	0.300	0.301	0.301	0.301	0.006
Max RMSE	0.011	0.302	0.300	0.007	0.300	0.324	0.303	0.304	**0.006**
Average RMSE	0.007	0.301	0.300	0.006	0.300	0.308	0.301	0.302	**0.006**
Std	0.0009	0.0003	1.3E-08	0.0003	**1.55E-15**	0.005	0.0005	0.0007	0.0001
DDM									
Parameters	ABC	CSA	CS	DE	DSA	GSA	HS	PSO	CMA-ES
RS (Ω)	0.011	2.0E-07	6.9E-05	0.012	1.9E-12	2.2E-02	1.2E-03	2.7E-01	8.7E-18
Rp (Ω)	16.81	5.0E+02	448.4	5.031	5.0E+02	2.4E+02	3.6E+02	4.3E+02	500
IL (A)	3.370	2.1E+00	2.164	3.4E+00	2.1E+00	2.1E+00	2.1E+00	9.3E-01	2.164
ID1 (A)	1.6E-07	1.6E-11	2.5E-10	4.8E-09	1.7E-15	1.0E-08	9.3E-09	6.6E-09	1.7E-21
ID2 (A)	1.3E-09	3.2E-11	7.1E-09	4.1E-08	2.4E-10	9.9E-09	7.2E-09	1.8E-09	2.4E-10
n1	1.369	1.000	1.000	2.454	1.988	1.781	1.261	1.558	1.988
n2	310.2	1.959	1.857	1.266	1	1.801	1.227	1.106	1
Min RMSE	0.004	0.300	0.300	0.004	0.300	0.301	0.300	0.300	**0.003**

(continued)

Table 4.11 (continued)

DDM

Parameters	ABC	CSA	CS	DE	DSA	GSA	HS	PSO	CMA-ES
Max RMSE	0.006	0.301	0.300	**0.004**	0.300	0.305	0.301	0.301	0.004
Average RMSE	0.005	0.300	0.300	0.004	0.300	0.302	0.301	0.301	**0.004**
Std	0.0002	3.3E-05	9.0E-08	8.7E-06	**7.9E-14**	0.0009	0.0001	0.0001	8.7E-06

Table 4.12 Solar cells module (multi-crystalline KC200GT) parameters estimation, mean and standard deviation for the single diode model (SDM) and double diode model (DDM) at 200 W/m^2

SDM									
Parameters	ABC	CSA	CS	DE	DSA	GSA	HS	PSO	CMA-ES
RS (Ω)	0.011	5.4E-08	1.6E-02	1.0E-02	0.014	0.017	0.001	1.7E-01	0.024
Rp (Ω)	147.1	499.9	167.8	201.0	339.9	2.2E+02	1.7E+02	429.5	24.31
IL (A)	1.680	2.164	1.675	1.676	1.675	1.411	1.680	1.678	1.680
ID (A)	3.7E-07	1.6E-11	7.5E-08	6.0E-07	1.6E-07	9.7E-06	6.3E-06	2.4E-06	4.5E-10
n	1.434	1.00	1.300	1.481	1.363	1.918	1.757	1.632	1
Min RMSE	0.006	0.300	0.300	**0.006**	0.300	0.301	0.300	0.301	0.006
Max RMSE	0.009	0.302	0.300	0.007	0.300	0.326	0.302	0.304	**0.006**
Average RMSE	0.007	0.301	0.300	0.006	0.300	0.308	0.301	0.302	**0.006**
Std	0.0007	0.0002	1.1E-08	0.0001	**5.3E-16**	0.005	0.0004	0.0008	0.0001
DDM									
Parameters	ABC	CSA	CS	DE	DSA	GSA	HS	PSO	CMA-ES
RS (Ω)	0.023	6.5E-08	0.024	0.022	0.024	1.6E-02	0.020	2.2E-02	2.4E-02
Rp (Ω)	257.6	499.1	23.34	51.2	3.6E+01	255.3	402.0	4.0E+02	24.31
IL (A)	1.676	2.164	1.680	1.6E+00	1.6E+00	1.6E+00	1.668	1.670	1.6E+00
ID1 (A)	3.4E-09	1.0E-11	4.5E-10	1.6E-09	6.6E-10	4.7E-11	6.8E-09	1.8E-09	1.4E-21
ID2 (A)	1.0E-09	1.6E-11	2.7E-09	8.5E-10	2.9E-10	7.8E-09	7.0E-09	2.4E-09	4.5E-10
n1	6.730	1.9E+00	1.00	1.062	1.018	1.645	1.199	1.067	1.869
n2	1.040	1.00	1.777	2.290	1.942	1.148	1.172	1.670	1
Min RMSE	0.004	0.300	0.300	0.004	0.300	0.301	0.300	0.300	**0.004**

(continued)

Table 4.12 (continued)

DDM									
Parameters	ABC	CSA	CS	DE	DSA	GSA	HS	PSO	CMA-ES
Max RMSE	0.005	0.301	0.300	**0.004**	0.300	0.306	0.301	0.301	0.004
Average RMSE	0.004	0.300	0.300	0.004	0.300	0.302	0.301	0.301	**0.004**
Std	0.0001	3.0E-05	7.5E-08	6.7E-06	**6.4E-14**	0.001	0.0001	0.0001	7.1E-06

Table 4.13 Wilcoxon test for the SDM and DDM for the mono-crystalline cell and the multi-crystalline module and SDM, DDM and TDM for the multi-crystalline cell at different irradiation conditions after the Bonferroni correction

C60 Mono-crystalline

IR		CMA-ES versus							
		ABC	CROW	CS	DE	DS	GSA	HS	PSO
1000	SDM	1.4E-14 ▲	1.4E-14 ▲	1.2E-14 ▲	1.2E-12 ▲	1.5E-13 ▲	1.3E-10 ▲	1.3E-14 ▲	1.4E-14 ▲
	DDM	1.0E-05 ▲	0.10087 ▼	1.5E-14 ▲	1.6E-12 ▲	5.2E-05 ▲	1.4E-10 ▲	1.1E-13 ▲	6.9E-08 ▲
800	SDM	1.6E-14 ▲	1.6E-14 ▲	1.3E-14 ▲	1.2E-14 ▲	1.0E-12 ▲	1.3E-14 ▲	1.5E-15 ▲	1.4E-14 ▲
	DDM	1.7E-05 ▲	0.17638 ▼	1.3E-14 ▲	1.5E-11 ▲	1.1E-07 ▲	1.4E-12 ▲	1.4E-15 ▲	7.2E-08 ▲
500	SDM	1.4E-14 ▲	1.4E-14 ▲	1.2E-14 ▲	1.2E-11 ▲	3.5E-13 ▲	1.4E-12 ▲	1.6E-16 ▲	1.4E-14 ▲
	DDM	1.2E-06 ▲	0.01286 ▲	1.2E-14 ▲	1.3E-14 ▲	1.7E-05 ▲	1.5E-11 ▲	1.4E-14 ▲	1.5E-08 ▲
300	SDM	1.4E-14 ▲	1.5E-08 ▲	1.4E-14 ▲	1.6E-12 ▲	2.2E-12 ▲	1.3E-13 ▲	1.5E-10 ▲	1.3E-14 ▲
	DDM	4.0E-09 ▲	1.4E-14 ▲	1.3E-14 ▲	1.3E-14 ▲	1.4E-14 ▲	1.2E-14 ▲	1.5E-14 ▲	1.3E-14 ▲

D6P100 Multi-crystalline

IR		CMA-ES versus							
		ABC	CROW	CS	DE	DS	GSA	HS	PSO
1000	SDM	2.8E-14 ▲	9.9E-06 ▲	1.9E-06 ▲	5.8E-04 ▲	1.7E-04 ▲	1.4E-14 ▲	1.4E-14 ▲	1.4E-14 ▲
	DDM	5.0E-04 ▲	1.2E-12 ▲	2.0E-12 ▲	1.5E-09 ▲	1.3E-11 ▲	1.4E-14 ▲	2.2E-12 ▲	2.2E-12 ▲
	TDD	8.6E-04 ▲	3.4E-04 ▲	6.9E-08 ▲	5.1E-05 ▲	7.9E-09 ▲	1.4E-14 ▲	2.9E-12 ▲	2.9E-12 ▲
500	SDM	4.0E-14 ▲	6.5E-04 ▲	1.2E-14 ▲	1.2E-04 ▲	1.5E-04 ▲	1.4E-14 ▲	1.4E-14 ▲	1.4E-14 ▲
	DDM	3.4E-05 ▲	7.9E-14 ▲	1.4E-14 ▲	1.6E-11 ▲	1.6E-13 ▲	6.7E-09 ▲	1.1E-12 ▲	1.1E-12 ▲
	TDDD	1.0E-14 ▲	1.0E-14 ▲	1.0E-14 ▲	1.0E-14 ▲	1.0E-14 ▲	1.0E-14 ▲	1.0E-14 ▲	1.0E-14 ▲

(continued)

Table 4.13 (continued)

Mod. kc200gt Mono-crystalline									
IR		CMA-ES versus							
		ABC	CROW	CS	DE	DS	GSA	HS	PSO
1000	SDM	2.0E-11 ▲	1.3E-14 ▲	2.3E-11 ▲	7.1E-04 ▲	6.5E-08 ▲	2.8E-14 ▲	3.2E-17 ▲	4.6E-14 ▲
	DDM	1.9E-14 ▲	1.6E-12 ▲	1.6E-14 ▲	2.7E-05 ▲	1.5E-06 ▲	1.5E-17 ▲	7.2E-14 ▲	1.2E-13 ▲
800	SDM	1.5E-11 ▲	1.3E-11 ▲	2.4E-11 ▲	4.0E-04 ▲	7.4E-08 ▲	2.7E-14 ▲	6.5E-15 ▲	8.9E-13 ▲
	DDM	2.8E-14 ▲	1.4E-14 ▲	1.4E-14 ▲	9.3E-05 ▲	2.1E-06 ▲	1.6E-13 ▲	1.8E-11 ▲	3.3E-12 ▲
600	SDM	6.3E-14 ▲	1.3E-11 ▲	2.4E-11 ▲	4.7E-04 ▲	5.4E-08 ▲	3.1E-14 ▲	7.3E-11 ▲	4.7E-11 ▲
	DDM	1.5E-14 ▲	3.4E-14 ▲	1.4E-14 ▲	1.1E-05 ▲	1.3E-06 ▲	1.7E-15 ▲	3.2E-12 ▲	2.9E-14 ▲
400	SDM	1.9E-12 ▲	7.4E-10 ▲	2.4E-11 ▲	2.7E-04 ▲	7.8E-08 ▲	2.9E-15 ▲	8.4E-14 ▲	3.4E-10 ▲
	DDM	1.4E-14 ▲	1.4E-14 ▲	1.4E-14 ▲	7.4E-05 ▲	1.5E-06 ▲	1.5E-16 ▲	2.2E-13 ▲	1.4E-08 ▲
200	SDM	8.1E-12 ▲	1.3E-13 ▲	2.4E-11 ▲	2.7E-04 ▲	6.7E-08 ▲	3.1E-10 ▲	4.1E-14 ▲	2.7E-14 ▲
	DDM	3.2E-14 ▲	6.4E-14 ▲	1.4E-14 ▲	2.1E-05 ▲	1.4E-06 ▲	1.8E-14 ▲	2.3E-18 ▲	2.5E-09 ▲

technique, and ▶ represents no significant difference between compared techniques. The n-value determined by Bonferroni correction was computed in $n = 0.00139$.

After analyses Table 4.13, we find that the CMA-ES is able to find better solution in almost all cases. These results reveal that exist statistical difference between the CMA-ES and the rest of the technique used for this study. In Fig. 4.4, the *I*-*V* characteristics between the measured data and the approximate model found by the CMA-ES are shown, where two different conditions are considered; in the condition A, the irradiation is set in 1000 W/m^2. For the condition B, the irradiation is 500 W/m^2. The D6P100 Multi-crystalline solar cell was used for the graphics for the SDM, DDM and TDM.

Figure 4.5 shows the absolute error curves of between the measures and the values determined by the CMA-ES under 1000 W/m^2 and 500 W/m^2 of irradiation for the SDM, DDM and TDM. In Fig. 4.6, the convergence rate of the evolutionary computation techniques as an important characteristic to evaluate the performance solving complex problems.

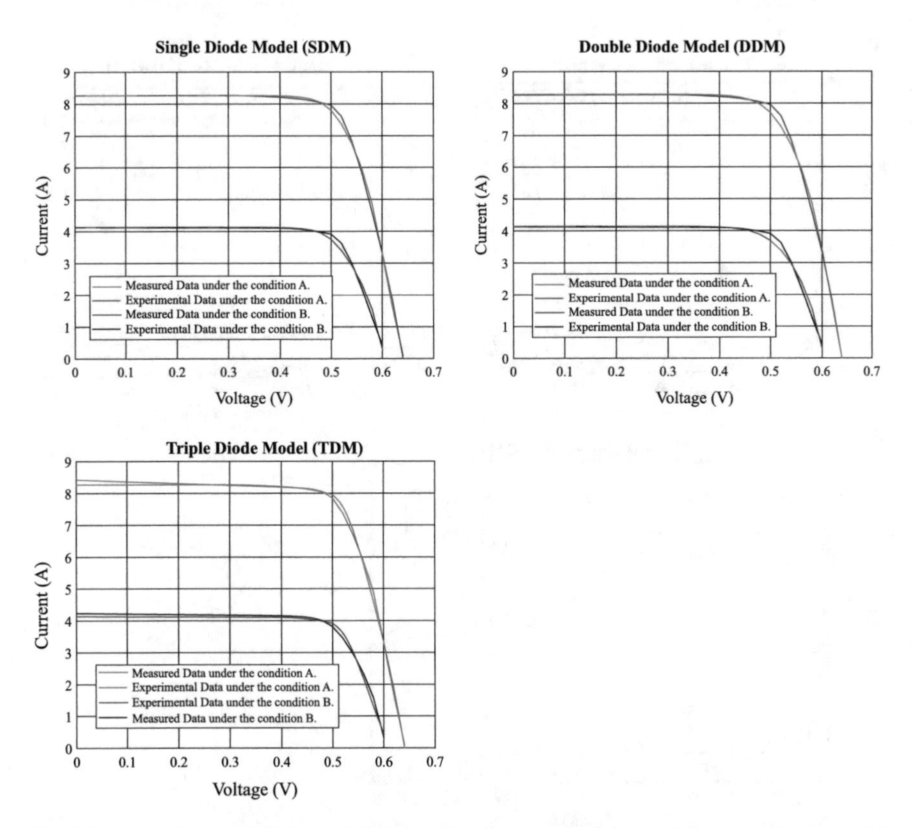

Fig. 4.4 Comparison of *I*-*V* characteristic between the measured data and the approximate model determined by CMA-ES for the D6P100 multi crystalline solar cell under two irradiation conditions: 1000 W/m^2 (condition A) and 500 W/m^2 (condition B)

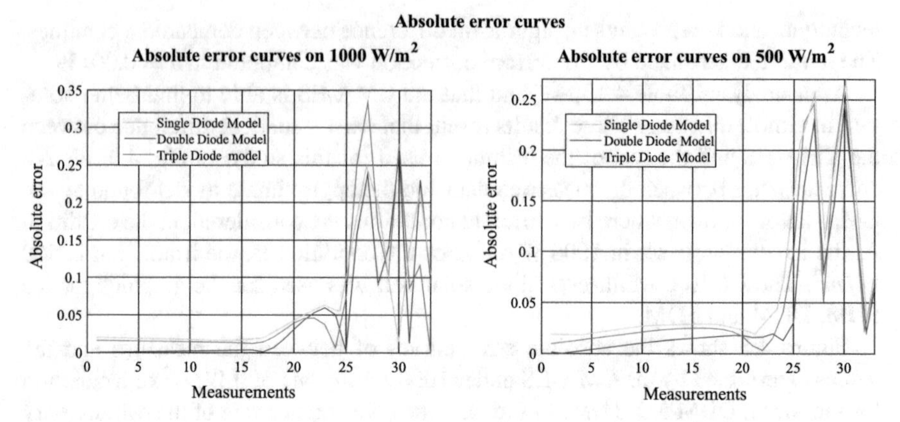

Fig. 4.5 Absolute error curves generated by the CMA-ES for the D6P100 multi-crystalline solar cell under two irradiation conditions: 1000 W/m^2 (condition A) and 500 W/m^2 (condition B) for the SDM, DDM and TDM

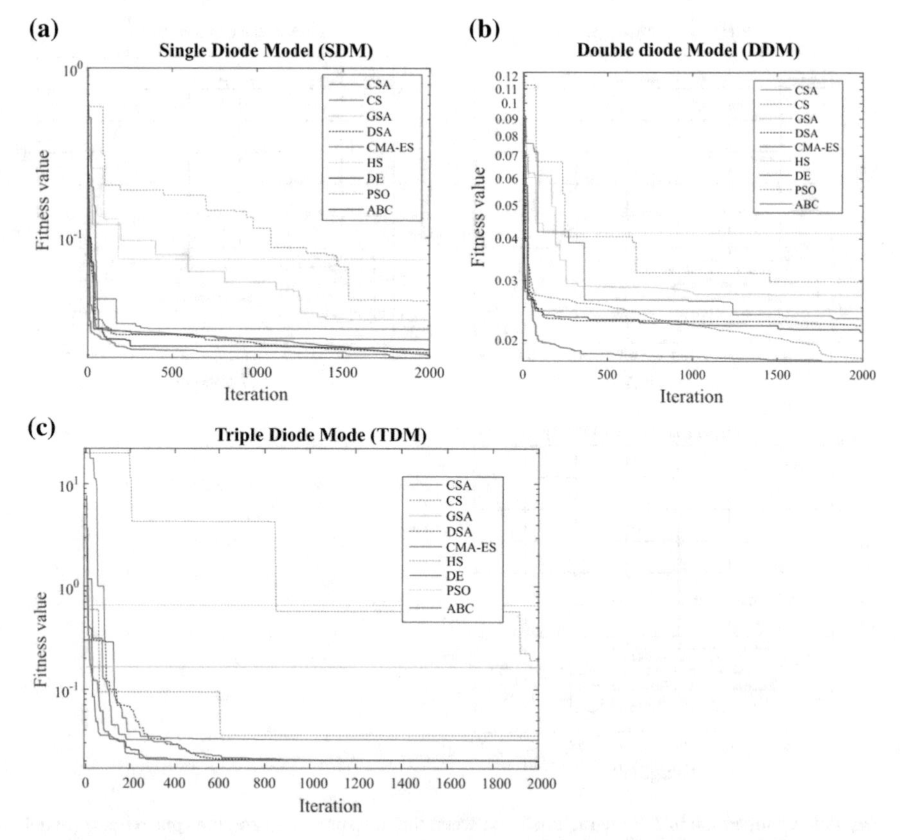

Fig. 4.6 Convergence evolution over iterations of the evolutionary computation techniques for the SDM **a**, DDM **b** and TDM **c** using the D6P100 Multi-crystalline solar cell

4.5 Conclusions

In this chapter a comparative study for solar cells parameter estimation is discussed. In the comparison different Evolutionary Computation Techniques are used, such as Artificial Bee Colony (ABC), Crow Search Algorithm (CSA), Cuckoo Search (CS), Differential Evolution (DE), Differential Search (DSA), Gravitational Search Algorithm (GSA), Harmony Search (HS), Particle Swarm Optimization (PSO) and Covariant Matrix Adaptation with Evolution Strategy (CMA-ES). The comparison was developed over three equivalent solar cell models, the Single Diode (SDM), Double Diode Model (DDM) and Three Diode Model (TDM) using a Mono-crystalline solar cell for the SDM and DDM, a Multi-crystalline solar cell for the SDM, DDM and TDM and a solar cell module for the SDM and DDM.

Since the solar cell parameter estimation becomes a complex task due their dimensionality, non-linearity and the multimodal error surface generated, the correct comparison of different techniques to determine which is able to perform better the solar cell parameter estimation. After compare the capabilities of the evolutionary computation techniques to determine the optimal parameters for the three diode models, the CMA-ES showed the best performance in almost all models for the different solar cells configurations. Moreover the CMA-ES have a better convergence rate through iterations over the rest of the algorithms.

In order to statistically corroborate the results found for each algorithm, a non-parametric test known as Wilcoxon test was developed as well as the typical statistical tests such as average, standard deviation minimum and maximum values, where the in most cases the CMA-ES obtained the best results from the set of algorithms. In the other hand, to avoid the error type 1 a Bonferroni correction was adopted, where a new significance value was computed, corroborating that the CMA-ES statistically outperforms the rest of the techniques used for the analysis.

References

1. S. Shafiee, E. Topal, When will fossil fuel reserves be diminished? Energy Policy **37**(1), 181–189 (2009)
2. Peer Review of Renewables 2017 Global Status Report—REN21. [Online]. Available: http://www.ren21.net/peer-review-renewables-2017-global-status-report/. Accessed 21 Mar 2017
3. C. Town, in *CIE42 Proceedings, 16–18 July 2012, Cape Town, South Africa © 2012 CIE & SAIIE*. pp. 16–18, July 2012
4. V. Quaschning, R. Hanitsch, Numerical simulation of current-voltage characteristics of photovoltaic systems with shaded solar cells. Sol. Energy **56**(6), 513–520 (1996)
5. G. Farivar, B. Asaei, Photovoltaic module single diode model parameters extraction based on manufacturer datasheet parameters, in *PECon2010—2010 IEEE International Conference on Power and Energy 2010*, vol. 2, pp. 929–934 (2010)
6. T.S. Babu, J.P. Ram, K. Sangeetha, A. Laudani, N. Rajasekar, Parameter extraction of two diode solar PV model using Fireworks algorithm. Sol. Energy **140**, 265–276 (2016)
7. V. Khanna, B.K. Das, D. Bisht, P.K. Singh, A three diode model for industrial solar cells and estimation of solar cell parameters using PSO algorithm. Renew. Energy **78**, 105–113 (2015)

8. T. Easwarakhanthan, J. Bottin, I. Bouhouch, C. Boutrit, Nonlinear minimization algorithm for determining the solar cell parameters with microcomputers. Int. J. Sol. Energy **4**, 1–12 (1986)
9. A. Ortiz-Conde, F.J. García Sánchez, J. Muci, New method to extract the model parameters of solar cells from the explicit analytic solutions of their illuminated I-V characteristics. Sol. Energy Mater. Sol. Cells **90**(3), 352–361 (2006)
10. A. Jain, A. Kapoor, Exact analytical solutions of the parameters of real solar cells using Lambert W-function. Sol. Energy Mater. Sol. Cells **81**(2), 269–277 (2004)
11. H. Saleem, S. Karmalkar, An analytical method to extract the physical parameters of a solar cell from four points on the illuminated J-V curve. Electron Device Lett. IEEE **30**(4), 349–352 (2009)
12. J. Appelbaum, A. Peled, Parameters extraction of solar cells—A comparative examination of three methods. Sol. Energy Mater. Sol. Cells **122**, 164–173 (2014)
13. J.A. Jervase, H. Bourdoucen, A. Al-Lawati, Solar cell parameter extraction using genetic algorithms. Meas. Sci. Technol. **12**(11), 1922–1925 (2001)
14. N. Moldovan, R. Picos, E. Garcia-Moreno, Parameter extraction of a solar cell compact model using genetic algorithms, in *Proceedings of the 2009 Spanish Conference on Electron Devices, CDE'09*, 2009, pp. 379–382
15. M. Zagrouba, A. Sellami, M. Bouaı, Identification of PV solar cells and modules parameters using the genetic algorithms: Application to maximum power extraction. Sol. Energy **84**(5), 860–866 (2010)
16. M. Ye, X. Wang, Y. Xu, Parameter extraction of solar cells using particle swarm optimization. J. Appl. Phys. **105**(9), 094502 (2009)
17. A. Khare, S. Rangnekar, A review of particle swarm optimization and its applications in solar photovoltaic system. Appl. Soft Comput (2013)
18. H. Qin, J.W. Kimball, Parameter determination of photovoltaic cells from field testing data using particle swarm optimization, in *2011 IEEE Power and Energy Conference at Illinois*, pp. 1–4 (2011)
19. E.E. Faculty, Implementation of artificial bee colony algorithm on maximum power point tracking for PV modules. Adv. Top. Electr. Eng. (ATEE). 1–4 (2013)
20. D. Oliva, E. Cuevas, G. Pajares, Parameter identification of solar cells using artificial bee colony optimization. Energy **72**, 93–102 (2014)
21. W. Gong, Z. Cai, Parameter extraction of solar cell models using repaired adaptive differential evolution. Sol. Energy **94**, 209–220 (2013)
22. K. Ishaque, Z. Salam, S. Mekhilef, A. Shamsudin, Parameter extraction of solar photovoltaic modules using penalty-based differential evolution. Appl. Energy **99**, 297–308 (2012)
23. K. Ishaque, Z. Salam, An improved modeling method to determine the model parameters of photovoltaic (PV) modules using differential evolution (DE). Sol. Energy **85**, 2349–2359 (2011)
24. A. Askarzadeh, A discrete chaotic harmony search-based simulated annealing algorithm for optimum design of PV/wind hybrid system. Sol. Energy (2013)
25. A. Askarzadeh, A. Rezazadeh, Parameter identification for solar cell models using harmony search-based algorithms. Sol. Energy (2012)
26. J. Ahmed, Z. Salam, A soft computing MPPT for PV system based on Cuckoo search algorithm, in *International Conference on Power Engineering, Energy and Electrical Drives*, pp. 558–562 (2013)
27. J. Ma, T.O. Ting, K.L. Man, N. Zhang, S.U. Guan, P.W.H. Wong, Parameter estimation of photovoltaic models via cuckoo search. J. Appl. Math. **2013**, 1–8 (2013)
28. A.M. Humada, M. Hojabri, S. Mekhilef, H.M. Hamada, Solar cell parameters extraction based on single and double-diode models: A review. Renew. Sustain. Energy Rev. **56**, 494–509 (2016)
29. R. Tamrakar, A. Gupta, A review: Extraction of solar cell modelling parameters. **3**(1) (2015)
30. D.S.H. Chan, J.R. Phillips, J.C.H. Phang, A comparative study of extraction methods for solar cell model parameters. Scopus (1986)
31. D. Karaboga, An idea based on honey bee swarm for numerical optimization, in *Tech. Rep. TR06, Erciyes Univ.*, no. TR06, p. 10 (2005)

32. R. Storn, K. Price, Differential evolution—A simple and efficient heuristic for global optimization over continuous spaces. J. Glob. Optim. 341–359 (1997)
33. Z.W. Geem, A new heuristic optimization algorithm: Harmony search. Simulation (2001)
34. E. Rashedi, H. Nezamabadi-pour, S. Saryazdi, GSA: A gravitational search algorithm. Inf. Sci. (Ny) **179**(13), 2232–2248 (2009)
35. X.S. Yang, S. Deb, Cuckoo search via Lévy flights, in *2009 World Congr. Nat. Biol. Inspired Comput. NABIC 2009—Proc.*, pp. 210–214 (2009)
36. P. Civicioglu, Transforming geocentric cartesian coordinates to geodetic coordinates by using differential search algorithm. Comput. Geosci. **46**, 229–247 (2012)
37. A. Askarzadeh, A novel metaheuristic method for solving constrained engineering optimization problems: Crow search algorithm. Comput. Struct. **169**, 1–12 (2016)
38. N. Hansen, A. Ostermeier, Adapting arbitrary normal mutation distributions in evolution strategies: the covariance matrix adaptation, in *Proceedings of IEEE International Conference on Evolutionary Computation*, pp. 312–317
39. N. Hansen, A. Ostermeier, Completely derandomized self-adaptation in evolution strategies. Evol. Comput. **9**(2), 159–195 (2001)
40. J. Kennedy, R. Eberhart, Particle swarm optimization, in *Neural Networks, 1995. Proceedings., IEEE International. Conference*, vol. 4, pp. 1942–1948 (1995)
41. P. Barthelemy, J. Bertolotti, D.S. Wiersma, A Lévy flight for light. Nature **453**(7194), 495–498 (2008)
42. A. Yona, T. Senjyu, T. Funabshi, H. Sekine, Application of neural network to 24-hours-ahead generating power forecasting for PV system. IEEJ Trans. Power Energy **128**(1), 33–39 (2008)
43. T. Hiyama, S. Kouzuma, T. Imakubo, Identification of optimal operating point of PV modules using neural network for real time maximum power tracking control. IEEE Trans. Energy Convers. **10**(2), 360–367 (1995)
44. E. Karatepe, M. Boztepe, M. Colak, Neural network based solar cell model. Energy Convers. Manage. **47**(9–10), 1159–1178 (2006)
45. K. Ishaque, Z. Salam, H. Taheri, Simple, fast and accurate two-diode model for photovoltaic modules. Sol. Energy Mater. Sol. Cells **95**(2), 586–594 (2011)
46. Y.-H. Ji, J.-G. Kim, S.-H. Park, J.-H. Kim, C.-Y. Won, C-language based PV array simulation technique considering effects of partial shading
47. H.-G. Beyer, Evolutionary algorithms in noisy environments: theoretical issues and guidelines for practice (1999)
48. L. Jun-hua, L. Ming, An analysis on convergence and convergence rate estimate of elitist genetic algorithms in noisy environments. Opt. Int. J. Light Electron Opt. **124**(24), 6780–6785 (2013)
49. T. Ma, H. Yang, L. Lu, Solar photovoltaic system modeling and performance prediction. Renew. Sustain. Energy Rev. **36**, 304–315 (2014)
50. K. Nishioka, N. Sakitani, Y. Uraoka, T. Fuyuki, Analysis of multicrystalline silicon solar cells by modified 3-diode equivalent circuit model taking leakage current through periphery into consideration. Sol. Energy Mater. Sol. Cells **91**(13), 1222–1227 (2007)
51. N.F.A. Hamid, N.A. Rahim, and J. Selvaraj, Solar cell parameters extraction using particle swarm optimization algorithm, in *2013 IEEE Conference on Clean Energy and Technology (CEAT)*, pp. 461–465 (2013)
52. F. Wilcoxon, *Breakthroughs in Statistics: Methodology and Distribution*, ed. by S. Kotz, N.L. Johnson (Springer, New York, NY, 1992), pp. 196–202
53. S. García, D. Molina, M. Lozano, F. Herrera, A study on the use of non-parametric tests for analyzing the evolutionary algorithms' behaviour: A case study on the CEC'2005 Special Session on Real Parameter Optimization. J. Heuristics **15**(6), 617–644 (2009)
54. Y. Hochberg, A sharper Bonferroni procedure for multiple tests of significance. Biometrika **75**(4), 800–802 (1988)
55. R.A. Armstrong, When to use the Bonferroni correction. Ophthalmic Physiol. Opt. **34**(5), 502–508 (2014)

Chapter 5
Gravitational Search Algorithm for Non-linear System Identification Using ANFIS-Hammerstein Approach

5.1 Introduction

Most of the real world systems have non-linear responses. Since the identification process of such systems is a difficult problem, the development of different modeling techniques has attracted the attention of the scientific in many fields of engineering. One typical example of such modeling techniques refers to Hammerstein models [1] which have proved their effectiveness to model difficult non-linear systems. They basically combine a non-linear adaptive system with a dynamic linear filter to develop an alternative model. The simplicity and easy definition of Hammerstein models have produced their use to represent a large number of complicated non-linear systems as solid oxide fuel cells [2], ultrasonic motors [3], power amplifier pre-distortion [4], control systems [5], PH process [6], just for mention a few.

The Hammerstein model identification is carried out considering two blocks: a static non-linear subsystem connected in cascade to a dynamic linear block. The static non-linear subsystem, considered as the crucial part in the identification, can be represented by an adaptive non-linear network as a multi-layer perceptron (MLP) [7], by a Functional link artificial neural network (FLANN) [8, 9] and by a Single hidden layer feed-forward neural network (SFLNs) [10]. On the other hand, the dynamic linear subsystem is normally described by an adaptive infinite impulse response (IIR) filter.

The Adaptive neuro-fuzzy inference system (ANFIS) [11] is a relevant adaptive non-linear network that has increased its popularity as a soft computing method. ANFIS is a hybrid system that integrates the learning capabilities of a neural network and the imprecise logic of a Takagi-Sugeno schema [12] Fuzzy inference system to produce input–output pairs with a high level of precision and generalization. According to its structure, ANFIS can be classified in a non-linear block and a linear system. Due to its unusual characteristics, ANFIS has been used in an large number of

E. Cuevas et al., *Recent Metaheuristics Algorithms for Parameter Identification*, Studies in Computational Intelligence 854, https://doi.org/10.1007/978-3-030-28917-1_5

applications including time-series prediction [13], estimation of wind speed profile [14], fault detection of an industrial steam turbine [15], representation of a non-linear liquid-level system [16], modeling of solid oxide fuel cell [17], and so on.

In the Hammerstein system identification, the parameter determination process is modified into a multi-dimensional problem where the weights of the non-linear system, as well as the parameters of the linear system, are considered as decision variables. Under these conditions, the difficulty of the problem tends to generate multimodal surfaces for which their cost functions are significantly complex to optimize. Most of the techniques used to identify systems consider deterministic approaches such as the gradient-based method or the least mean square (LMS) algorithm to determine the parameters of the Hammerstein model.

As another option to deterministic methods, the Hammerstein model identification has also been conducted through evolutionary computing techniques. They have shown to deliver better models than those based on deterministic methods regarding accuracy and robustness [18]. Some examples of these strategies used in model identification involve the Particle Swarm Optimization (PSO) [19], Locust Search (LS) [20] and Differential Evolution (DE) [21], just to mention a few. Although these techniques perform interesting results, they present important limitations: They generally obtain sub-optimal solutions as a result of the inadequate balance between exploration and exploitation in their search strategies.

On the other hand, the Gravitational Search Algorithm (GSA) [22] is a recent evolutionary computation technique that is based on the laws of gravitation. In GSA, the operators are devised in such a way that they mimic the gravitational process in its particles. Different from most of the existent evolutionary techniques, GSA achieves better performance in high multimodal problems, avoiding important flaws as the premature convergence to sub-optimal solutions [23, 24]. Such peculiarities have motivated its use to solve many engineering problems in energy [25], image processing [26] and machine learning [27], and so on.

In this chapter, a non-linear system identification technique based on a Hammerstein model is conducted. Under this scheme, a Hammerstein system is formed through the adaptation of an ANFIS model. Under this assumption, the non-linear and linear ANFIS sub-systems correspond to the non-linear element and the dynamic linear block of a Hammerstein model. Taking into account the similitude between the two models, the presented approach combines the excellent modeling capabilities of ANFIS and the interesting multi-modal capacities of the GSA to determine the parameters of a Hammerstein model, which are coded in the ANFIS structure. To confirm the effectiveness of the suggested scheme, its modeling accuracy is compared with other popular evolutionary computing methods for identification models based on numerical simulations on different complex cases.

5.2 Background

5.2.1 *Hybrid ANFIS Models*

The ANFIS model is a hybrid computational intelligence architecture generated by combining the learning capability of neural networks and inference capabilities of fuzzy logic. The ANFIS model is used to model and to identify many systems [13–17]. In its learning stage, an optimization algorithm is used to determine the optimal parameters of ANFIS. Two different classes of methods are commonly applied for this purpose: gradient-based and evolutionary [28]. Since gradient-based techniques frequently achieve sub-optimal solutions in the estimation of the ANFIS parameters, evolutionary techniques are recently preferred to estimate them. Evolutionary computation methods are good competitors to solve difficult problems; they perform better in global search spaces than gradient-based techniques [28]. Their optimization strategies enable the simultaneous exploration and exploitation for obtaining optimal solutions.

Many methods that combine ANFIS and evolutionary methods are reported in the literature for modeling and identifying systems. Some examples are mentioned below.

In [29], Sarkheyli et al. applied a modified Genetic Algorithm (GA) to determine the ANFIS parameters. Wei, in [30], combined a hybrid model based on GA and ANFIS to determine the stock price in Taiwan. Rini et al. [31] analyzed the interpretability and efficiency of the ANFIS modeling capabilities when the PSO algorithm is employed as a training method. Wang and Ning, in [32], predicted time series behavior by coupling PSO and ANFIS. Liu et al. [33] determined the parameters of an ANFIS by employing an improved Quantum-behaved Particle Swarm Optimization (QPSO). In [34], Catalão et al. generated a hybrid model based on the PSO, wavelet transform and ANFIS for short-term electricity cost estimation. Suja Priyadharsini et al. [35] coupled an Artificial Immune System (AIS) with the ANFIS system for modeling the elimination of artifacts in Electroencephalography (EEG) signals. In [36], Gunasekaran and Ramaswami represented the future index value of the National Stock Exchange (NSE) of India by applying an ANFIS model trained with an AIS technique. Asadollahi-Baboli [37] introduced a predictor system of chemical particles combining a modified Ant Colony Optimization (ACO) algorithm with ANFIS structure while the prediction accuracy is estimated through shuffling cross-validation (SCV) technique. In [38], Teimouri and Baseri introduced an ANFIS scheme to determine the parameter correlation in an electrical discharge machining (EDM) method. In the proposal, the Ant Colony Optimization (ACO) technique is employed as a learning algorithm for the ANFIS structure. Varshini et al. [39] used the Bacterial Foraging Optimization Algorithm (BFOA) to determine the ANFIS parameters for modeling of different proposes. In [40], Bhavani et al. introduced a model based on ANFIS and BFOA for the control of load frequency in a competitive electricity market. Karaboga and Kaya [28] proposed a modified Artificial Bee Colony (ABC) algorithm for training the ANFIS model. In [41], Azarbad et al. suggested an ANFIS structure to model the Global Positioning System (GPS) data. The proposal combines Radial Basis Functions (RBF) and ABC for improving the learning process. Rani and Sankaragomathi [42] designed a PID controller by the combination of ANFIS and the cuckoo search (CS). In [43], Chen and Do represented the student educational performance by using ANFIS and CS as a training process. Khazraee et al. [44] applied ANFIS and the DE algorithm to model a reactive batch distillation process. In [45], Zangeneh et al. combinate the DE and a gradient descent algorithm to train an ANFIS system to identify non-linear systems. In the proposal, while the antecedent parameters are optimized by DE, the consequent elements are determined with a gradient-based technique. Afifi et al. [46] suggested a model that predicts mobile malware presence. The proposal contemplates the union between the ANFIS model and DE. All methods consider the ANFIS scheme as a black box model while the evolutionary techniques are used to improve its modeling as training algorithms. Different from such procedures, in our proposal, the ANFIS structure is employed to represent a plant with a Hammerstein model. Considering the high similarity between the ANFIS and Hammerstein models, the parameters of the Hammerstein system are efficiently coded in the ANFIS architecture. Hence, the training stage in the ANFIS composition corresponds to the identification of the Hammerstein model. Under these circumstances, the resulting ANFIS system can be represented as a gray box system where their parameters match with the parameters of the Hammerstein model.

On the other hand, the implementation of evolutionary algorithms to the identification process of Hammerstein models has been previously reported in the state-of-art. Some cases involve the studies performed by Cui et al. [7], Tang et al. [8], Gotmare et al. [10], Duwa [19] and Mete et al. [21], where some evolutionary methods are applied as optimization techniques to determine Hammerstein systems. Nevertheless, in such procedures, the evolutionary techniques are employed in combination with other computing schemes that operate as black box models without the probability of directly examine the inner parameters. In these procedures, it is frequently used evolutionary techniques that generally obtain sub-optimal solutions as a consequence of their bad search strategy in multimodal optimization problems.

5.2.2 *Adaptive Neuro-fuzzy Inference System (ANFIS)*

ANFIS [11] considers an adaptive Takagi-Sugeno fuzzy system. According to this structure, ANFIS can be distributed in a non-linear system and a linear block. The non-linear system includes the antecedent of the fuzzy inference model, while the linear block describes the linear functions of the consequent (see Fig. 5.1). The ANFIS model includes five layers. Each layer has different purposes which are detailed below.

Layer 1: Every node i in this layer defines the order to which an input u satisfies the linguistic concept described by A_i.

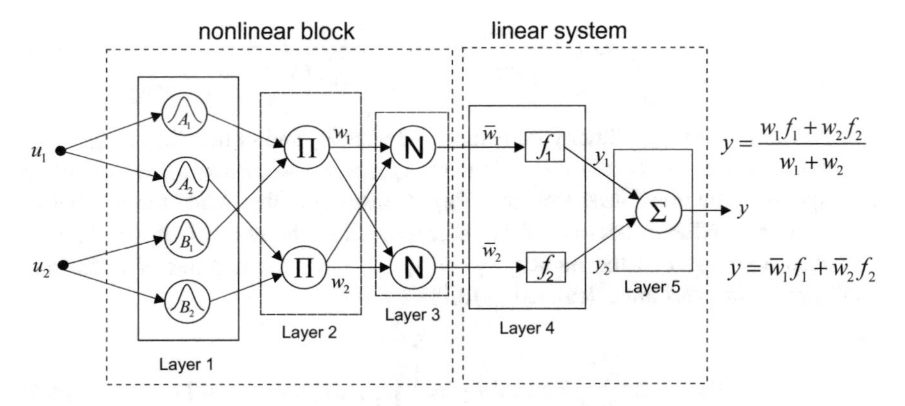

Fig. 5.1 ANFIS structure

$$O_i^1 = \mu_{A_i}(u) \tag{5.1}$$

The correspondence is computed by a membership function represented as follows:

$$\mu_{A_i}(u) = e^{-\frac{(u-\sigma_i)}{(2s_i)^2}} \tag{5.2}$$

where $\{\sigma_i, s_i\}$ describes the flexible parameters of the membership function. The setting of this parameter is known as premise parameters.

Layer 2: Here, the output O_i^2 ($O_i^2 = w_i$) of each node i $\left(\prod\right)$ describes the firing force of its corresponding rule.

Layer 3: Each node i (N) in this layer computes the normalized firing force expressed as follows:

$$O_i^3 = \bar{w}_i = \frac{w_i}{\sum_{j=1}^{h} w_j}, \tag{5.3}$$

where h describes the number of nodes in this layer.

Layer 4: Each node i in this layer corresponds to an adaptive processing unit that performs a linear function described as follows:

$$O_i^4 = \bar{w}_i f_i = \bar{w}_i (p_i u_1 + q_i u_2 + r_i) \tag{5.4}$$

where $\{p_i, q_i, r_i\}$ describe the adjustable parameters that define the function response. The parameter setting in this layer is assigned as the consequence elements. Generally, f_i is a function that can be represented by any linear function as long as it can properly represent the response of the system within the fuzzy area defined by the antecedent of rule i.

Layer 5: This node $\left(\sum\right)$ computes the overall output y as the sum of all incoming signals:

$$y = O_i^5 = \sum_i \bar{w}_i \cdot f_i = \frac{\sum_i w_i \cdot f_i}{\sum_i w_i} \tag{5.5}$$

ANFIS distributes its inner parameters P_T linear P_L and in non-linear P_{NL}. The linear parameters P_L represent the elements $\{p_i, q_i, r_i\}$ while $\{\sigma_i, s_i\}$ the elements correspond to the non-linear elements P_{NL}. Considering that *Nu* is the number of inputs of the ANFIS model and FM_q corresponds to the number of membership functions associated to the input q ($q = 1, \ldots, Nu$), the number of linear and nonlinear parameters are calculated as follows:

$$P_{NL} = 2 \cdot \sum_{q=1}^{Nu} FM_q, \quad P_L = \left[\prod_{q=1}^{Nu} FM_q \right] \cdot (Nu + 1) \tag{5.6}$$

Hence, the number of adaptive parameters of the ANFIS model is $P_T = P_{NL} + P_L$. They are represented as the vector $\mathbf{E}_{ANFIS}$ described as follows:

$$\mathbf{E}_{ANFIS} = \left\{ e_1^{ANFIS}, \ldots, e_{P_T}^{ANFIS} \right\} \tag{5.7}$$

ANFIS employs a hybrid training technique that combines the steepest descent method for P_{NL} and the least-squares technique for P_L. This hybrid training process performs good results; nevertheless; they usually obtain sub-optimal solutions as a consequence of its poor search strategy in multi-modal functions.

5.2.3 Gravitational Search Algorithm (GSA)

The GSA is an optimization technique based on gravitational phenomena, introduced by Rashedi et al. [22] in 2009. GSA is an optimization technique that is increasing its demand due to its interesting performance in difficult applications, avoiding completely sub-optimal solutions in multi-modal problems. Each solution of the GSA mimics a mass that has a certain gravitational force which allows attracting other masses. GSA is a population technique that considers N different candidate solutions or masses, where each mass symbolizes a d-dimensional vector expressed as follow:

$$\mathbf{m}_i(t) = \left\{ m_i^1, \ldots m_i^d \right\} \quad i = 1, \ldots, N \tag{5.8}$$

In GSA, the gravity force pulling from mass i to mass j in the h variable ($h \in 1, \ldots, d$), at time *t*, is described as follows:

$$F_{ij}^h(t) = G(t) \frac{M_i(t) \cdot M_j(t)}{R_{ij}(t) + \varepsilon} \left(m_j^h(t) - m_i^h(t) \right), \tag{5.9}$$

where $G(t)$ describes the gravitational constant, M_i and M_j are the gravitational mass of candidates i and j, respectively. R_{ij} expresses the Euclidian distance between the ith and jth candidates, and ε is a small constant. To preserve the balance between exploration and exploitation, GSA adjusts the gravity strength $G(t)$ between solutions through the optimization process. Under such conditions, the final force operating over the candidate i is represented as follows

$$F_i^h(t) = \sum_{j=1,\, j \neq i}^{N} rand \cdot F_{ij}^h(t), \tag{5.10}$$

where *rand* expresses an arbitrary number in the interval [0, 1]. The acceleration of the candidate i is described as follows:

$$a_i^h(t) = \frac{F_i^h(t)}{M_i(t)}, \tag{5.11}$$

Hence, a new position and velocity of each candidate solution i are calculated as follows:

$$\begin{aligned} v_i^h(t+1) &= rand \cdot v_i^h(t) + a_i^h(t) \\ m_i^h(t+1) &= m_i^h(t) + v_i^h(t) \end{aligned} \tag{5.12}$$

The Gravitational masses are calculated at each time t by applying the cost function which measures the quality of the candidates. Hence, the new gravitational masses are computed as follows:

$$p_i(t) = \frac{f(m_i(t)) - worst(t)}{best(t) - worst(t)} \tag{5.13}$$

$$M_i(t) = \frac{p_i(t)}{\sum_{l=1}^{N} p_l(t)}, \tag{5.14}$$

where $f(\cdot)$ represents to the objective function, *worst*(t) and *best*(t) is the worst and best fitness values respectively.

5.3 Hammerstein Model Identification by Applying GSA

A Hammerstein method basically connects a non-linear static system NS in cascade with a dynamic linear system LS to represent a non-linear plant P. Following this procedure; the non-linearity is represented by the static function NS while the linear system LS describes the dynamic response.

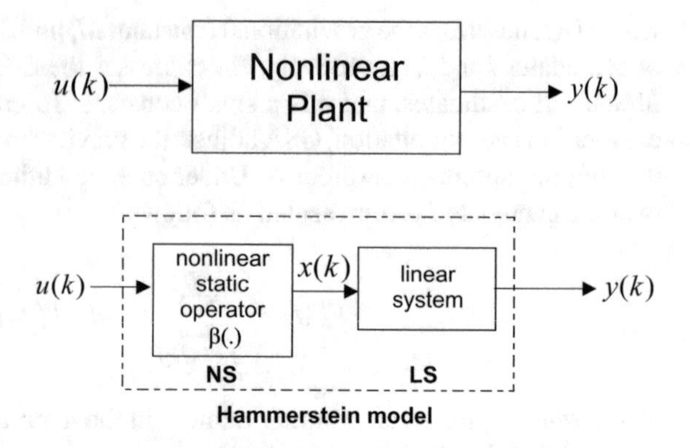

Fig. 5.2 Graphic illustration of the Hammerstein system

The non-linear static system **NS** takes an input $u(k)$ and gives an output $x(k)$ which is converted by a non-linear function $\beta(\cdot)$ such that $x(k) = \beta(u(k))$. Under these conditions, $u(k)$ of NS describes the input of the Hammerstein model while the input of the linear system LS of the Hammerstein structure is the output $x(k)$ of NS. Figure 5.2 presents a graphical illustration of the Hammerstein model.

The input–output approximation of the linear system **LS** in the Hammerstein structure is defined as adaptable IIR filter described as follows:

$$\begin{aligned}
y(k) &= \frac{B(z)}{A(z)}x(k-1) + \frac{C(z)}{A(z)}\delta(k) \\
A(z^{-1}) &= 1 + a_1 z^{-1} + \ldots + a_P z^{-P} \\
B(z^{-1}) &= b_0 + b_1 z^{-1} + \ldots + b_L z^{-L-1} \\
C(z^{-1}) &= c_0 + c_1 z^{-1} + \ldots + c_S z^{-S-1}
\end{aligned} \tag{5.15}$$

where P, L and S are the polynomial order of $A(z)$, $B(z)$ and $C(z)$, respectively. In the linear system represented in Eq. (5.15), $x(k)$ and $y(k)$ express the input and output of the model while $\delta(k)$ is the noise signal that represents the unmodeled consequences.

In this chapter, a non-linear system identification approach based on the Hammerstein scheme is presented. In the suggested approach, the model is approximated through the adaptation of an ANFIS model, taking advantage of the similitude between them. The presented approach is shown in Fig. 5.3. The structure of ANFIS is an adaptable Takagi-Sugeno fuzzy scheme. According to the ANFIS structure, it can be classified into a non-linear system and a linear block. The non-linear system includes the antecedent of the fuzzy inference system, while the linear block describes the linear functions of the consequent. The input of the linear system represents the output of the non-linear part $x(k)$. In the suggested ANFIS method, the value of $x(k)$ that describes the Hammerstein non-linearity is divided in the signals $\bar{w}_i$ provided by layer 3 (Fig. 5.1). Hence, to maintain the coherence to the Hammerstein system,

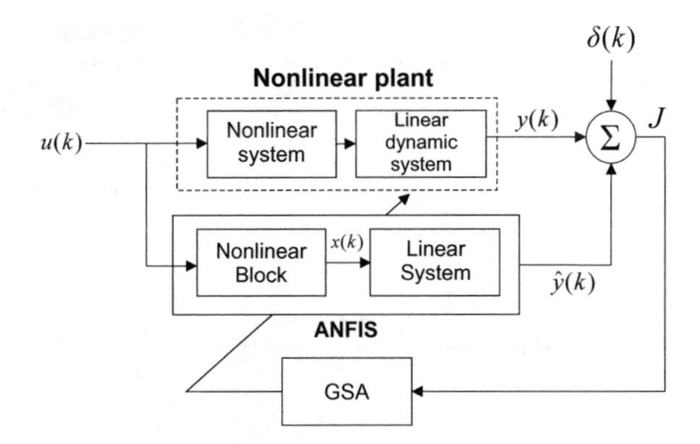

Fig. 5.3 Graphic illustration of the identification process regarding the presented scheme

the linear block of the ANFIS scheme performs the adaptable IIR f_i of the ANFIS system are represented as follows:

$$
\begin{aligned}
&f_i(k) = \frac{B(z)}{A(z)}\bar{w}_i(k-1) \\
&A(z^{-1}) = 1 + a_1 z^{-1} + \ldots + a_P z^{-P} \\
&B(z^{-1}) = b_0 + b_1 z^{-1} + \ldots + b_L z^{-L-1} \\
&f_i(k)A(z) = \bar{w}_i(k-1)B(z) \\
&f_i(k)\left[1 + a_1 z^{-1} + \ldots + a_P z^{-P}\right] = \bar{w}_i(k-1)\left[b_0 + b_1 z^{-1} + a_L z^{-L-1}\right] \\
&f_i(k) + a_1 f_i(k-1) + \ldots + a_P f_i(k-P) = b_0\bar{w}_i(k-1) + b_1\bar{w}_i(k-2) + \ldots + b_L\bar{w}_i(k-L-2) \\
&f_i(k) = b_0\bar{w}_i(k-1) + b_1\bar{w}_i(k-2) + \ldots + b_L\bar{w}_i(k-L-2) - a_1 f_i(k-1) - \ldots - a_P f_i(k-P)
\end{aligned} \tag{5.16}
$$

Hence, every function f_i of the ANFIS model describes an IIR filter that appropriately defines the dynamic behavior of the plant in the fuzzy region defined by the antecedent of rule i. Figure 5.4 presents the structure of the ANFIS linear model.

Under this scheme, the suggested identification technique consists of an ANFIS system (MISO) with two inputs and one output. The input system includes $u(k)$ and a delayed signal of $u(k)$ represented as $u(k-Q)$ (where Q is the maximum input delay). The output $y(k)$ is the response of the system. Every ANFIS input is characterized by two membership functions, regarding two different Takagi-Sugeno rules. Figure 5.5 illustrates the suggested ANFIS identification model. In this chapter, the input delay Q is considered as 2.

In the presented ANFIS scheme, $x(k)$ represents the static Hammerstein nonlinearity distributed in the $\bar{w}_1$ and $\bar{w}_2$, through this, $x(k)$ can be modeled as follows:

$$x(k) = \bar{w}_1 + \bar{w}_2, \tag{5.17}$$

After the performed adaptation from the original ANFIS scheme, in the linear model for every consequent function f_i, the linear parameters described in Sect. 5.2.2 have been modified. The new linear parameters P'_L are computed as follows:

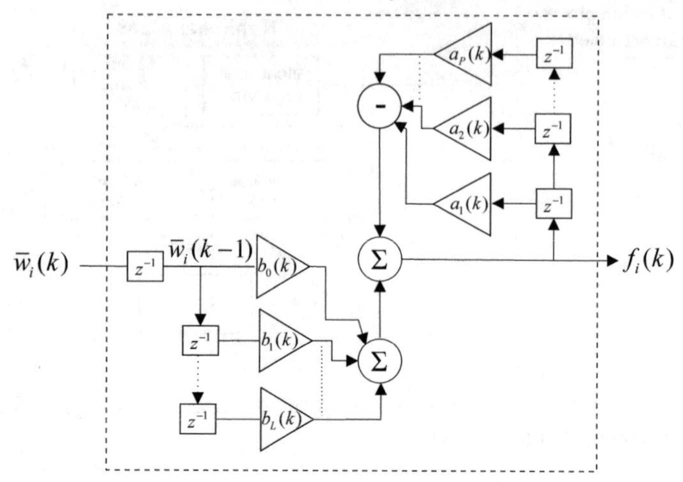

Fig. 5.4 Structure of the ANFIS linear model

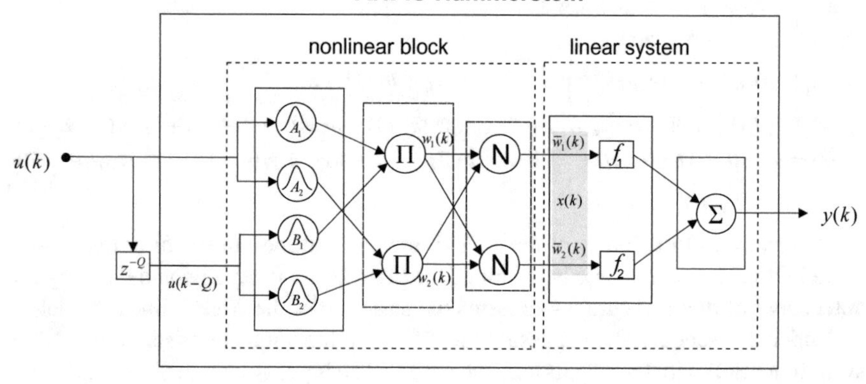

Fig. 5.5 The presented ANFIS identification model

$$P'_L = 2 \cdot (P + L + 1); \tag{5.18}$$

where P and $L + 1$ is the polynomial order of the transference functions which perform the IIR filter. In this chapter, every function is modeled considering $P = 3$ and $L = 2$.

For this reason, the number of adaptive parameters P'_T in the presented ANFIS-Hammerstein approach is $P'_T = P_{NL} + P'_L$. Hence, the new parameter vector $\mathbf{E}'_{ANFIS}$ is determined as follows:

$$\mathbf{E}'_{ANFIS} = \left\{ e_1^{ANFIS}, \ldots, e_{P'_T}^{ANFIS} \right\} \tag{5.19}$$

In the presented scheme, the identification process is carried out finding the best parameters of $\mathbf{E}'_{ANFIS}$ that present the optimal possible similarity between the actual model and the Hammerstein alternative system. For this reason, the Hammerstein parameters describe the dimensions of every solution (mass) for the identification.

To measure the quality of the solutions in the identification, the Mean squared error (MSE) criterion is considered. The MSE index, expressed by J, evaluates the similarity between the response $y(k)$ generated by the actual plant regarding the estimation $\hat{y}(k)$ produced by the Hammerstein model. Hence, the quality of every solution is evaluated considering the following model:

$$J = \frac{1}{NS} \sum_{k=1}^{NS} [y(k) - \hat{y}(k)]^2, \tag{5.20}$$

where NS symbolizes the input samples analyzed in the identification process, to estimate the entire set of parameters $\mathbf{E}'_{ANFIS}$, the GSA is employed. For this reason, the objective function J is used to measure the performance of every parameter set. Handled by the values of the objective function, the group of solutions is defined through the GSA technique to determine the optimal solution.

The values of Q, P and L of the presented method are set to maintain agreement with similar approaches reported in the state-of-art [7–10, 47]. Considering that in such works the number of delays is 2, in our proposal $Q = 2$. Furthermore, the values of $P = 3$ and $L = 2$ to model the polynomial order of $A(z)$ and $B(z)$, in the presented scheme we have used the same configuration.

5.4 Experimental Analysis

The objective of this section is to perform the modeling results of the presented ANFIS-GSA approach in different numeric identification problems. In the analysis, a set of 7 tests are considered to evaluate the accuracy of the presented scheme. Regarding the configuration, all the analyzed plants hold the NARMAX structure (non-linear autoregressive moving average model with exogenous inputs). Nonetheless, considering their nonlinear properties, they are grouped into the following categories:

(A) Hammerstein systems with polynomial non-linearities (Test 1 and Test 4). In a model that keeps a polynomial non-linearity, the relationship between the input and output variables is represented as an *nth* degree polynomial regarding the input. The Hammerstein systems with a polynomial non-linearity are complex to approximate considering that the algorithms must properly represent the turning points of the model [48]. Turning points are locations where the system behavior significantly varies from sloping downwards to sloping upwards.

(B) Hammerstein systems with non-linearities represented with linear piecewise functions (Test 2). Complex non-linear responses are usually modeled through linear piecewise functions. For this reason, the approximation of Hammerstein systems with linear piecewise functions signifies a difficult task, since its global system response must carefully approximate for every one of the single behaviors [49].

(C) Hammerstein systems with trigonometric non-linearities (Test 3 and Test 5). In a plant that keeps a trigonometric non-linearity, the relation between the input and output variables is represented as a trigonometric function of the input. The approximation of these models under Hammerstein system is considered a complex task because trigonometric functions generate a bounded output regarding high or low input values [50]. This fact is significantly difficult due to this non-linearity cannot be proportionally estimated by the learning technique during the approximation.

(D) Hammerstein systems based on real industrial problems (Test 6). Non-linear models that represent realistic engineering schemes are very interesting due to their employment context. They are generally difficult models whose approximation symbolizes a challenge from the view of any optimization technique [51]. In this chapter, we use the real non-linear heat exchanger model to identify a Hammerstein scheme.

The ANFIS-GSA results generated from the identification process are compared to those produced by the incorporation of the ANFIS structure and other popular evolutionary techniques. The experimental results are verified by statistical analysis.

In the test study, the presented ANFIS-GSA scheme is applied in the Hammerstein structure identification of 7 plants while the results are compared regarding with those generated by the DE [52], the PSO techniques [53] and the original ANFIS model [11]. The DE and PSO techniques are selected, because they have the largest number of references regarding the most popular scientific databases ScienceDirect [54], SpringerLink [55] and IEEE-Xplore [56], regarding the last ten years. In the tests, the number of elements N is set in 50. To reduce the random outcomes, every test is executed for 30 independent runs. In the comparative analysis, a fixed number of iterations is determined as a stop criterion. Hence, each execution of a system consists of 1000 iterations. This stop criterion has been selected to maintain adaptability with similar approaches reported in the literature [7–10, 47]. All tests are executed on a Pentium dual-core computer with 2.53-GHz and 8-GB RAM under MATLAB 8.3.

The parameters setting of the techniques used for the comparison are described below:

1. **DE** [52]: Its parameters set as $CR = 0.5$ and $F = 0.2$.
2. **PSO** [53]: The parameters setting are $c_1 = 2$ and $c_2 = 2$; and the weight factor decreases linearly from 0.9 to 0.2.
3. **Original ANFIS** [11]: Two inputs, two membership functions in the range of $[-1, 1]$ for Tests 1, 3, 4 and 5 whereas $[-3, 3]$ for Test 2.
4. **GSA** [22]: $G_t = G_O e^{-\alpha \frac{t}{T}}$, $\alpha = 20$, $G_O = 100$ and $T = 1000$.

The parameters configuration are selected because they reach the best optimization performance according to their reported references [11, 22, 52, 53, 57].

For the identification process, every technique achieves a solution for the parameter set $\mathbf{E}'_{ANFIS}$ at each execution. This estimation corresponds to the structure of the identified Hammerstein model with the lowest J value (Eq. 5.20). In the experimentation, four performance indexes are employed for the comparative study: the worst approximation value (W_J), the best approximation value (B_J), the mean approximation value (μ_J) and the standard deviation (σ_J). The first three indexes evaluate the precision of the identification between the estimated model and the actual plant whereas the latter evaluates the variability or robustness in the approximation. The worst approximation value (W_J) represents the lower estimation that corresponds to the largest value of J (Eq. 5.20) achieved by every technique regarding a set of 30 independent runs. The best approximation value (B_J) symbolizes the first-rate archived which is the lowest value of J reached in 30 runs. The mean approximation value (μ_J) determines the averaged estimation achieved by each technique in 30 independent executions. The Mean value just describes the average value of J over 30 values generated by each technique. The standard deviation (σ_J) is a value that measures the distribution of a set regarding the J data values. σ_J indicates the variability of the evaluated models.

5.4.1 Test I

This approximation problem is selected from [8, 58]. It analyzes the following system:

$$y(k) = \frac{B(z)}{A(z)}x(k-1) + \frac{C(z)}{A(z)}\delta(k), \tag{5.21}$$

where

$$\begin{aligned} A(z) &= 1 + 0.8z^{-1} + 0.6z^{-2}, \\ B(z) &= 0.4 + 0.2z^{-1}, \\ C(z) &= 1, \\ x(k) &= u(k) + 0.5u^3(k) \end{aligned} \tag{5.22}$$

In the test, the input signal values $u(k)$ used for this model Eq. (5.22) are arbitrary components uniformly distributed between the range $[-1, 1]$. Furthermore, in the test, a noisy signal $\delta(k)$ is added as a white Gaussian distribution with a signal to noise ratio (SNR) of 70 dB. For this reason, the presented ANFIS-GSA scheme, the DE [52], the PSO technique [53] and the original ANFIS model [11] are employed to the parameter identification $\mathbf{E}'_{ANFIS}$ of the system. All techniques are arbitrarily initialized in the first iteration. The experimental results achieved from 30 runs are shown in Table 5.1. It summarizes reports B_J, W_J and μ_J solutions in terms of J collected through the 30 runs. In the table, it is also presented the σ_J values estimated

Table 5.1 Experimental results of the Test 1

		B_J	W_J	μ_J	σ_J
EX 1	DE	7.94E−04	2.80E−02	2.85E−03	3.08E−03
	PSO	1.13E−03	9.23E−02	2.19E−02	2.34E−02
	ANFIS	3.11E−03	5.73E−03	4.30E−03	5.10E−04
	ANFIS-GSA	2.92E−04	1.56E−02	1.03E−03	2.19E−03

in the 30 executions. According to Table 5.1, the ANFIS-GSA scheme presents better performance than DE, PSO and the original ANFIS techniques regarding their μ_J values. This evidence indicates that the presented approach produces more precise system estimations than the other evolutionary techniques DE and PSO, while the original ANFIS model gets the inferior performance. In Table 5.1, is also shown that the ANFIS-GSA obtains the best approximation result (B_J) regarding 30 independent runs. On the other hand, the worst determination is produced by the original ANFIS model, this represents the estimation with the higher performance concerning the other techniques. Concerning the standard deviation value (σ_J), it is the original ANFIS model which shows the smallest interval in relation to the other evolutionary algorithms. This fact reflexes the performance of the gradient-based approaches in multi-modal problems. For this reason, the original ANFIS model converges faster (lower σ_J values), but getting a sub-optimal solution (higher μ_J values).

In Fig. 5.6 is shown the ideal static non-linearity $x(k)$ and the estimations generated by the ANFIS-GSA, DE, PSO, and original ANFIS model. From this figure, it can appreciate that the presented ANFIS-GSA approach performs the best approximation regarding the other techniques. To visualize the precision of each method, in Fig. 5.6 is incorporated a zoom window that represents the model response with more detail.

To assess the performance of the approximation process, the Hammerstein system in Eq. (5.22) and the identified models are undergone using the same input signal $u(k)$ defined below:

$$u(k) = \begin{cases} \sin\left(\frac{2\pi k}{250}\right) & 0 < k \leq 500 \\ 0.8\ \sin\left(\frac{2\pi k}{250}\right) + 0.2\sin\left(\frac{2\pi k}{25}\right) & 500 < k \leq 700 \end{cases} \tag{5.23}$$

Figure 5.7 presented the output $y(k)$ of every technique. In the figure, it is obvious that each method performs distinct depending on the segment of the model response. Nevertheless, the ANFIS-GSA scheme archives better similarity regarding the true system. Finally, Fig. 5.8 presents the convergence comparison of the objective function J for the original ANFIS model, DE, PSO, and the ANFIS-GSA scheme. From the figure, it is confirmed that the ANFIS-GSA scheme reaches faster the final J values in contrast with its competitors.

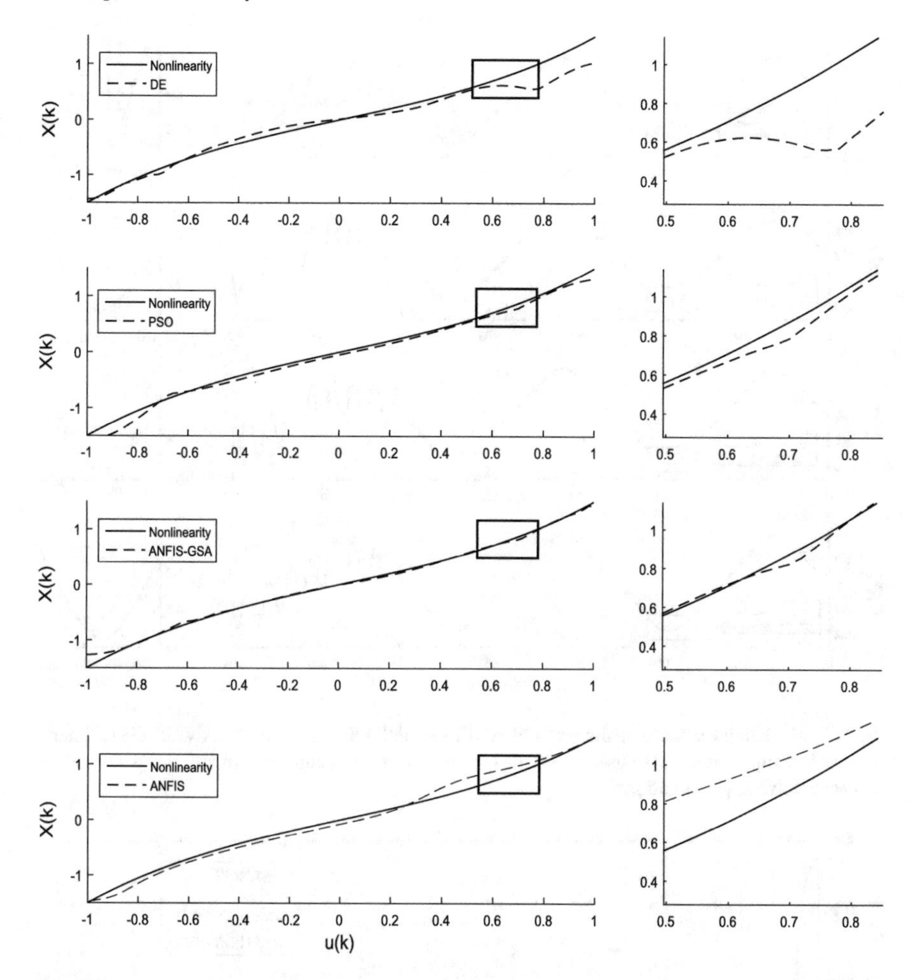

Fig. 5.6 Static non-linearity x(k) and the approximations generated by the DE, PSO, ANFIS-GSA, and the original ANFIS for the system of Eq. (5.22)

5.4.2 Test II

The non-linear system analyzed in this test is selected from [8]. This system is similar to the employed in Test I, but the input $u(k)$ and output of the static non-linearity $x(k)$ in the model are described by a set of linear piecewise functions such as is described below:

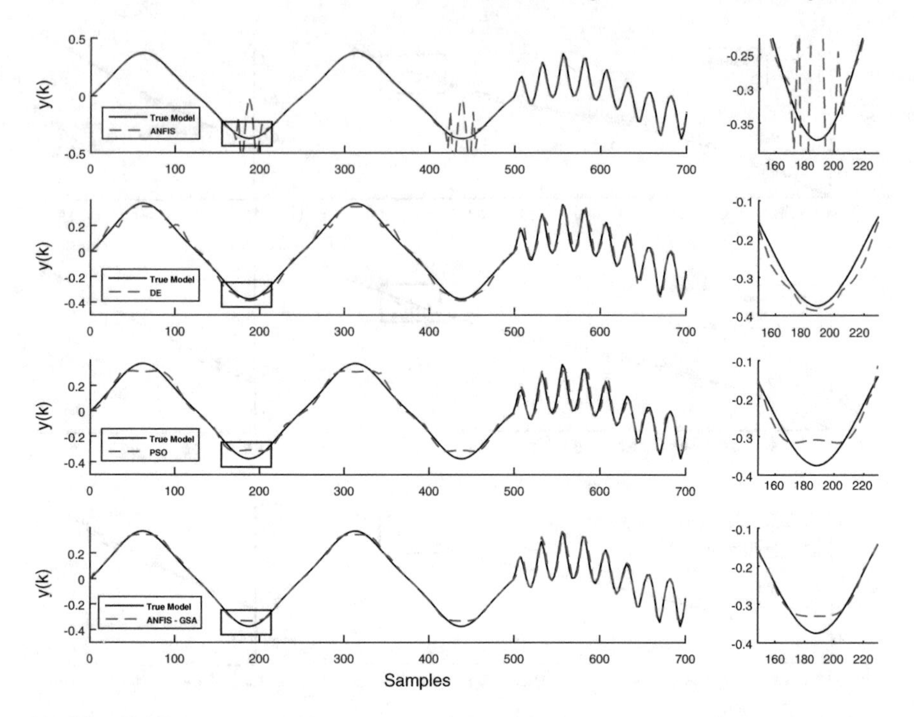

Fig. 5.7 Modeling response of the original ANFIS model, DE, PSO and the ANFIS-GSA scheme for Test I. **a** Input signal $u(k)$ used in the test, **b** output of the actual system and the generated by their respective approximations

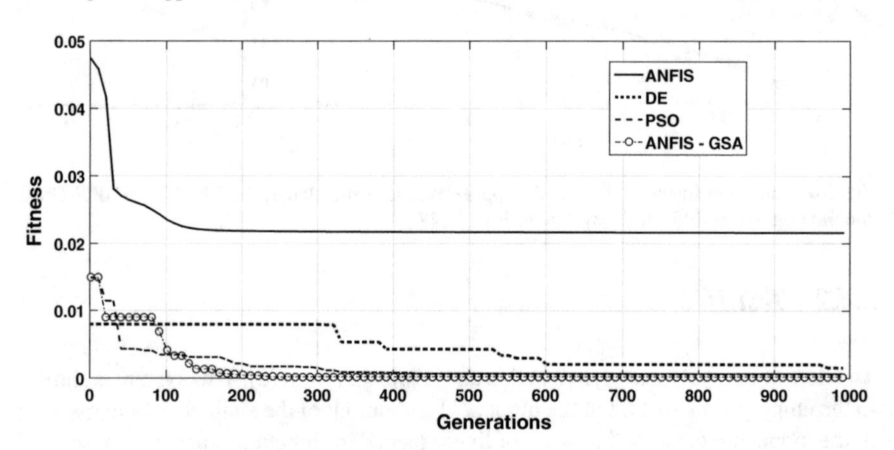

Fig. 5.8 The performance of the objective function generated through the identification process by DE, PSO, the ANFIS-GSA scheme and the original ANFIS model for Test I

Table 5.2 Experimental results of the Test 2

		B_J	W_J	μ_J	σ_J
EX 2	DE	5.99E−03	7.67E−02	1.68E−02	9.57E−03
	PSO	6.78E−03	9.02E−02	3.54E−02	2.09E−02
	ANFIS	1.01E−02	1.04E−02	1.03E−02	4.23E−05
	ANFIS-GSA	2.02E−03	9.26E−03	4.59E−03	1.64E−03

$$x(k) = \begin{cases} -2.0 & \text{if } -3.0 \leq u(k) < -1.8 \\ \frac{u(k)}{0.6} + 1 & \text{if } -1.8 \leq u(k) < -0.6 \\ 0 & \text{if } -0.6 \leq u(k) < 0.6 \\ \frac{u(k)}{0.6} - 1 & \text{if} 0.6 \leq u(k) < 1.8 \\ 2 & \text{if} 1.8 \leq u(k) < 3.0 \end{cases}, \tag{5.24}$$

Furthermore, in this test, the input $u(k)$ to the non-linear system as well as to the adaptive model is an arbitrary signal uniformly distributed in the range [−3, 3]. Considering this setting, the ANFIS-GSA scheme, DE, PSO, and the original ANFIS model are employed to the parameters identification $\mathbf{E}'_{ANFIS}$ of the system. Table 5.2 shows the approximation results after estimate the parameters. In Table 5.2, the ANFIS-GSA scheme shows the best results in terms of μ_J, B_J and W_J values. This evidence suggests that the ANFIS-GSA scheme provides better system estimation than the other techniques concerning the objective function J. The performance variations between ANFIS-GSA, DE, and PSO can be associated with their diverse capabilities for handling multi-modal functions. In contrast with the original ANFIS model generates minimal variation in its approximation (lower σ_J values), it converges consecutively in sub-optimal solutions (higher μ_J values). For this reason, the ANFIS-GSA scheme holds the most accurate approximation with low variability.

Figure 5.9 contrasts the non-linearity $x(k)$ of the real Hammerstein model considered in this test and the non-linearities generated by the four techniques after the estimation. Figure 5.9 presents the complexity to correctly determine the set of linear piecewise functions with the approximated systems. Under such modeling aspects, the ANFIS-GSA scheme achieves the best visual approximation regarding the other algorithms. Figure 5.10 illustrates the notable non-linear identification capabilities of the ANFIS scheme trained by the GSA. The figure presents the generated output of the approximated system regarding all the methods used for the comparative study, considering as input $u(k)$ the signal described in Eq. (5.23). In this figure, it is shown that the ANFIS-GSA scheme achieves the best approximation results regarding its competitors. The accuracy of the estimation is observed from the zoom window found in the right of Fig. 5.10. From these figures, it is clear that the ANFIS-GSA scheme provides the lower ripples in the approximation.

Furthermore, Fig. 5.11 shows the convergence comparison of the original ANFIS model, DE, PSO, and the ANFIS-GSA scheme regarding the objective function J. In Fig. 5.11, it is shown that the ANFIS-GSA scheme determines faster and better solutions than its competitors.

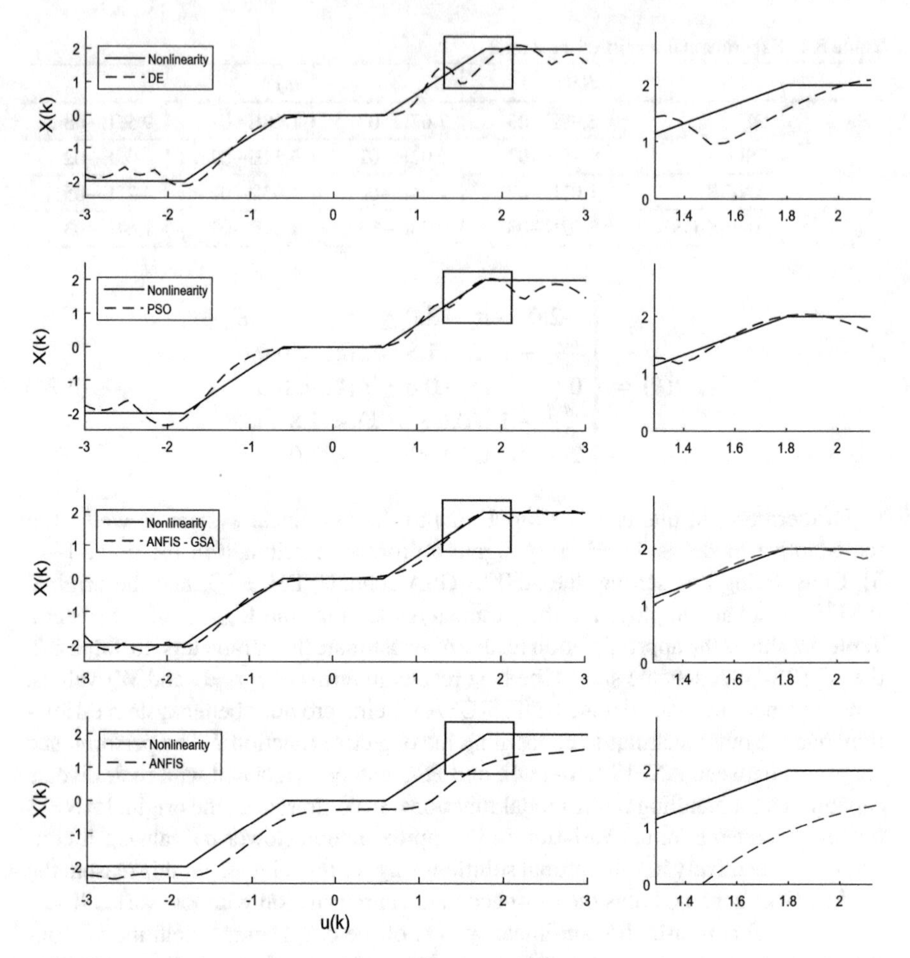

Fig. 5.9 Static non-linearity x(k) and the estimates generated by DE, PSO, the ANFIS-GSA scheme and the original ANFIS model for system Eqs. (5.22)–(5.24)

5.4.3 Test III

In this test, the non-linear system to be approximated is described as follows:

$$\begin{aligned} A(z) &= 1 + 0.9z^{-1} + 0.15z^{-2} + 0.02z^{-3}, \\ B(z) &= 0.7 + 1.5z^{-1}, \\ C(z) &= 1, \\ x(k) &= 0.5\sin^3(\pi u(k)) \end{aligned} \tag{5.25}$$

This system is taken from [47]. In the test, the input signal $u(k)$ employed in the approximation is represented by arbitrary numbers uniformly distributed in the range

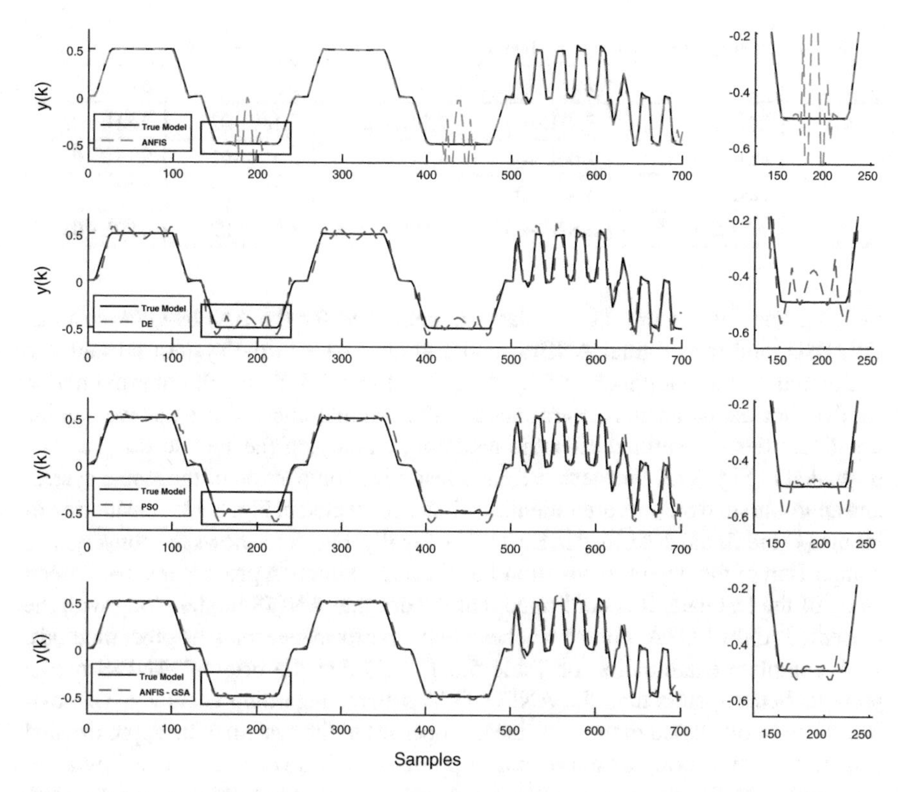

Fig. 5.10 Approximation results of the original ANFIS model, DE, PSO, and the ANFIS-GSA scheme for Test II. **a** Input signal u(k) considered in the test, **b** output of the actual model and the generated by the respective approximation

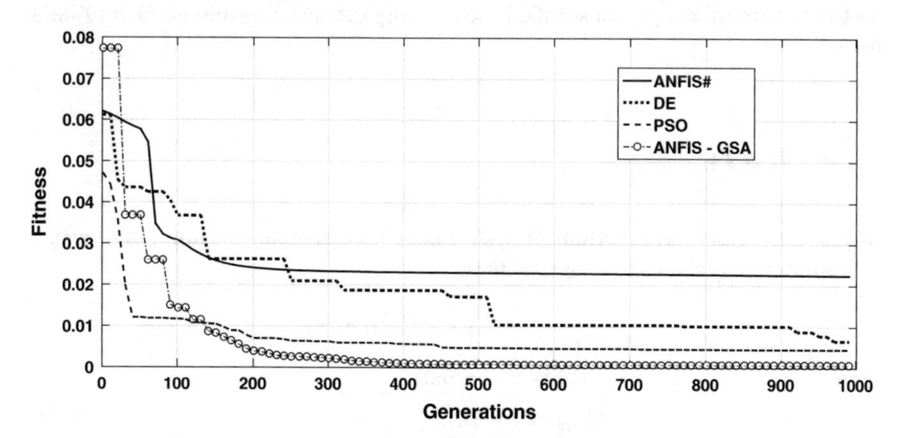

Fig. 5.11 The performance of the objective function generated through the approximation process by DE, PSO, the ANFIS-GSA scheme and the original ANFIS model for Test II

Table 5.3 Experimental results of the Test 3

		B_J	W_J	μ_J	σ_J
EX 3	DE	7.01E−03	2.21E−02	1.18E−02	2.81E−03
	PSO	1.06E−02	9.94E−02	3.81E−02	2.17E−02
	ANFIS	5.66E−03	2.84E−01	4.01E−02	1.51E−06
	ANFIS-GSA	5.82E−03	3.59E−01	1.07E−02	4.92E−03

[−1, 1]. For this reason, the test data are employed for the ANFIS-GSA scheme, DE, PSO, and the original ANFIS model to approximate the system through the estimation of the parameters vector $\mathbf{E}'_{ANFIS}$. Table 5.3 shows the approximation results after the parameter determination. The original non-linearity, as well as the non-linearities determined after the modeling employing the four techniques, are presented in Fig. 5.12. Furthermore, the illustrative comparison of the actual system and approximated outputs after identification is presented in Fig. 5.13, considering as input $u(k)$ the signal described in Eq. (5.23). Finally, Fig. 5.14 shows the convergence comparison of the objective function J as the approximation process results. After a study of the test data, it is confirmed that the original ANFIS model along with the presented ANFIS-GSA scheme reaches a better performance than the other methods.

A complete examination of Table 5.3 reveals that the original ANFIS model presents better results than the ANFIS-GSA scheme regarding accuracy. This evidence can be attributed to the non-linear properties of the system to be approximated (Eq. 5.25). Considering its non-linear response (Fig. 5.12) can be modeled by a linear composite of Gaussians, an objective function J with one minimum is generated. For this reason, it is well known that gradient-based methods, as back-propagation employed by the original ANFIS model, are a better option than evolutionary techniques to determine optimal solutions concerning the accuracy and number of iterations [59].

5.4.4 Test IV

In this test, the transfer functions $A(z)$, $B(z)$ and $C(z)$ that model the non-linear system is represented by the following structure:

$$\begin{aligned} A(z) &= 1 - 1.5z^{-1} + 0.7z^{-2}, \\ B(z) &= z^{-1} + 0.5z^{-2}, \\ C(z) &= 1 - 0.5z^{-1}, \end{aligned} \tag{5.26}$$

The non-linearity system taken from [47] is defined as:

$$x(k) = u(k) + 0.5u^2(k) + 0.3u^3(k) + 0.1u^4(k) \tag{5.27}$$

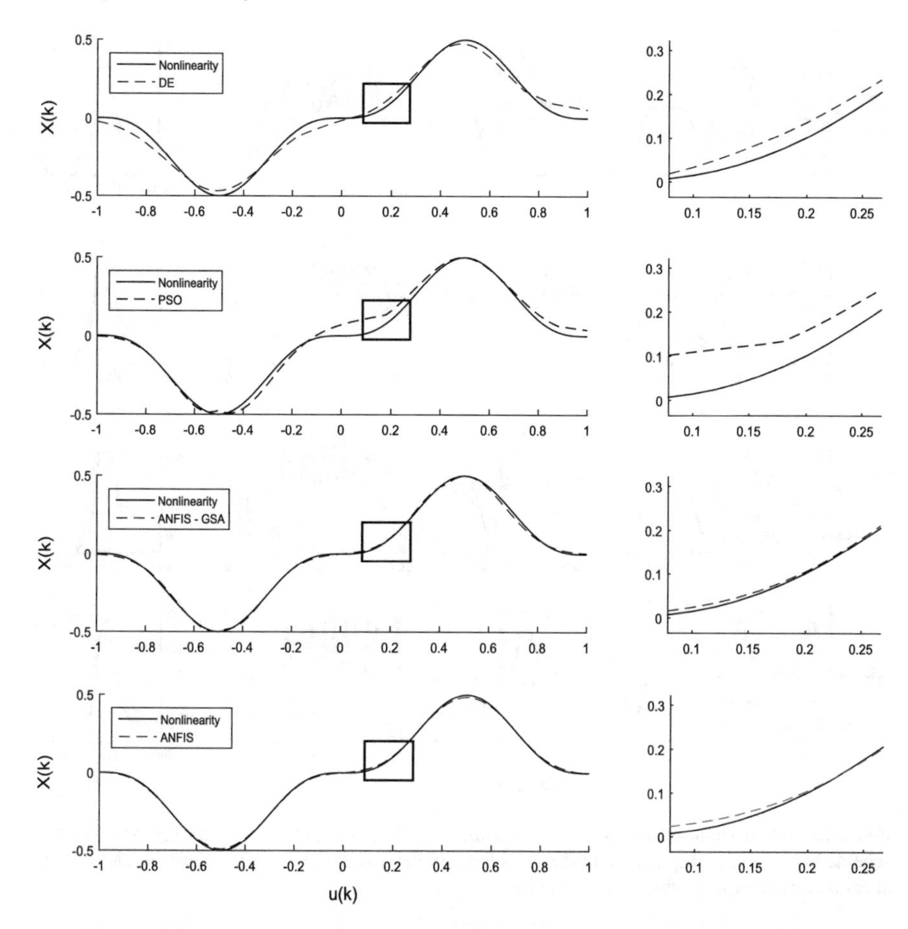

Fig. 5.12 Static non-linearity x(k) and the estimation generated by DE, PSO, the ANFIS-GSA scheme and the original ANFIS model for system Eq. (5.25)

The input u(k) employed in the approximation is a uniformly distributed arbitrary signal in the range $[-1, 1]$. Related to the other tests, the parameter vector $\mathbf{E}'_{ANFIS}$ is estimated by applying PSO, DE, the ANFIS-GSA scheme, and the original ANFIS model to minimize J. In this test, the polynomial order of the non-linear section is high, many turning points must be determined. As a result, a multi-modal response in J is assumed. Table 5.4 shows the results after the approximation process. According to these results, the ANFIS-GSA scheme achieves the lowest μ_J value, symbolizing that this technique generates the best system approximation. On the other hand, the approximation produced by the DE maintains a comparable performance, considering the variations of the μ_J values among ANFIS-GSA scheme and DE are short. Table 5.4 further shows that the ANFIS-GSA scheme offers the smallest discrepancy among the Best (B_J) and the Worst (W_J) system approximation. This evidence shows that the ANFIS-GSA scheme produces approximation with the lowest vari-

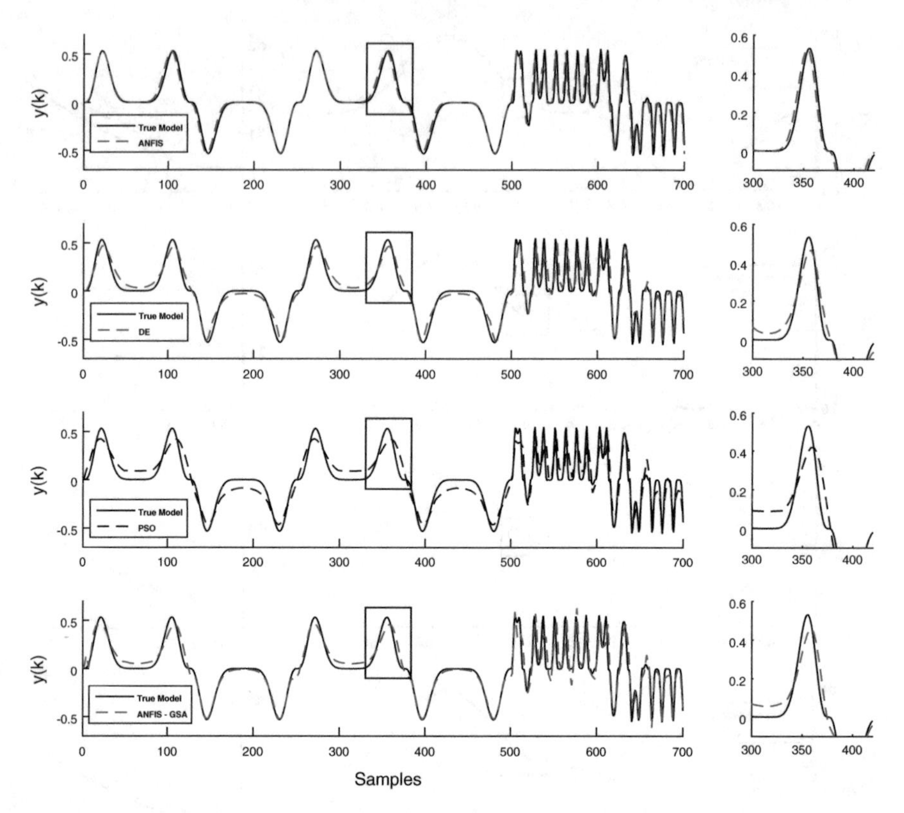

Fig. 5.13 Approximation results of the original ANFIS model, DE, PSO and the ANFIS-GSA scheme for Test III. **a** Input signal u(k) considered in the test, **b** output of the actual model and the generated by their respective approximation

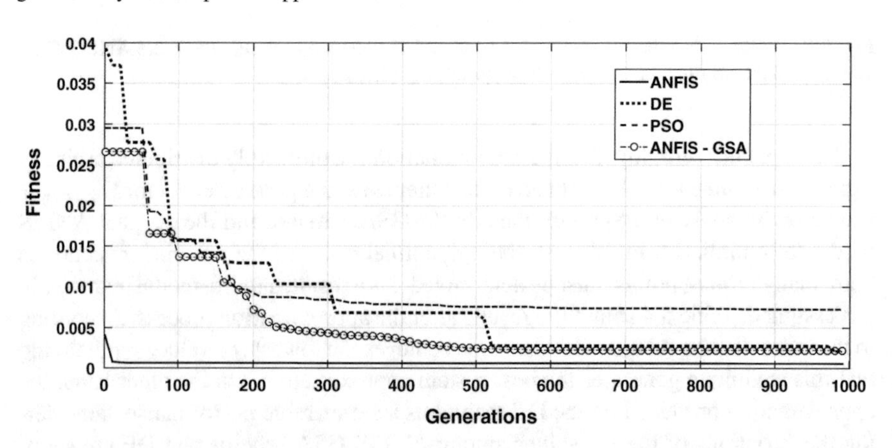

Fig. 5.14 Performance of the objective function generated during the approximation process by DE, PSO, the ANFIS-GSA scheme and the original ANFIS model for Test III

Table 5.4 Experimental results of Test 4

		B_J	W_J	μ_J	σ_J
EX 4	DE	2.02E−03	1.42E−02	3.58E−03	1.36E−03
	PSO	2.08E−03	8.98E−02	1.38E−02	1.69E−02
	ANFIS	6.50E−03	7.44E−01	7.75E−03	6.18E−02
	ANFIS-GSA	1.65E−03	5.21E−03	2.48E−03	9.01E−04

ation. This robustness is also valued in terms of the standard deviation values (σ_J), where the ANFIS-GSA scheme and the DE reach the smallest values. In this test, it is remarkable, that the original ANFIS model reaches the inferior σ_J concerning other techniques. This fact is described due to the multi-modality of the objective function J. Under such circumstances, the numerous local minima increase the localization of many sub-optimal solutions at each run, allowing a big variation in the ANFIS model approximation.

The non-linearities of the actual model, as well as the non-linearities determined by the approximation process, are shown in Fig. 5.15. From the figure, it is shown that the ANFIS-GSA scheme and DE offer the best graphical illustration regarding the actual non-linearity. Moreover, the comparative study of the actual system and the approximated models are presented in Fig. 5.16, considering the input $u(k)$ as the signal described in Eq. (5.23). From the figure, it is shown that the approximation accuracy of the ANFIS-GSA scheme is better regarding the other techniques. This fact is illustrated in the small zoom windows presented on the right side of the figure. Finally, Fig. 5.17 presents the convergence performance of the objective function J in the approximation process. After a study of Fig. 5.17, it can be seen that the presented ANFIS-GSA scheme presents good convergence qualities comparable with the other techniques.

5.4.5 Test V

In this test, we examine the system taken from [9] which is modeled as follows:

$$
\begin{aligned}
&A(z) = 1 + 0.3z^{-1} + 0.65z^{-2} - 0.35z^{-3},\\
&B(z) = 0.125 - 0.6z^{-1},\\
&C(z) = 1,\\
&x(k) = 0.5\sin^3(\pi u(k)) - \frac{2}{u^3(k) + 2} - 0.1\cos(4\pi u(k)) + 1.125
\end{aligned} \tag{5.28}
$$

In this analysis, the input signal $u(k)$ used to the non-linear system is considered as an arbitrary number uniformly distributed in the range $[-1, 1]$. Under such circumstances, a set of test data is generated. With the data, the ANFIS-GSA scheme, DE,

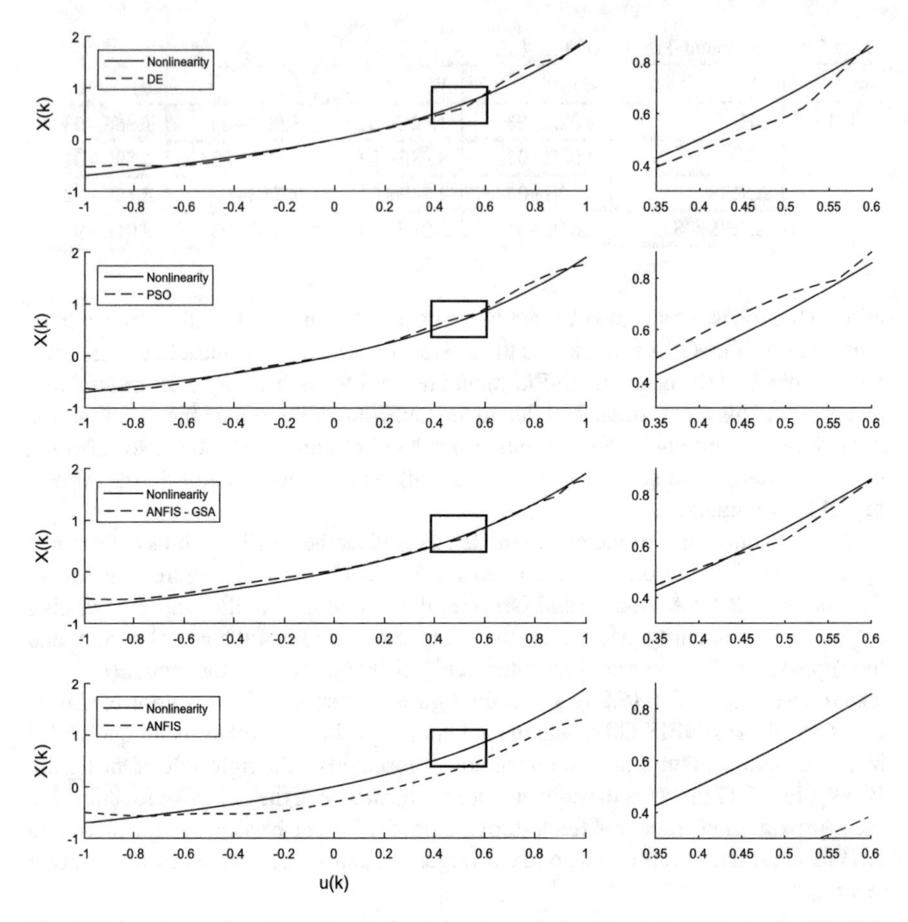

Fig. 5.15 Static non-linearity $x(k)$ and the estimation generated by DE, PSO, the ANFIS-GSA scheme and the original ANFIS model for system Eqs. (5.25)–(5.26)

PSO, and the original ANFIS model are employed to estimate the parameters $\mathbf{E}'_{ANFIS}$ of the system. In this test, since the trigonometric representation of the non-linear segment includes a complicated structure, a high multi-modal objective function J is generated. The test results achieved from 30 independent runs are shown in Table 5.5. According to Table 5.5, the ANFIS-GSA scheme produces better performance than DE, PSO and the original ANFIS model regarding B_J, W_J, μ_J and standard deviation. The accuracy variations between the ANFIS-GSA scheme, DE and PSO can be associated with their distinct capabilities for handling multi-modal objective functions. In this analysis, comparable to Test IV, the original ANFIS model achieves a large σ_J value due to the multi-modality of the objective function J. Figure 5.18 presents the ideal static non-linearity $x(k)$ and the estimation generated by ANFIS-GSA scheme, DE, PSO, and the original ANFIS model. An examination of Fig. 5.18 confirms that

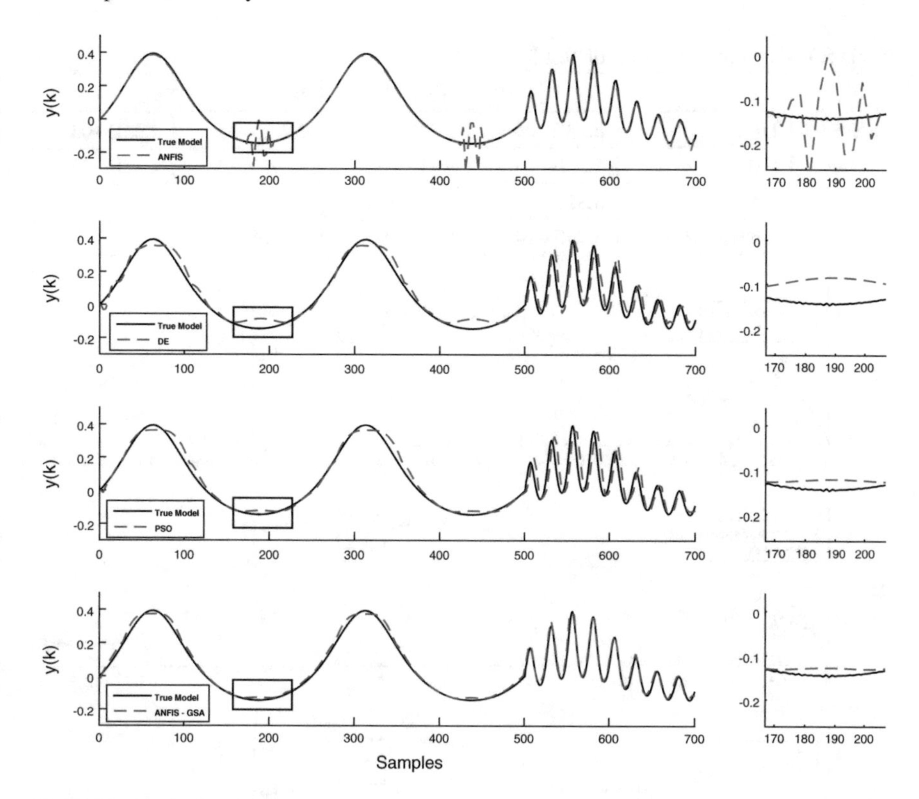

Fig. 5.16 Approximation results of the original ANFIS model, DE, PSO and the ANFIS-GSA scheme for Test IV. **a** Input signal $u(k)$ considered in the test, **b** output of the actual model and the generated by their respective approximations

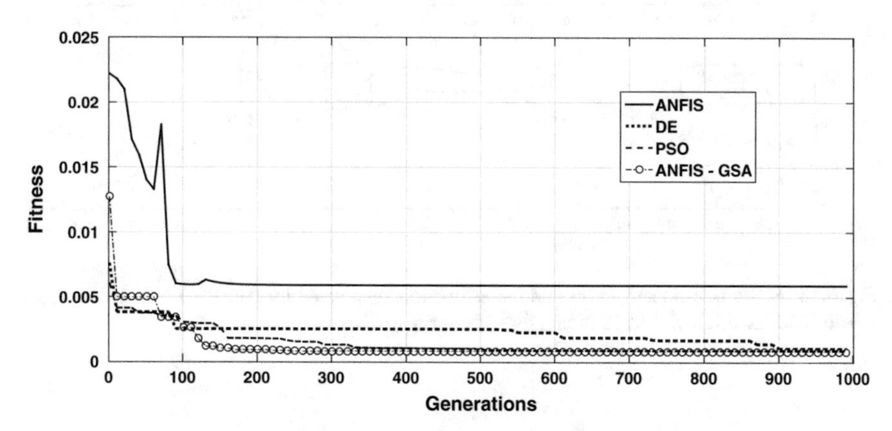

Fig. 5.17 Performance of the objective function generated during the approximation process by DE, PSO, the ANFIS-GSA scheme and the original ANFIS model for Test VI

Table 5.5 Experimental results of Test 5

		B_J	W_J	μ_J	σ_J
EX 5	DE	8.53E−04	9.34E−03	2.28E−03	1.05E−03
	PSO	1.33E−03	9.71E−02	1.29E−02	1.81E−02
	ANFIS	3.82E−03	5.58E−03	4.68E−03	3.80E−04
	ANFIS-GSA	3.79E−04	5.05E−03	1.16E−03	1.01E−04

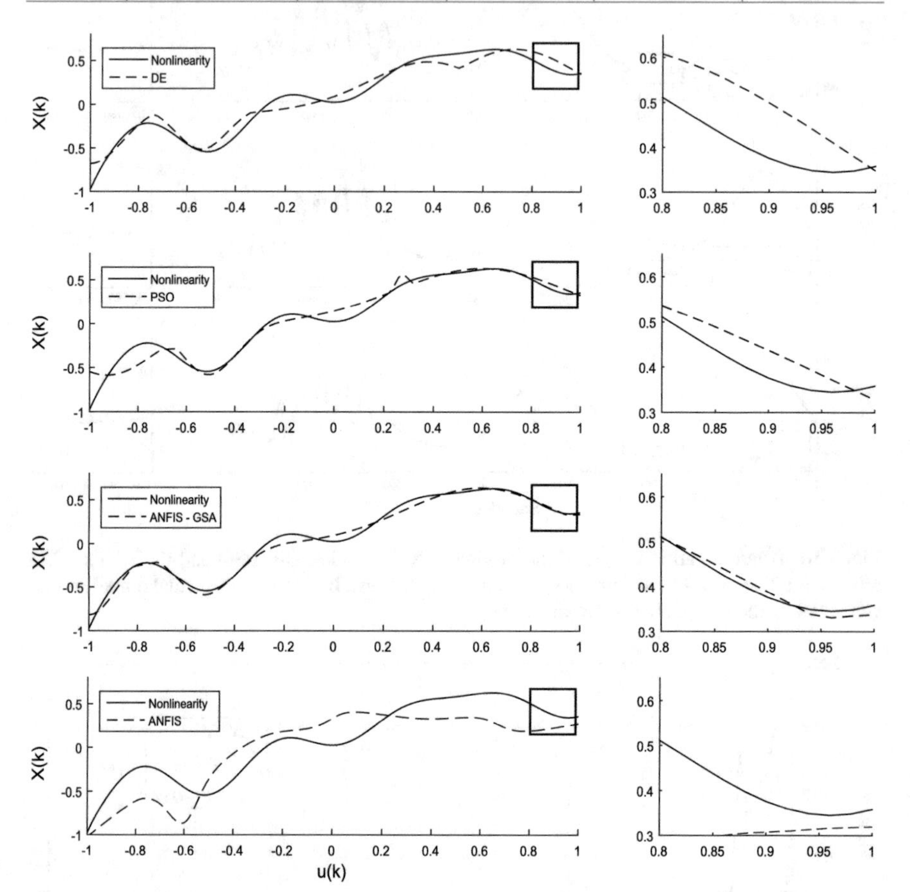

Fig. 5.18 Static non-linearity $x(k)$ and the estimations generated by DE, PSO, the ANFIS-GSA scheme and the original ANFIS model for system Eq. (5.28)

the ANFIS-GSA scheme achieves the best graphical illustration regarding the actual non-linearity.

To validate the accuracy of the approximation, the input signal $u(k)$, described in Eq. (5.29), is used as the signal test.

$$u(k) = \begin{cases} \sin(\frac{2\pi k}{250}) & 0<k \leq 250 \\ 0.8\sin(\frac{2\pi k}{250}) + 0.2\sin(\frac{2\pi k}{25}) & 250<k \leq 800 \end{cases} \quad (5.29)$$

In Fig. 5.19, the performance comparative analysis of the actual system and the approximated models are presented. From the figure, it can be seen that the ANFIS-GSA scheme archives the best graphical illustration identification between all approaches. On the other hand, Fig. 5.20 presents the convergence qualities of objective function J for the original ANFIS model, DE, PSO, and the ANFIS-GSA scheme. After a study of the convergence analysis, it is confirmed that the ANFIS-GSA scheme reaches the best accuracy between all techniques.

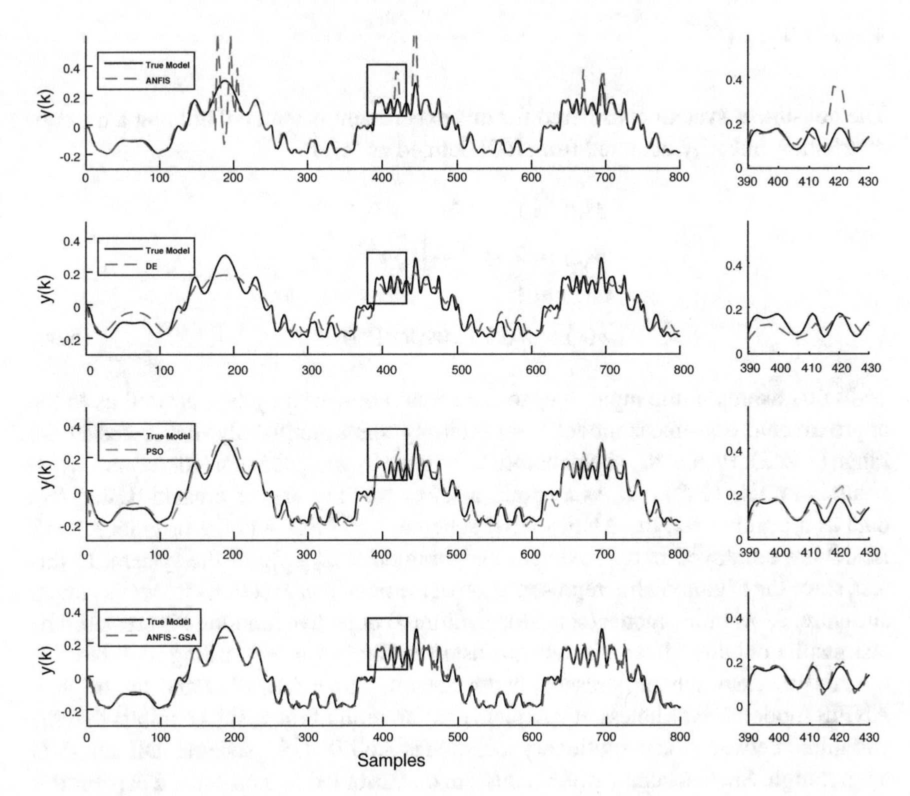

Fig. 5.19 Approximation results of the original ANFIS model, DE, PSO, and the ANFIS-GSA scheme for Test V. **a** Input signal $u(k)$ considered in the test, **b** output of the actual model and the produced by their respective approximation

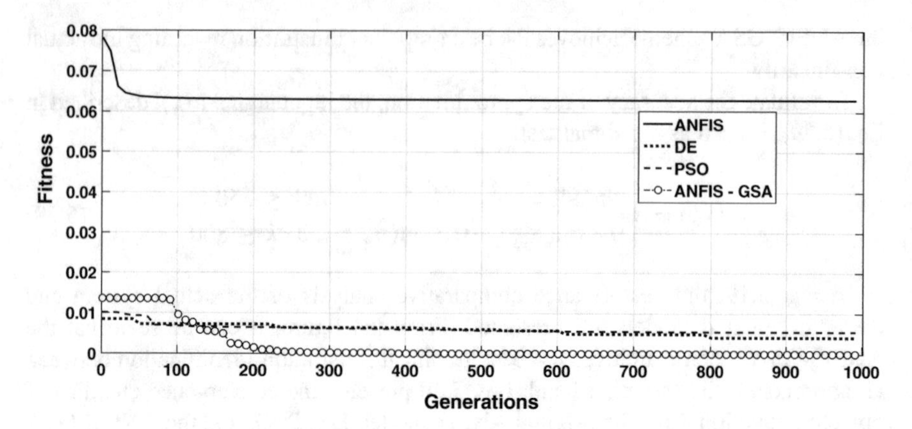

Fig. 5.20 Performance of the objective function generated through the approximation process by DE, PSO, the ANFIS-GSA scheme and the original ANFIS model for Test V

5.4.6 Test VI

The non-linear system considered for this experiment is a model holding a trigonometric non-linearity obtained from [60] defined as follows:

$$
\begin{aligned}
A(z) &= 1 + 0.9z^{-1} + 0.7z^{-2}, \\
B(z) &= 0.4z^{-1} - 0.5z^{-2}, \\
C(z) &= 1, \\
x(k) &= u(k) \cdot \cos(\pi u(k))
\end{aligned}
\tag{5.30}
$$

In this example, the input $u(k)$ to the actual non-linear system as well as to the approximated non-linear model is an arbitrary signal uniformly distributed in the range $[-2, 2]$. Hence, the signal before to be used by models is considered as a white Gaussian noise of 50 dB. As a result, a set of training data is created. Using this data as a training set, the ANFIS-GSA scheme, DE, PSO, and the original ANFIS model are employed to approximate the parameters $\mathbf{E}'_{ANFIS}$ of the system. In this test, since the trigonometric representation of the non-linear section includes a simple structure, a low multi-modal (some local minima) objective function J is created. The test results obtained from 30 runs are listed in Table 5.6. According to Table 5.6, the ANFIS-GSA scheme presents better accuracy than DE, PSO and the original ANFIS model. Nevertheless, a complete examination of the results exhibits that the variations between the evolutionary techniques ANFIS-GSA scheme, DE and PSO are not high. Such accuracy differences can be attributed to their related capabilities for solving low multi-modal objective problems.

Figure 5.21 presents the actual static non-linearity $x(k)$ and the estimation generated by ANFIS-GSA scheme, DE, PSO, and the original ANFIS model. Figure 5.21, shows that all the techniques perform a good visual estimation, but the ANFIS-GSA

Table 5.6 Experimental results of Test 6

		B_J	W_J	μ_J	σ_J
EX 6	DE	1.40E−03	3.98E−02	8.10E−03	5.60E−03
	PSO	7.22E−03	4.29E−02	5.60E−03	6.90E−03
	ANFIS	5.30E−03	5.60E−01	12.5E−03	7.32E−05
	ANFIS-GSA	5.14E−04	4.50E−03	2.30E−03	1.20E−03

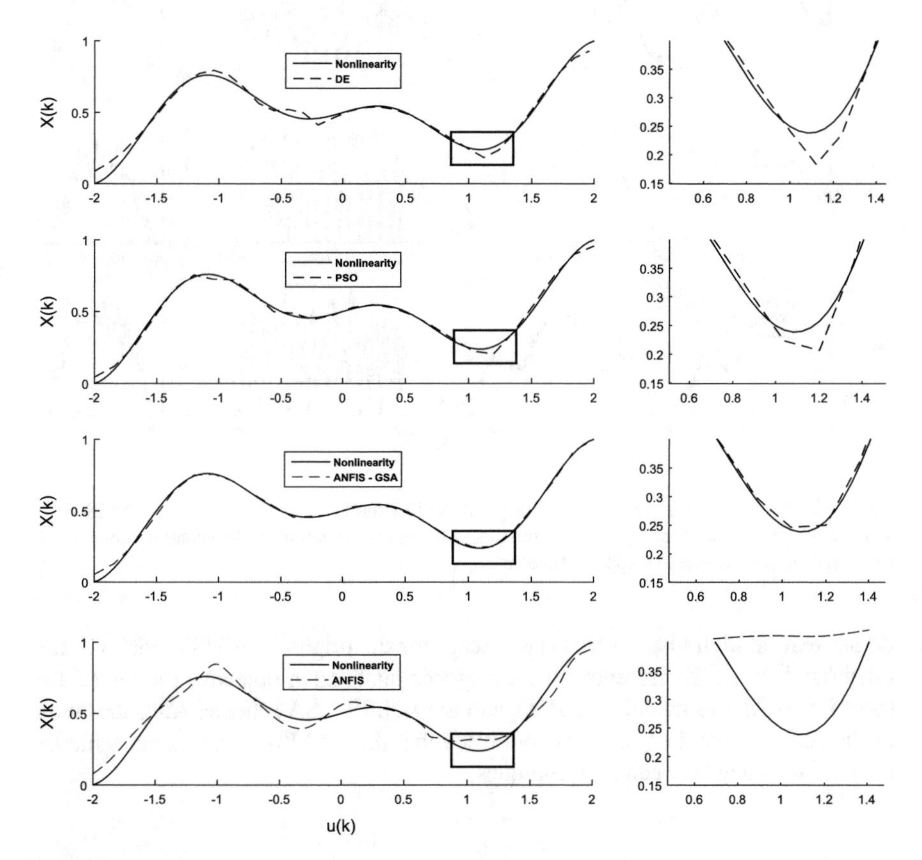

Fig. 5.21 Static non-linearity $x(k)$ and the estimations generated by DE, PSO, the ANFIS-GSA scheme and the original ANFIS model for system Eq. (5.30)

scheme performs the best similarity. For this test, the original ANFIS model holds the worst estimation, as a consequence of its insufficiency to handle multi-modal problems. In Fig. 5.22, the comparison between the actual system and the approximated models are presented, considering as input $u(k)$ the signal represented in Eq. (5.23). Figure 5.22 verify the same results that those exposed to Fig. 5.21. From the figure, it can be seen that the ANFIS-GSA scheme reaches the best graphical illustration of the estimation between all techniques. The graphs also point that they

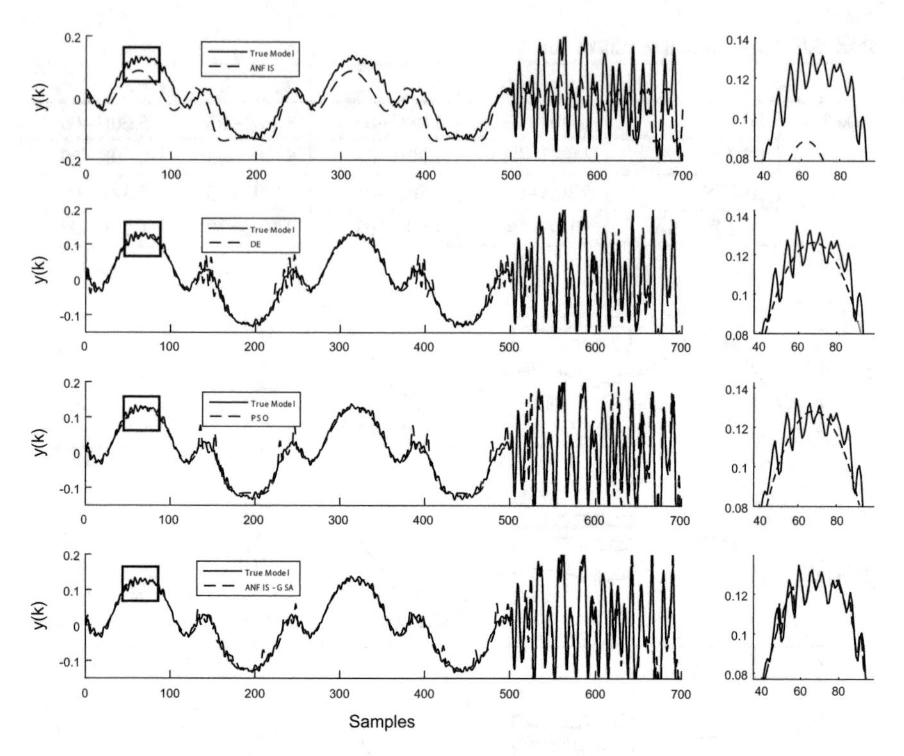

Fig. 5.22 Approximation results of the original ANFIS model, DE, PSO, and the ANFIS-GSA scheme for Test VI. **a** Input signal $u(k)$ considered in the test, **b** output of the actual model and the generated by their respective approximations

do not exhibit high visual variations, except for the original ANFIS model. On the other hand, Fig. 5.23 presents the convergence qualities of objective function J for the original ANFIS model, DE, PSO, and the ANFIS-GSA scheme. After the study of the convergence analysis, it is confirmed that the ANFIS-GSA scheme achieves the best accuracy between all techniques.

5.4.7 *Test VII*

The non-linear Hammerstein system studied in this test is presented to model the real non-linear dynamics of a heat exchanger system. This practical problem selected from [51]. Hence, the non-linear dynamics of this model is defined as follows:

$$A(z) = 1 - 1.608z^{-1} + 0.6385z^{-2},$$
$$B(z) = 0.207z^{-1} - 0.1764z^{-2},$$
$$C(z) = 1,$$

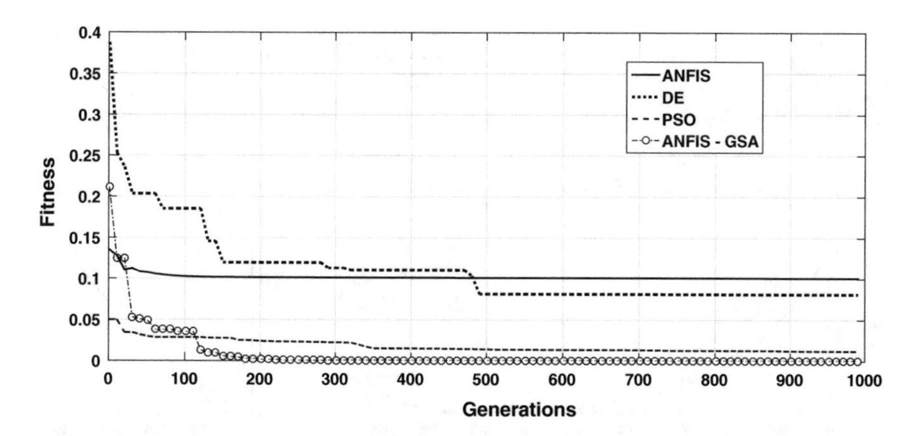

Fig. 5.23 Performance of the objective function generated during the estimation process by DE, PSO, the ANFIS-GSA scheme and the original ANFIS model for Test VI

$$x(k) = -31.549x(n) + 41.732x^2(n) - 24.201x^3(n) + 68.634x^4(n) \tag{5.31}$$

As can be seen, the characteristics of this system are comparable to those present in Test 4. In this analysis, the input of the non-linear system as well as to the adaptive model is an arbitrary signal uniformly distributed in the range $[-1, 1]$. Under this context, the ANFIS-GSA scheme, DE, PSO, and the original ANFIS model are used to estimate the parameters $\mathbf{E}'_{ANFIS}$ of the system. As a result of the training process, the results of the different techniques of the objective function J (Eq. 5.20) are described in Table 5.7. After an examination of Table 5.7, it is clear that the ANFIS-GSA scheme offers the best accuracy in terms of B_J, W_J, μ_J and σ_J values. This evidence indicates that the ANFIS-GSA scheme gets the most precise system identification for the model regarding the other techniques. The original ANFIS model maintains the worst accuracy of the whole set of methods due to the multi-modality.

Figure 5.24 presents the comparative analysis of the real non-linearity generated by the Hammerstein system with the approximated models generated by the ANFIS-GSA scheme, DE, PSO, and the original ANFIS model. According to this figure, it is easy to see that the ANFIS-GSA scheme offers the best graphics illustration of the approximations regarding the true model. Although the ANFIS-GSA scheme

Table 5.7 Experimental results of Test 7

		B_J	W_J	μ_J	σ_J
EX 7	DE	2.74E−02	1.30E+00	4.25E−01	3.11E−01
	PSO	2.90E−03	3.13E−01	6.49E−02	1.55E−01
	ANFIS	1.54E+00	1.54E+00	1.54E+00	3.14E−04
	ANFIS-GSA	2.62E−03	6.74E−02	7.63E−03	7.03E−03

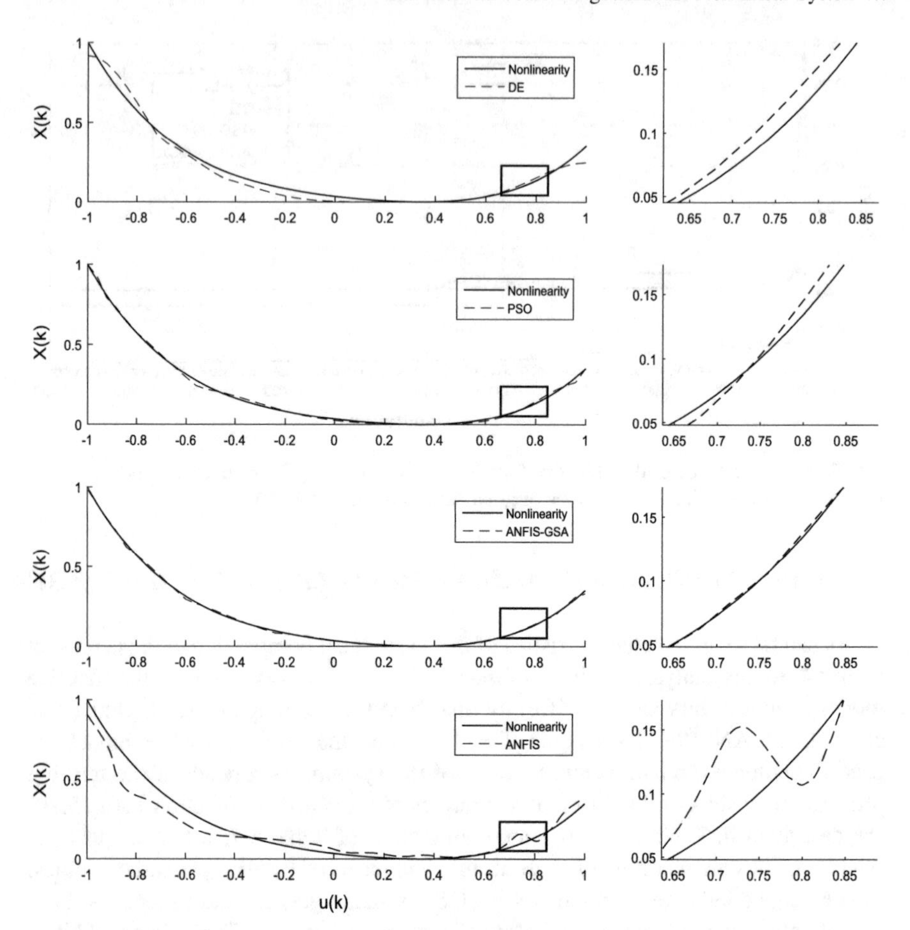

Fig. 5.24 Static non-linearity $x(k)$ and the estimation generated by DE, PSO, the ANFIS-GSA scheme and the original ANFIS model for system Eq. (5.31)

preserves the best relationship, the other approaches such as PSO and DE have competitive performance. Different from the other techniques, the original ANFIS model holds the worst performance due to its problems with multi-modal optimization problems.

In Fig. 5.25, the accuracy study of the actual model and the identified systems are presented. The input considered for the responses is the signal represented for Eq. (5.23). The graphs of Fig. 5.25 validate the results found for Fig. 5.24. From this figure, it is shown that the ANFIS-GSA scheme provides the best visual estimation between all techniques. On the other hand, Fig. 5.26 presents the convergence qualities of objective function J for the original ANFIS model, DE, PSO, and the ANFIS-GSA scheme. After a study of the performance results, it is confirmed that the ANFIS-GSA scheme achieves the best accuracy regarding other techniques.

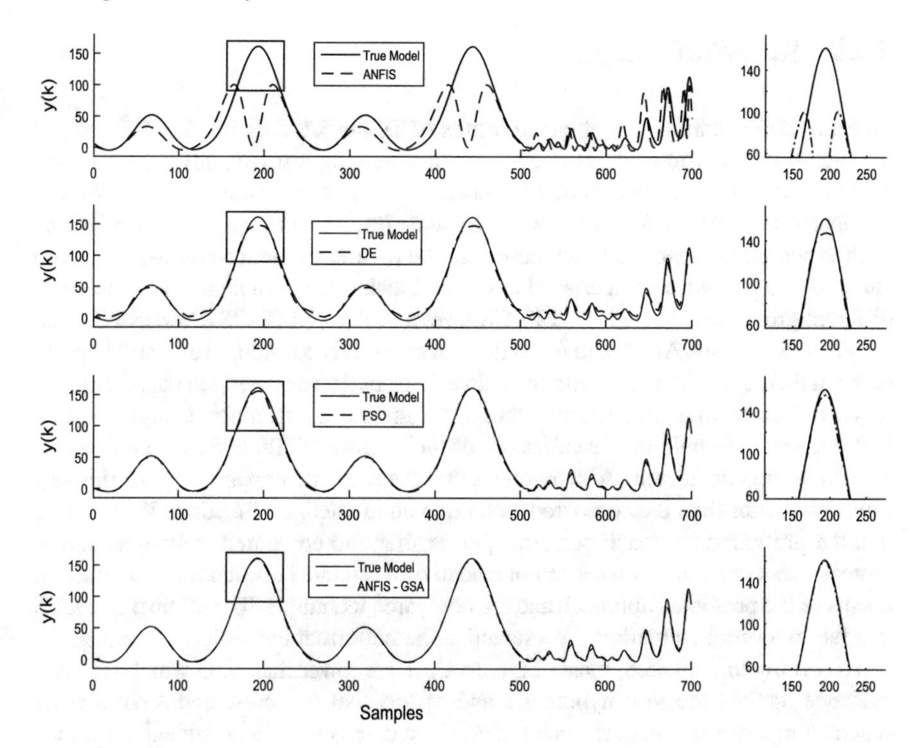

Fig. 5.25 Approximation results of the original ANFIS model, DE, PSO, and the ANFIS-GSA scheme for Test VII. **a** Input signal $u(k)$ considered in the test, **b** output of the actual model and the produced by their respective approximations

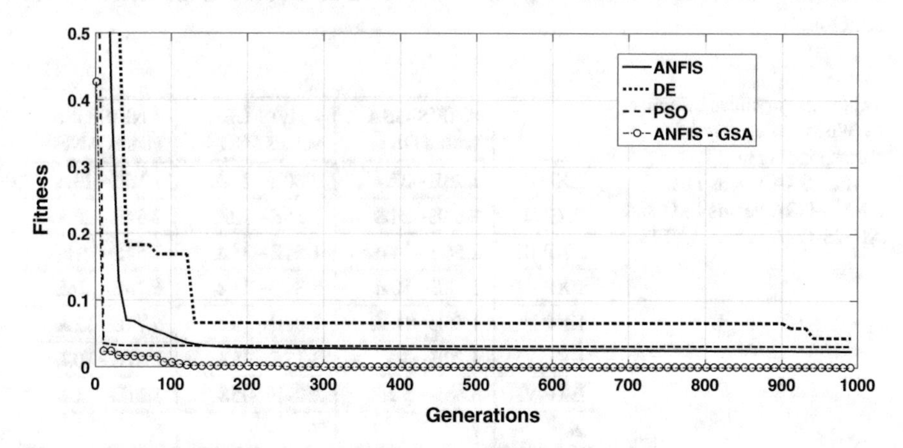

Fig. 5.26 Performance of the objective function generated during the approximation process by DE, PSO, the ANFIS-GSA scheme and the original ANFIS model for Test VII

5.4.8 Statistical Study

To statistically examine the obtained results of Tables 5.1, 5.2, 5.3, 5.4, 5.5, 5.6 and 5.7, a non-parametric analysis known as the Wilcoxon test [61, 62] is carried out. It allows determining the differences between two related approaches. The study is conducted using the 5% (0.005) of significance level over the μ_J data provided by each experimental test. Table 5.8 describes the p-values produced by the Wilcoxon study for the pair-wise comparison between the techniques. Under this scheme, three different groups are created: ANFIS-GSA versus DE, ANFIS-GSA versus PSO and ANFIS-GSA versus ANFIS. In the Wilcoxon study, it is considered as a null hypothesis that there is no difference enough between the two techniques. On the other hand, it is accepted as an alternative hypothesis that there is a difference enough between both methods. To help the visualization of the analysis of Table 5.8, the figures ▲, ▼, and ► are considered. ▲ symbolizes that the presented approach performs significantly better than the compared technique on the defined function. ▼ expresses that the presented approach performs poorer than the compared technique, and ► indicates that the Wilcoxon test is not able to differentiate between the experimental results of the presented approach and the compared technique. The number of events that fall in a certain situation is presented at the bottom of the table.

According to Table 5.8, most of the p-values are lower than 0.05 which is strong evidence against the null hypothesis and means that the presented ANFIS-GSA scheme outperforms the other techniques. Such data is statistical evidence that they have not happened by chance (i.e. due to the normal noise included in the process). On exception, according to the results from Table 5.6, in Test 3 there is no statistical evidence of the difference between the original ANFIS model and the ANFIS-GSA scheme.

Table 5.8 *p*-values provided by Wilcoxon's analysis corresponding to ANFIS-GSA versus DE, ANFIS-GSA versus PSO and ANFIS-GSA versus ANFIS

	ANFIS-GSA versus DE	ANFIS-GSA versus PSO	ANFIS-GSA versus ANFIS
EXP I	1.28E−22▲	2.60E−22▲	1.16E−23▲
EXP II	2.64E−24▲	1.56E−30▲	2.54E−28▲
EXP III	2.56E−14▲	1.51E−34▲	1.54E−01►
EXP IV	2.60E−10▲	8.89E−21▲	6.74E−10▲
EXP V	1.20E−10▲	7.88E−08▲	6.74E−12▲
EXP VI	4.20E−4▲	3.17E−05▲	1.07E−10▲
EXP VII	6.45E−5▲	2.29E−05▲	3.11E−11▲
▲	7	7	6
►	0	0	1
▼	0	0	0

5.5 Conclusions

Most of the real-life systems have a non-linear response. For this reason, their identification process is considered a difficult problem, the development of novel identification techniques has attracted the attention in many fields of research. One typical case of such modeling systems is t Hammerstein models that have demonstrated their effectiveness in the identification of difficult non-linear systems.

In this chapter, a non-linear system identification technique using the Hammerstein model is presented [63]. In the presented approach, a non-linear plant is approximated by the adaptation of an ANFIS model, using the similitude among it and the Hammerstein system. To determine the parameters of the system, the presented technique employs a nature-inspired approach knows as the GSA. Different from the recent optimization methods, GSA achieves a better performance in difficult multimodal problems, avoiding the premature convergence to sub-optimal solutions.

To confirm the capability and robustness of our method, the identification scheme is evaluated analyzing an experimental set of 7 complicated tests. To evaluate the performance of the presented technique, other famous evolutionary algorithms as the PSO, DE and the original ANFIS model have been used to compare its training capabilities. The obtained results are statistically validated using the Wilcoxon test, confirming that the presented approach performs better than the other optimization techniques in terms of its solution quality for almost all experiments.

The Hammerstein identification using evolutionary computation techniques allows several interesting subjects on which more study would be useful. Further areas involve: (1) The description of novel Hammerstein benchmark systems that provide an appropriate assessment of identification methods. Nowadays, the benchmark systems reflect the most common non-linearities, under these circumstances, it is necessary the addition of more complexities now known in traditional non-linear systems to Hammerstein models. (2) The use of multi-objective optimization approaches to enhance the accurate identification of non-linear and linear blocks. The presented approach in this chapter studies the identification system as a problem with a single objective function (minimize Eq. 5.20). Considering the multi-modal nature of the problem, the identification process can be considered to two distinct objectives. The first objective concerns to the approximation of the non-linear system while the second objective represents the quality solution of the determination of the linear filter. (3) The development of a novel single objective function that includes structural components to optimize not just the ANFIS model parameters, but also its structure.

References

1. L. Piroddi, M. Farina, M. Lovera, Black box model identification of nonlinear input–output models: a Wiener–Hammerstein benchmark. Control Eng. Pract. **20**(11), 1109–1118 (2012)

2. F. Jurado, A method for the identification of solid oxide fuel cells using a Hammerstein model. J. Power Sources **154**(1), 145–152 (2006)
3. X. Zhang, Y. Tan, Modelling of ultrasonic motor with dead-zone based on Hammerstein model structure. J. Zhejiang Univ. Sci. A **9**(1), 58–64 (2005)
4. P. Gilabert, G. Montoro, E. Bertran, On the Wiener and Hammerstein models for power amplifier predistortion. Asia-Pac. Microw. Conf. Proc. APMC **2**(2), 5–8 (2005)
5. H. Huo, X. Zhu, W. Hu, H. Tu, J. Li, J. Yang, Nonlinear model predictive control of SOFC based on a Hammerstein model. J. Power Sources **185**(1), 38–344 (2008)
6. J.G. Smith, S. Kamat, K.P. Madhavan, Modeling of pH process using wavenet based Hammerstein model. J. Process Control **17**(6), 551–561 (2007)
7. M. Cui, H. Liu, Z. Li, Y. Tang, X. Guan, Identification of Hammerstein model using functional link artificial neural network. Neurocomputing **142**, 419–428 (2014)
8. S.J. Nanda, G. Panda, B. Majhi, Improved identification of Hammerstein plants using new CPSO and IPSO algorithms. Expert Syst. Appl. **37**(10), 6818–6831 (2010)
9. A. Gotmare, R. Patidar, N.V. George, Nonlinear system identification using a cuckoo search optimized adaptive Hammerstein model. Expert Syst. Appl. **42**(5), 2538–2546 (2015)
10. Y. Tang, Z. Li, X. Guan, Identification of nonlinear system using extreme learning machine based Hammerstein model. Commun. Nonlinear Sci. Numer. Simul. **19**(9), 3171–3183 (2014)
11. J.S.R. Jang, ANFIS: adaptive-network-based fuzzy inference system. IEEE Trans. Syst. Man Cybern. **23**(3), 665–685 (1993)
12. T. Takagi, M. Sugeno, Fuzzy identification of systems and its applications to modeling and control. IEEE Trans. Syst. Man Cybern. SMC **15**, 116–132 (1985)
13. C. Vairappan, H. Tamura, S. Gao, Z. Tang, Batch type local search-based adaptive neuro-fuzzy inference system (ANFIS) with self-feedbacks for time-series prediction. Neurocomputing **72**(7–9), 1870–1877 (2009)
14. M. Mohandes, S. Rehman, S.M. Rahman, Estimation of wind speed profile using adaptive neuro-fuzzy inference system (ANFIS). Appl. Energy **88**(11), 4024–4032 (2011)
15. K. Salahshoor, M. Kordestani, M.S. Khoshro, Fault detection and diagnosis of an industrial steam turbine using fusion of SVM (support vector machine) and ANFIS (adaptive neuro-fuzzy inference system) classifiers. Energy **35**(12), 5472–5482 (2010)
16. S.N. Engin, J. Kuvulmaz, V.E. Ömurlü, Fuzzy control of an ANFIS model representing a nonlinear liquid-level system. Neural Comput. Appl. **13**(3), 202–210 (2004)
17. X.J. Wu, X.J. Zhu, G.Y. Cao, H.Y. Tu, Nonlinear modeling of a SOFC stack based on ANFIS identification. Simul. Model. Pract. Theory **16**(4), 399–409 (2008)
18. I. Zelinka, A survey on evolutionary algorithms dynamics and its complexity—mutual relations, past, present and future. Swarm Evol. Comput. **25**, 2–14 (2015)
19. H.N. Al-Duwaish, Nonlinearity structure using particle swarm optimization. Arab. J. Sci. Eng. **36**(7), 1269–1276 (2011)
20. E. Cuevas, A. González, D. Zaldívar, M. Pérez-Cisneros, An optimisation algorithm based on the behaviour of locust swarms. Int. J. Bio-Inspir. Comput. **7**(6), 402–407 (2015)
21. S. Mete, S. Ozer, H. Zorlu, System identification using Hammerstein model optimized with differential evolution algorithm. AEU—Int. J. Electron. Commun. **70**(12), 1667–1675 (2016)
22. E. Rashedi, H. Nezamabadi-Pour, S. Saryazdi, GSA: a gravitational search algorithm. Inf. Sci. **179**(13), 2232–2248 (2009)
23. F. Farivar, M.A. Shoorehdeli, Stability analysis of particle dynamics in gravitational search optimization algorithm. Inf. Sci. **337–338**, 25–43 (2016)
24. S. Yazdani, H. Nezamabadi-Pour, S. Kamyab, A gravitational search algorithm for multimodal optimization. Swarm Evol. Comput. **14**, 1–14 (2014)
25. A. Yazdani, T. Jayabarathi, V. Ramesh, T. Raghunathan, Combined heat and power economic dispatch problem using firefly algorithm. Front. Energy **7**(2), 133–139 (2013)
26. V. Kumar, J.K. Chhabra, D. Kumar, Automatic cluster evolution using gravitational search algorithm and its application on image segmentation. Eng. Appl. Artif. Intell. **29**, 93–103 (2014)

27. W. Zhang, P. Niu, G. Li, P. Li, Forecasting of turbine heat rate with online least squares support vector machine based on gravitational search algorithm. Knowl.-Based Syst. **39**, 34–44 (2015)
28. D. Karaboga, E. Kaya, An adaptive and hybrid artificial bee colony algorithm (aABC) for ANFIS training. Appl. Soft Comput. **49**, 423–436 (2016)
29. A. Sarkheyli, A.M. Zain, S. Sharif, Robust optimization of ANFIS based on a new modified GA. Neurocomputing **166**, 357–366 (2015)
30. L.Y. Wei, A GA-weighted ANFIS model based on multiple stock market volatility causality for TAIEX forecasting. Appl. Soft Comput. **13**(2), 911–920 (2013)
31. D.P. Rini, S.M. Shamsuddin, S.S. Yuhaniz, Particle swarm optimization for ANFIS interpretability and accuracy. Soft Comput. **20**(1), 251–262 (2016)
32. J.S. Wang, C.X. Ning, ANFIS based time series prediction method of bank cash flow optimized by adaptive population activity PSO algorithm. Information **6**(3), 300–313 (2015)
33. P. Liu, W. Leng, W. Fang, Training ANFIS model with an improved quantum-behaved particle swarm optimization algorithm. Math. Probl. Eng. **2013** (2013)
34. J.P.S. Catalão, H.M.I. Pousinho, V.M.F. Mendes, Hybrid wavelet-PSO-ANFIS approach for short-term electricity prices forecasting. IEEE Trans. Power Syst. **26**(1), 137–144 (2011)
35. S. Suja Priyadharsini, S. Edward Rajan, S. Femilin Sheniha, A novel approach for the elimination of artefacts from EEG signals employing an improved Artificial Immune System algorithm. J. Exp. Theor. Artif. Intell. **21**, 1–21 (2015)
36. M. Gunasekaran, K.S. Ramaswami, A fusion model integrating ANFIS and artificial immune algorithm for forecasting Indian stock market. J. Appl. Sci. **11**(16), 3028–3033 (2011)
37. M. Asadollahi-Baboli, In silico prediction of the aniline derivatives toxicities to *Tetrahymena pyriformis* using chemometrics tools. Toxicol. Environ. Chem. **94**(10), 2019–2034 (2012)
38. R. Teimouri, H. Baseri, Optimization of magnetic field assisted EDM using the continuous ACO algorithm. Appl. Soft Comput. **14**, 381–389 (2014)
39. G.S. Varshini, S.C. Raja, P. Venkatesh, Design of ANFIS controller for power system stability enhancement using FACTS device, in *Power Electronics and Renewable Energy Systems* (Springer, India, 2015), pp. 1163–1171
40. M. Bhavani, K. Selvi, L. Sindhumathi, Neuro fuzzy load frequency control in a competitive electricity market using BFOA tuned SMES and TCPS, in *Swarm, Evolutionary, and Memetic Computing* (Springer International Publishing, 2014), pp. 373–385
41. M. Azarbad, H. Azami, S. Sanei, A. Ebrahimzadeh, New neural network-based approaches for GPS GDOP classification based on neuro-fuzzy inference system, radial basis function, and improved bee algorithm. Appl. Soft Comput. **25**, 285–292 (2014)
42. M.A.F. Rani, B. Sankaragomathi, Performance enhancement of PID controllers by modern optimization techniques for speed control of PMBL DC motor. Res. J. Appl. Sci. Eng. Technol. **10**(10), 1154–1163 (2015)
43. J.F. Chen, Q.H. Do, A cooperative cuckoo search–hierarchical adaptive neuro-fuzzy inference system approach for predicting student academic performance. J. Intell. Fuzzy Syst. **27**(5), 2551–2561 (2014)
44. S.M. Khazraee, A.H. Jahanmiri, S.A. Ghorayshi, Model reduction and optimization of reactive batch distillation based on the adaptive neuro-fuzzy inference system and differential evolution. Neural Comput. Appl. **20**(2), 239–248 (2011)
45. A.Z. Zangeneh, M. Mansouri, M. Teshnehlab, A.K. Sedigh, Training ANFIS system with DE algorithm, in *2011 Fourth International Workshop on Advanced Computational Intelligence (IWACI)*, IEEE, Oct 2011, pp. 308–314
46. F. Afifi, N.B. Anuar, S. Shamshirband, K.K.R. Choo, DyHAP: dynamic hybrid ANFIS-PSO approach for predicting mobile malware. PLoS ONE (2016). https://doi.org/10.1371/journal.pone.0162627
47. A. Gotmare, R. Patidar, N.V. George, Nonlinear system identification using a cuckoo search optimized adaptive Hammerstein model. Expert Syst. Appl. **42**, 2538–2546 (2015)
48. X. Gao, X. Ren, C. Zhu, C. Zhang, Identification and control for Hammerstein systems with hysteresis non-linearity. IET Control Theory Appl. **9**(13), 1935–1947 (2015)

49. J. Voros, Identification of Hammerstein systems with time-varying piecewise-linear characteristics. IEEE Trans. Circuits Syst. II Express Briefs **52**(12), 865–869 (2005)
50. W. Greblicki, Non-parametric orthogonal series identification of Hammerstein systems. Int. J. Syst. Sci. **20**(12), 2355–2367 (1989)
51. E. Eskinat, S.H. Johnson, W.L. Luyben, Use of Hammerstein models in identification of non-linear systems. AIChE J. **37**(2), 255268 (1991)
52. R. Storn, K. Price, *Differential Evolution—A Simple and Efficient Adaptive Scheme for Global Optimisation Over Continuous Spaces*. Technical Report TR-95-012 (ICSI, Berkeley, CA, 1995)
53. J. Kennedy, R. Eberhart, Particle swarm optimization, in *Proceedings of the 1995 IEEE International Conference on Neural Networks*, vol. 4 (1995), pp. 1942–1948
54. http://www.sciencedirect.com
55. http://www.springerlink.com
56. http://ieeexplore.ieee.org
57. E. Cuevas, Block-matching algorithm based on harmony search optimization for motion estimation. Appl. Intell. **39**(1), 165–183 (2013)
58. T. Hachino, K. Deguchi, H. Takata, Identification of Hammerstein model using radial basis function networks and genetic algorithm, in *2004 5th Asian Control Conference*, vol. 1 (2004), pp. 124–129
59. R. Salomon, Evolutionary algorithms and gradient search: similarities and differences. IEEE Trans. Evol. Comput. **2**(2), 45–55 (1998)
60. T. Hatanaka, K. Uosaki, M. Koga, Evolutionary computation approach to block oriented non-linear model identification, in *5th Asian Control Conference*, vol. 1, Malaysia, 2004, pp. 90–96
61. F. Wilcoxon, Individual comparisons by ranking methods. Biometrics **1**, 80–83 (1945)
62. S. Garcia, D. Molina, M. Lozano, F. Herrera, A study on the use of non-parametric tests for analyzing the evolutionary algorithms' behaviour: a case study on the CEC'2005 special session on real parameter optimization. J. Heurist. **15**, 617–644 (2009)
63. E. Cuevas, P. Díaz, O. Avalos, D. Zaldívar, M. Pérez-Cisneros, Nonlinear system identification based on ANFIS-Hammerstein model using gravitational search algorithm. Appl. Intell. **48**(1), 182–203 (2018)

Chapter 6
Fuzzy Logic Based Optimization Algorithm

6.1 Introduction

There are many real-world problems which are solved better by humans than deterministic systems. This may be due to our unique thinking abilities and complex cognitive processing. Although some processes can be complex, under such circumstances, humans undertake them by using simple rules of thumb obtained from their expertise.

Fuzzy logic [1] is a useful discipline for a variety of challenging purposes since it provides a convenient design for constructing systems through the use of human experience. In the fuzzy logic scheme, it is asked to write down a collection of rules to manipulate certain process. Then, we incorporate them into a fuzzy inference system that emulates the decision-making process [2]. Under such circumstances, the partitioning of the system into regions is an important characteristic of a fuzzy system [3]. For each region, the inherent characteristics of the system can be easily modeled using a rule that connects the region with some actions [4]. Typically, a fuzzy representation consists of a set of rules, where the information available is understandable and easily legible. The fuzzy modeling methodology has been widely employed in many fields such as pattern recognition [5, 6], control [7, 8] and image processing [9, 10].

Recently, many optimization algorithms based on stochastic principles have been introduced with impressive results. Such methods are inspired by our scientific knowledge of biological or social systems, which at some level of abstraction can be represented as optimization processes [11]. These methods imitate the social behavior of the bird flocking and the fish schooling in the Particle Swarm Optimization (PSO) method [12], the collective behavior of bee colonies in the Artificial Bee Colony (ABC) technique [13], the improvisation process that happens when a musician seeks for a better state of harmony in the Harmony Search (HS) [14], the characteristics of the bat behavior in the Bat Algorithm (BAT) method [15], the mating behavior of firefly insects in the Firefly (FF) method [16], the social behaviors of spiders in the Social Spider Optimization (SSO) [17], the properties of animal

E. Cuevas et al., *Recent Metaheuristics Algorithms for Parameter Identification*, Studies in Computational Intelligence 854, https://doi.org/10.1007/978-3-030-28917-1_6

behavior in a group in the Collective Animal Behavior (CAB) [18] and the emulation of the differential and regular evolution in species in the Differential Evolution (DE) [19] and Genetic Algorithms (GA) [20], respectively.

On the other hand, the union of fuzzy systems with metaheuristic algorithms has lately attracted the study in the Computational Intelligence society. As a result, a novel class of systems known as Evolutionary Fuzzy Systems (EFSs) [21, 22] has emerged. These procedures essentially consider the automatic generation and the tuning of fuzzy systems during a training process based on a metaheuristic method. The EFSs approaches described in the literature can be divided into two classes [21, 22]: tuning and learning.

In a tuning approach, a metaheuristic algorithm is employed to adjust the parameters of an existent fuzzy system, without altering its rule base. Some examples of tuning in EFSs involve the calibration of fuzzy controllers [23, 24], the adaptation of type-2 fuzzy models [25] and the enhancement of precision in fuzzy models [26, 27]. In learning, the rule base of a fuzzy system is produced by a metaheuristic algorithm, so that the final fuzzy system has the potential to precisely reproduce the modeled system. There are many models of learning in EFSs, which consider diverse kinds of problems such as the selection of fuzzy rules with membership functions [28, 29], rule generation [30, 31] and determination of the entire fuzzy structure [32–34].

The introduced approach cannot be contemplated as EFSs method, since the fuzzy system, used as an optimizer, is not automatically generated or tuned by a learning procedure. On the opposite, its design is based on expert observations obtained directly from the optimization task. Consequently, the amount of rules and its configuration are fixed, during its operation. Furthermore, in a common EFSs scheme, a metaheuristic algorithm is used to locate an optimal base rule for a fuzzy system to an evaluation function. Contrary to such strategies, in our proposal, a fuzzy system is applied to obtain the optimum value of an optimization problem. Therefore, the proposed Fuzzy system directly operates like any other metaheuristic algorithm directing the optimization strategy implemented in its rules.

A metaheuristic algorithm is formulated as a high-level methodology that involves a set of guidelines and procedures to produce an optimization strategy. In this chapter, we explain how the fuzzy logic scheme can be employed to develop algorithms for optimization problems. Contrary to traditional metaheuristic approaches where the center is on the construction of evolutionary operators that imitate natural or social behavior, in our proposal, we concentrate on getting an intuitive perception of how to conduct an adequate search strategy to model it directly into a fuzzy system.

Although sometimes unnoticed, it is well recognized that human experience plays an essential role in optimization techniques. It must be admitted that metaheuristic approaches use the human experience to tune their corresponding parameters or to choose the suitable algorithm for a specific problem [35]. Under such conditions, it is crucial to ask the following questions: How much of the benefit may be attached to the use of a certain metaheuristic methodology? How much should be connected to its intelligent heuristic tuning or selection? Also, if we employ the use of human experience during the entire design process, can we reach higher performance?

The utility of fuzzy logic for the development of optimization methods presents numerous advantages. (A) Generation. Traditional metaheuristic methodologies follow complex natural or social behaviors. Such as reproduction, which requires the mathematical modeling of partially-known and non-characterized behaviors, which are sometimes even hidden [36]. Hence, it is distinctly difficult to perfectly model even rather simple metaphors. On the other hand, fuzzy logic presents a simple and well-known technique for constructing systems via the use of human experience [37]. (B) Transparency. The metaphors used by metaheuristic approaches point to algorithms that are challenging to understand from an optimization viewpoint. Hence, the metaphor cannot be immediately understood as a harmonious search strategy [36]. On the other hand, fuzzy logic generates completely interpretable models whose content expresses the search strategy as humans can conduct it [38]. (C) Improvement. Once conceived, metaheuristic methods keep the identical procedure to produce candidate solutions. Combining changes to improve the quality of prominent solutions is complicated and critically injures the conception of the original metaphor [36]. As human specialists interact with an optimization task, they get a better comprehension of the accurate search strategies that allow determining the optimal solution. As a result, new rules are collected so that their incorporation in the existing rule base increases the quality of the original search strategy. Under the fuzzy logic scheme, new rules can be quickly joined to a previously existent system. The extension of such rules gives the capabilities of the original system to be prolonged [39].

In this chapter, a proposal to implement the human experience to develop optimization strategies is introduced. In the proposed methodology, a Takagi-Sugeno Fuzzy inference system [40] is employed to generate a specific search strategy created by a human expert. Hence, the number of rules and its configuration simply depend on the specialist experience without considering any knowledge rule process. Each fuzzy rule describes an expert point of view that models the circumstances under which candidate solutions are adjusted to obtain the optimal position. To demonstrate the performance and robustness of the proposed methodology, a comparison to other well-known optimization methods is presented. The study considers several numerical test functions that are commonly encountered in the metaheuristic optimization literature. The numerical results demonstrate a higher performance of the proposed method in contrast to existing optimization procedures.

This chapter is designed as follows: In Sect. 6.2, the essential features of fuzzy logic and the various reasoning models are included. In Sect. 6.3, the proposed methodology is shown. Section 6.4 it is presented the computational procedure of FLOA. Section 6.5 explains the characteristics of the proposed methodology. In Sect. 6.6 the experimental and the comparative studies are presented. Finally, in Sect. 6.7, conclusions are described.

6.2 Reasoning Models

This section presents a brief introduction to the principal fuzzy logic concepts. The discussion particularly examines the Takagi-Sugeno Fuzzy inference model [40].

6.2.1 Fuzzy Logic Concepts

A fuzzy set (A) [1] is considered as a generalization of the Boolean or Crisp set, which is determined within a universe of discourse X. A is a linguistic description which describes the fuzzy set into the word A. Such a word represents how a human expert recognizes the variable X in connection to A. The fuzzy set (A) is defined by a membership function $\mu_A(x)$ which affords a measure of degree of similarity of an element x from X to the fuzzy set A. It also takes values within the range [0, 1], such as:

$$\mu_A(x) : X \to [0, 1] \tag{6.1}$$

Hence, a general variable x_c can be considered using multiple fuzzy sets $\{A_1^c, A_2^c, \ldots, A_m^c\}$, each one modeled by a membership function $\{\mu_{A_1^c}(x_c), \mu_{A_2^c}(x_c), \ldots, \mu_{A_m^c}(x_c)\}$.

A fuzzy system is a reasoning model based on the theories of fuzzy logic. It incorporates three conceptual components: a rule base, which comprises a collection of fuzzy rules; a database, which determines the membership functions related by the fuzzy rules; and a reasoning mechanism, which operates the inference system. There are two different inference fuzzy systems: Mamdani [41] and Takagi-Sugeno (TS) [40].

The fundamental distinction among the two inference models is in the consequent segment of the fuzzy systems. In the Mamdani model, all of the construction of the fuzzy system has linguistic variables and fuzzy sets. Nevertheless, the consequent segment of the TS model consists of mathematical formulations. Contrary to Mamdani construction, the TS model presents computational efficiency and mathematical integrity in the rules [42]. Hence, to reach higher modeling exactness considering several rules, the TS fuzzy model is a suitable candidate that gets better models when the rules are exposed as functional relationships represented in several local behaviors [42, 43]. Since the available information for the construction of the fuzzy system examined in our approach includes functional, local behaviors, the TS inference model has been employed in this work.

6.2.2 *The Takagi-Sugeno (TS) Fuzzy Model*

TS fuzzy inference systems permit us to express complex nonlinear systems by disintegrating the input space into several local operations, each of which is designed by a simplistic regression model [3].

The principal element of a TS fuzzy system is the collection of its K fuzzy rules. Since they implement the human experience that describes the performance of the current task. Each rule expressed by R^i describes the input variables to a consequent. A typical TS fuzzy rule is broken into two pieces: Antecedent (I) and consequent (II), which are defined as follows:

$$R^i : \overbrace{\text{IF } x_1 \text{ is } A_p^1 \text{ and } x_2 \text{ is } A_q^2, \ldots, \text{ and } x_n \text{ is } A_r^n}^{\text{I}} \text{ Then } \underbrace{y_i = g_i(\mathbf{x})}_{\text{II}} \; i = 1, 2, \ldots, K, \tag{6.2}$$

where $\mathbf{x} = [x_1, x_2, \ldots, x_n]^T$ is the n-dimensional input and y_i describes the output rule. $g(\mathbf{x})$ is a function that can be represented by any function as long as it can properly explain the operation of the system inside the fuzzy region designated by the antecedent of rule i. In Eq. 6.2, r, p and q expresses one fuzzy set which displays the role of variables x_1, x_2 and x_n, respectively.

6.2.2.1 Antecedent (I)

The antecedent segment is considered as a simple preposition of the form "x_e is A_d^e". Such a preposition, is mathematically modeled by the membership function $\mu_{A_d^e}(x_e)$, which presents the degree of similarity among x_e and the fuzzy set A_d^e. Since the antecedent segment is concatenated by applying the and operator, the degree of accomplishment of the antecedent $\beta_i(\mathbf{x})$ is determined using a t-norm operator such as:

$$\beta_i(\mathbf{x}) = \min\Big(\mu_{A_p^1}(x_1), \mu_{A_q^2}(x_2), \ldots, \mu_{A_r^n}(x_n)\Big) \tag{6.3}$$

6.2.2.2 Consequent (II)

$g_i(\mathbf{x})$ is a function which can be represented by any function that can properly represent the role of the system within the fuzzy region designated by the antecedent of rule i.

6.2.2.3 Inference Model in the TS Model

The final output y of a TS fuzzy system is produced as the union of the local behaviors, and can be perceived as the weighted mean of the consequent segments:

$$y = \frac{\sum_{i=1}^{K} \beta_i(\mathbf{x}) \cdot y_i}{\sum_{i=1}^{K} \beta_i(x)} \tag{6.4}$$

where $\beta_i(\mathbf{x})$ is the level of accomplishment of the ith rule's antecedent segment while y_i is the output of the consequent representation. Figure 6.1 exhibits the fuzzy reasoning procedure for a TS fuzzy inference system considering two rules. In the figure, it is considered two variables (x_1,x_2) and only two membership functions (I and II). Hence, it should be clear that the meaning of fuzzy logic systems mirrors the divide and conquer approach. Consequently, the antecedent segment of a fuzzy rule determines a local fuzzy area, while the consequent segment defines the role inside the region.

6.3 The Proposed Method

Because there is no particular solution for many types of difficult problems, human specialists usually follow a trial and error strategy to solve them. Under this manner, humans get knowledge as the information obtained by the interaction with the problem. In general, a fuzzy system is a representation that imitates the decisions and behavior of a human that has specific knowledge and experience in a precise

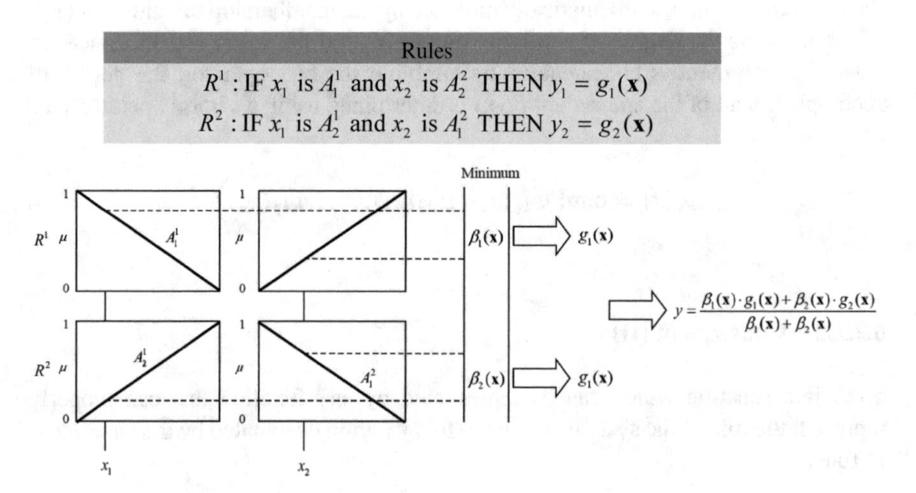

Fig. 6.1 TS fuzzy model

discipline. Accordingly, a fuzzy system is then considered to be capable of representing the role of a target system. For instance, if the target system is a human laborer in charge of a chemical reaction process, then the fuzzy system fits a fuzzy controller that can adjust the chemical process. Likewise, if the target system is a person who is closely related to optimization algorithms and decision making processes, then the fuzzy reasoning matches a fuzzy expert system that can obtain the optimal solution to a specific optimization problem as if the search strategy were carried by the human expert. In this chapter, we introduce a methodology for imitating human search tactics in algorithmic composition. In this part, the fuzzy optimization proposal is explained in detail. First, each component of the fuzzy system is defined; then, the entire computational system is exhibited.

Given a set of circumstances, an expert contributes a report of how to manage an optimization policy for finding the optimal solution to a general problem applying natural communication. Then, the aim is to catch this linguistic definition and mirror it into a fuzzy system. The linguistic description provided by the expert is split into two parts: (A) linguistic variables and (B) rule base formulation.

(A) Linguistic instances specify the behavior in which a human specialist understands the details of a particular variable in terms of its related values. One example is the velocity that could be recognized as flat, moderate and large. (B) Rule base formulation involves the building process of a set of IF-THEN relationships. Each connection exposes the circumstances under which specific procedures are accomplished. Typically, a fuzzy inference system consists of a set of rules joins fuzzy areas to actions. Under this context, the addition of any rule to the behavior of the fuzzy system will be modified considering the operating zone.

6.3.1 Optimization Mechanism

From the previously published literature, most of the optimization approaches have been created to reach the global solution for complex nonlinear optimization problems considering box constraints in the following form [44]:

$$\begin{aligned} &\text{maximize } f(\mathbf{x}), \quad \mathbf{x} = (x_1, \ldots, x_n) \in \mathbb{R}^n \\ &\text{subject to } \mathbf{x} \in \mathbf{X} \end{aligned} \tag{6.5}$$

where $f : \mathbb{R}^n \rightarrow \mathbb{R}$ is a nonlinear function whereas $\mathbf{X} = \{\mathbf{x} \in \mathbb{R}^n | l_i \leq x_i \leq u_i, i = 1, \ldots, n\}$ is a bounded feasible search space, constrained by the lower (l_i) and upper (u_i) limits.

To determine the optimal solution from the optimization problem presented in Eq. 6.5, a population of N candidate solutions $\mathbf{P}^k(\{\mathbf{p}_1^k, \mathbf{p}_2^k, \ldots, \mathbf{p}_N^k\})$, evolves from an initial state ($k = 0$) to a maximum number of generations ($k = Maxgen$) [45]. At first, the method starts constructing the set of N promissory solutions with values that are uniformly distributed within the range of the predefined lower (l_i) and upper

(u_i) boundaries. For each generation, a collection of evolutionary operations are then applied over the population $\mathbf{P}^k$ to produce the new population $\mathbf{P}^{k+1}$. In the population, an individual $\mathbf{p}_i^k$ ($i \in [1, \ldots, N]$) resembles to a n-dimensional vector $\{p_{i,1}^k, p_{i,2}^k, \ldots, p_{i,n}^k\}$ where the dimensions describe the decision variables of the optimization problem to be solved. The quality of the solution $\mathbf{p}_i^k$ is computed by considering an objective function $f(\mathbf{p}_i^k)$ whose value matches the fitness value of $\mathbf{p}_i^k$. As the optimization process grows, the best individual $\mathbf{g}$ marked so-far is conserved. Since it describes the prevailing best solution.

In this study, the human experience is modelled in the rule base of a TS Fuzzy inference system, as a result, the implemented fuzzy system reveals the requirements under which candidate solutions from are evolved to produce new positions.

6.3.1.1 Linguistic Variables Characterization (A)

To create a fuzzy inference system from human experience, it is essential the characterization of the linguistic variables and the representation of a rule set. A linguistic variable is formed by using membership functions. They express functions which attach a numerical value to a subjective observation of the variable. The amount and the form of the membership functions that display a certain linguistic variable depend on the application context [46]. Hence, in order to keep the idea of the fuzzy system as manageable as possible, we designate each linguistic variable by handling only two membership functions [47]. One example is the velocity which could be determined by the membership functions: low and high. Such a membership function is mutually exclusive or disjoint. Hence, if $\mu_L = 0.7$, then $\mu_H = 0.3$. Considering that the linguistic variable velocity has a numerical value within the interval [0, 100] revolutions per minute (rpm), and are designated regarding to the membership functions depicted in Fig. 6.2.

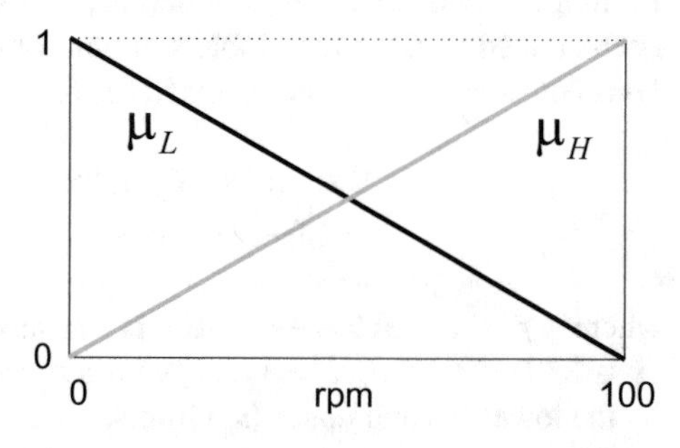

Fig. 6.2 Example of membership functions which represents a linguistic variable

6.3.1.2 Rule Base Formulation (B)

Many optimization methodologies can be expressed by adopting human experience. In this part, a simplistic search mechanism is expressed regarding the basic measurements of the optimization method. Hence, the simplistic search strategy is to relocate candidate solutions to explore regions of the space where it is expected to obtain the optimal solution. Since the values of the fitness function are known by the positions defined by the candidate solutions, the positions with better probabilities of designating latent solutions are those found near the best ones in terms of its fitness value.

Considering this, a simple search procedure could be expressed by the subsequent rules:

1. (Attraction) IF the distance from $\mathbf{p}_i^k$ to g is short AND $f(\mathbf{p}_i^k)$ is good THEN $\mathbf{p}_i^k$ is moved towards g
This rule describes the condition where the candidate solution $\mathbf{p}_i^k$ is relocated to the best candidate solution to improve its fitness value. Since the fitness values of $\mathbf{p}_i^k$ and g are better than other members of $\mathbf{P}^k$, the region among $\mathbf{p}_i^k$ and g keeps likely solutions that could improve g. With this movement, it is expected to explore the unexplored region among $\mathbf{p}_i^k$ and g. To depict how each rule operates. Figure 6.3 displays a simplistic case that shows the conditions under which rules are performed. In the example, a population $\mathbf{P}^k$ of five candidate solutions is analyzed (Fig. 6.3a). In the case of rule 1, as it is presented in Fig. 6.3b), the candidate solution $\mathbf{p}_5^k$ that achieved the requirements is attracted to g
2. (Repulsion) **IF** the distance from $\mathbf{p}_i^k$ to **g** is short **AND** $f(\mathbf{p}_i^k)$ is bad **THEN** $\mathbf{p}_i^k$ is moved away from **g**
Even if the distance among $\mathbf{p}_i^k$ and **g** is short, the proof reveals that there are no good solutions between them. Consequently, the enhancement of $\mathbf{p}_i^k$ is examined in the reverse direction of **g**. A visual illustration of this behavior is displayed in Fig. 6.3c)
3. (Refining) **IF** the distance from $\mathbf{p}_i^k$ to **g** is large **AND** $f(\mathbf{p}_i^k)$ is good **THEN** $\mathbf{p}_i^k$ is refined
In this rule, a good solution $\mathbf{p}_i^k$ which is located far from **g** is improved by searching within its region. The purpose is to promote the quality of competitive solutions which have been found. Such a scenario is exhibit in Fig. 6.3d) where the original solution $\mathbf{p}_2^k$ is interchanged by a new position $\mathbf{p}_2^{k+1}$ which is randomly produced within the vicinity of $\mathbf{p}_2^k$
4. (Substitution) **IF** the distance from $\mathbf{p}_i^k$ to **g** is large **AND** $f(\mathbf{p}_i^k)$ is bad **THEN** a new position is randomly chosen
This rule describes the condition in Fig. 6.3e where the solution $\mathbf{p}_4^k$ is bad and so away from **g** that it is better to substitute it by other solution, randomly produced within the search space **X**

Each rule is a representation of a linguistic rule which holds only linguistic knowledge. As linguistic expressions are not clear descriptions of the values that they express, linguistic rules are not accurate. They express only conceptual ideas concerning how to produce a good optimization procedure according to the human viewpoint. Under such circumstances, it is important to define the purpose of their linguistic representations from a computational point of view.

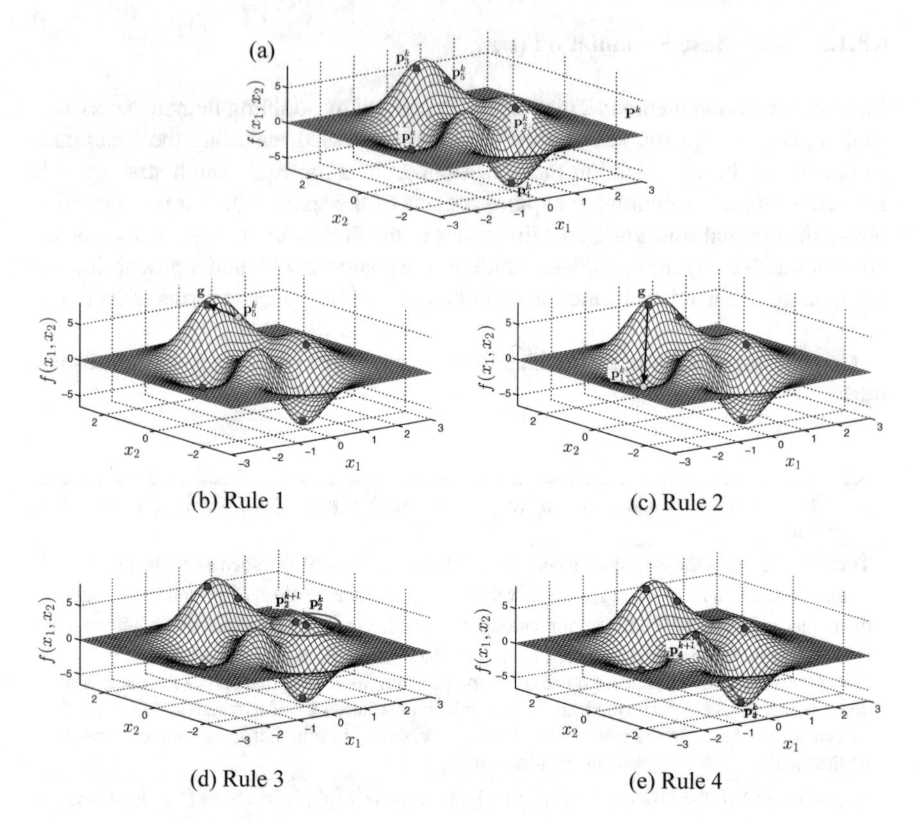

Fig. 6.3 Graphical example that exposes the required conditions under which rules are executed. **a** Actual configuration of the candidate solution population $\mathbf{P}^k$, **b** rule 1, **c** rule 2, **d** rule 3 and **e** rule 4

6.3.1.3 Implementation of the TS Fuzzy System

In this segment, we will examine the implementation of the skillful understanding related to the optimization process.

(I) *Membership functions and its antecedents*

From the aforementioned rules, two different linguistic variables are studied, the distance from de candidate solution $\mathbf{p}_i^k$ to the best global solution $\mathbf{g}$ ($D(\mathbf{p}_i^k, \mathbf{g})$) and its fitness value $\left(f\left(\mathbf{p}_i^k\right)\right)$. Hence, $D(\mathbf{p}_i^k, \mathbf{g})$ is designated by two functions: short and large. On the other hand, $f\left(\mathbf{p}_i^k\right)$ is formed by the functions good and bad. Figure 6.4 graphically presents the fuzzy membership functions for the linguistic variables.

The distance $D(\mathbf{p}_i^k, \mathbf{g})$ is defined as the Euclidian distance $\left\|\mathbf{g} - \mathbf{p}_i^k\right\|$. Hence, as it is exposed in Fig. 6.4a, complementary functions produce the relative distance $D(\mathbf{p}_i^k, \mathbf{g})$: short (**S**) and large (**L**). Their values range from 0 and $d_{\max}$, where $d_{\max}$ represents the maximum distance within the search space $\mathbf{X}$ and it is defined as follows:

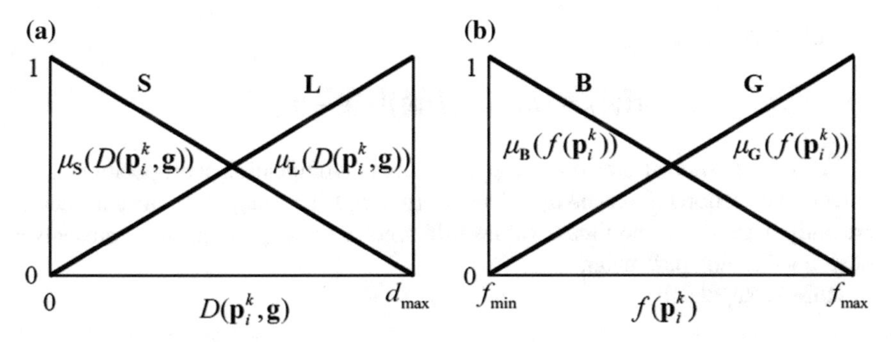

Fig. 6.4 Membership functions for distance $D(\mathbf{p}_i^k, \mathbf{g})$ (**a**) and for the fitness $f(\mathbf{p}_i^k)$ (**b**)

$$d\text{max} = \sqrt{\sum_{s=1}^{d} (u_s - l_s)^2} \tag{6.6}$$

where d represents the dimensionality of the search space $\mathbf{X}$. For the case of $f(\mathbf{p}_i^k)$, two different functions produce its relative values: bad (**B**) and good (**G**). Their values are within $f\text{min}$ and $f\text{max}$. These values represent the minimum and maximum fitness values ben so far. Hence, they can be formulated as follows:

$$f\text{min} = \min_{\substack{i \in \{1,2,\ldots,N\} \\ k \in \{1,2,\ldots,gen\}}} \left(f(\mathbf{p}_i^k)\right) \text{ and } f\text{max} = \max_{\substack{i \in \{1,2,\ldots,N\} \\ k \in \{1,2,\ldots,gen\}}} \left(f(\mathbf{p}_i^k)\right) \tag{6.7}$$

From Eq. 6.7, it is quite clear that $f\text{max} = f(\mathbf{g})$. If a new minimum value or maximum value of $f(\mathbf{p}_i^k)$ is detected, it replaces the previously values of $f\text{min}$ or $f\text{max}$. Figure 6.4(b) demonstrates the membership functions that represent $f(\mathbf{p}_i^k)$.

Taking into account the functions defined in Fig. 6.4, the degree of achievement of the antecedent segment $\beta_w(\mathbf{x})$ for each rule ($w \in [1, 2, 3, 4]$) is then, defined in Table 6.1.

(II) *Consequents or Actions*

Consequents or Actions are mathematical functions that can be designated by any function that describes properly the desired performance of the inference system within the fuzzy region specified by the antecedent segment. The consequents of the four rules are formed by considering the following behaviors.

Table 6.1 Degree of achievement of the antecedent segment $\beta_w(\mathbf{x})$ for each rule ($w \in [1, 2, 3, 4]$)

Rule	Degree of fulfilment $\beta_w(\mathbf{x})$
1	$\beta_1(\mathbf{p}_i^k) = \min\left(\mu_S(D(\mathbf{p}_i^k, \mathbf{g})), \mu_G(f(\mathbf{p}_i^k))\right)$
2	$\beta_2(\mathbf{p}_i^k) = \min\left(\mu_S(D(\mathbf{p}_i^k, \mathbf{g})), \mu_B(f(\mathbf{p}_i^k))\right)$
3	$\beta_3(\mathbf{p}_i^k) = \min\left(\mu_L(D(\mathbf{p}_i^k, \mathbf{g})), \mu_G(f(\mathbf{p}_i^k))\right)$
4	$\beta_4(\mathbf{p}_i^k) = \min\left(\mu_L(D(\mathbf{p}_i^k, \mathbf{g})), \mu_B(f(\mathbf{p}_i^k))\right)$

Bold data represent the best obtained values

Rule 1. Attraction

$$At(\mathbf{p}_i^k) = \left| f\text{max} - f(\mathbf{p}_i^k) \right| \cdot (\mathbf{g} - \mathbf{p}_i^k) \cdot \alpha_1 \tag{6.8}$$

Considering this rule, the rule $At(\mathbf{p}_i^k)$ produces a modification of the position in the attraction direction influenced by the vector $(\mathbf{g} - \mathbf{p}_i^k)$. The magnitude of such vector, crucially depends on the fitness values difference among $\mathbf{g}$ and $\mathbf{p}_i^k$. α_1 represents a predefined tuning parameter.

Rule 2. Repulsion

$$Rep(\mathbf{p}_i^k) = \left| f\text{max} - f(\mathbf{p}_i^k) \right| \cdot (\mathbf{g} + \mathbf{p}_i^k) \cdot \alpha_2 \tag{6.9}$$

where α_2 corresponds to a predefined tuning factor.

Rule 3. Refining

$$Ref(\mathbf{p}_i^k) = \left| f\text{max} - f(\mathbf{p}_i^k) \right| \cdot \mathbf{v} \cdot \gamma \tag{6.10}$$

where $\mathbf{v} = \{v_1, v_2, \ldots, v_d\}$ is considered as a random vector where each dimension corresponds to a random number within $[-1, 1]$ whereas γ corresponds to a predefined tuning factor. Under this rule, $Ref(\mathbf{p}_i^k)$ produces a random position within the bounds $\pm \left| f_{\max} - f(\mathbf{p}_i^k) \right|$.

Rule 4. Substitution

$$Ran(\mathbf{p}_i^k) = \mathbf{r} \tag{6.11}$$

where $\mathbf{r} = \{r_1, r_2, \ldots, r_d\}$ is considered as a random vector where each component r_u corresponds to a number between $[l_u, u_u]$.

(III) Inference of the TS model.

The entire modification of the position $\Delta\mathbf{p}_i^k$ in the TS fuzzy system, is buit as the concatenation of the local behaviors obtained by the aforementioned rules, and can be formulated as the weighted mean of the consequents:

$$\Delta\mathbf{p}_i^k = \frac{At(\mathbf{p}_i^k) \cdot \beta_1(\mathbf{p}_i^k) + Rep(\mathbf{p}_i^k) \cdot \beta_2(\mathbf{p}_i^k) + Ref(\mathbf{p}_i^k) \cdot \beta_3(\mathbf{p}_i^k) + Ran(\mathbf{p}_i^k) \cdot \beta_4(\mathbf{p}_i^k)}{\beta_1(\mathbf{p}_i^k) + \beta_2(\mathbf{p}_i^k) + \beta_3(\mathbf{p}_i^k) + \beta_4(\mathbf{p}_i^k)} \tag{6.12}$$

Once $\Delta\mathbf{p}_i^k$ has been computed, the newest position $\mathbf{p}_i^{k+1}$ is computed as follows:

$$\mathbf{p}_i^{k+1} = \mathbf{p}_i^k + \Delta\mathbf{p}_i^k \tag{6.13}$$

6.4 Computational Procedure

The proposed fuzzy methodology, has been implemented as an iterative process in which many actions are executed. Such actions can be compiled in the form of pseudo-code in Algorithm 6.1. The proposed method uses as input data the number of candidate solutions (*N*), the maximum number of iterations, and the tuning parameters $\alpha_1, \alpha_2, \gamma$. Similar to other metaheuristic algorithms, in the first step (line 2), the algorithm begins assembling the set of *N* candidate solutions with values that are uniformly distributed between the limits of the search space. These promissory solutions correspond to the first population $\mathbf{P}^0$. After initialization, the best element $\mathbf{g}$ is selected (line 3). Next, for each particle $\mathbf{p}_i^k$ its relation (distance) to the best solution $\mathbf{g}$ is computed (line 6).

With $D(\mathbf{p}_i^k, \mathbf{g})$ and $f(\mathbf{p}_i^k)$, the search approach executed in the fuzzy system is employed (lines 7–9). Under such facts, the antecedents (line 7) and consequents (line 8) are calculated while the final movement $\Delta\mathbf{p}_i^k$ is achieved as a result of the operation made by the TS model (line 9). Eventually, the new position $\mathbf{p}_i^{k+1}$ is updated (line 10). Once the new population $\mathbf{P}^{k+1}$ is obtained as a result of the iterative procedure of lines 6–10, the best value $\mathbf{g}$ is updated (line 12). This sequence is renewed until the maximum number the generations has been reached.

Algorithm 6.1. Fuzzy Logic Optimization Algorithm (FLOA) pseudo-code

1.	**Input:** *N*, *Maxgen*, $\alpha_1, \alpha_2, \gamma$, *k*=0.	
2.	$\mathbf{P}^k$ ←**Initialize**(*N*);	
3.	$\mathbf{g}$ ←**SelectBestParticle**($\mathbf{P}^k$);	
4.	**while** *k*<=*Maxgen* **do**	
5.	**for** (*i*=1;*i*>*N*;*i*++)	
6.	$D(\mathbf{p}_i^k, \mathbf{g})$←**CalculateTheDistancetoTheBest** ($\mathbf{p}_i^k, \mathbf{g}$);	
7.	[$\beta_1,\beta_2,\beta_3,\beta_4$]←**EvaluateAntecedents**($D(\mathbf{p}_i^k, \mathbf{g}), f(\mathbf{p}_i^k)$);	Fuzzy System
8.	[*At*,*Rep*,*Ref*,*Ran*] ←**EvaluateConsequents**($\mathbf{p}_i^k$,$\mathbf{g}$, $f(\mathbf{p}_i^k)$);	
9.	$\Delta\mathbf{p}_i^k$ ←**InferenceTS**($\beta_1,\beta_2,\beta_3,\beta_4$,*At*,*Rep*,*Ref*,*Ran*);	
10.	$\mathbf{p}_i^{k+1} \leftarrow \mathbf{p}_i^k + \Delta\mathbf{p}_i^k$	
11.	end **for**	
12.	$\mathbf{g}$ ←**SelectBestParticle**($\mathbf{P}^{k+1}$);	
13.	$k \leftarrow k+1$	
14.	end **while**	
15.	**Output**: $\mathbf{g}$	

6.5 Discussion

In this sub-section, many characteristics of the proposed method are discussed. First, the sub-Sect. 6.5.1, introduces the operations of the optimization process are analyzed. Second, the sub-Sect. 6.5.2 analysis the modelling properties of the proposed approach.

6.5.1 *Optimization Algorithm*

A metaheuristic search algorithm is formulated as a high-level independent methodology that comprises of a set of rules and procedure to construct an optimization strategy. In the proposed method, a fuzzy system is produced based on expert experience over the optimization field. The fuzzy system employed, performs several fuzzy logic operations to obtain a new candidate solution $\mathbf{p}_i^{k+1}$ from the current solution $\mathbf{p}_i^k$. During this process, the following steps are involved:

1. Computation of the degree of membership among the input information $(D(\mathbf{p}_i^k, \mathbf{g}), f(\mathbf{p}_i^k))$ and the fuzzy sets.
2. Calculation of the degree of proportion for each rule based on the degree of achievement $\beta_w(\mathbf{x})$ in the antecedent segment of the rule.
3. Evaluation of the consequent segment for each rule: *At*, *Rep*, *Ref*, *Ran*.
4. Generation of the new solutions $\mathbf{p}_i^{k+1}$ based on the weighted mean of the consequent functions, as it is followed to the TS model.

Under such conditions, the produced fuzzy system is applied over the solutions from the population $\mathbf{P}^k$, to produce the new population $\mathbf{P}^{k+1}$. This procedure is executed until the maximum number of iterations has been reached.

6.5.2 *Modeling Characteristics*

Metaheuristic techniques are extensively applied to solving difficult optimization problems. Such methodologies have been generated by a combination of deterministic patterns and randomness, simulating the behavior of natural or social systems. Most of the metaheuristic approaches split the individual behavior into many processes which show no coupling among them [11, 36].

In the proposed algorithm, the formed fuzzy system creates a complex optimization strategy. This construction is performed by an amount of IF-THEN rules, each of which defines the local behavior of the model. In particular, the rules communicate the circumstances under which new places are explored. To calculate a new solution $\mathbf{p}_i^{k+1}$, the consequent actions of all rules are combined. In this mode, all the actions are performed in the computation of a certain solution $\mathbf{p}_i^{k+1}$, but with different weight levels. By linking local behaviors, fuzzy systems can model complex behaviors. An interesting example of such modeling aspects is rule 1 and rule 2. If these rules are separately analyzed, the attraction and repulsion actions conducted by the functions are entirely deterministic. However, when all rules are considered, rule 3 and rule 4 add randomness.

6.6 Experimental Study

A representative set of 19 benchmark functions has been adopted to measure the performance of FLOA. These test functions describe the base functions from the latest competition on single objective optimization at CEC2015 [48]. Tables B1, B2 and B3 in Appendix B show the test functions applied in the experimental study. These functions are grouped into three different classes: Unimodal (Table B1), multimodal (Table B2) and Hybrid (Table B3) benchmark functions. In the tables, n designates the dimension in which the function is accomplished, $f(\mathbf{x}^*)$ describes the optimal value of the function in the location $\mathbf{x}^*$ and S is the established search space.

The principal aim of this section is to manifest the performance of the proposed algorithm over numerical optimization problems. Furthermore, the numerical results of our method are compared with some traditional optimization algorithms. The results of the proposed algorithm are validated by a statistical analysis of the experimental data.

The experimental results are split into two sections. In the first one, the performance of the proposed method is evaluated with respect to its tuning parameters. In the second one, the performance of the proposed fuzzy method is compared to six traditional evolutionary algorithms.

6.6.1 Performance Evaluation Considering the Tuning Parameters

The parameters of the proposed rules, α_1, α_2 and γ influence the expected performance of the proposed algorithm. In this section, we examine the behavior of the proposed method regarding the different configurations of these parameters. All the experiments have been performed on a Pentium dual core computer with 2.53-GHz and 4-GB RAM under MATLAB 8.3. For the purpose of simplicity, just the functions f_1–f_{14} have been considered in the study. In the experiments, all the functions run with a dimension n = 30. As an initial condition, the parameter are configured to their default values, such as: $\alpha_1 = 1.4$, $\alpha_2 = 0.05$ and $\gamma = 0.005$. Then, the three parameters are evaluated each at a time, while the remaining are fixed to their corresponding default values. To minimize the random effect, each test function is evaluated separately a total of 10 events. As a termination criterion, the maximum number of generations is set to 1000. For this experiment, the population size N is fixed to 50 individuals.

In the first step, the performance of the proposed emethod is analyzed examining different values for α_1. In the study, the values of α_1 vary from 0.6 to 1.6 while the values of α_2 and γ maintain fixed at 0.05 and 0.005. In the experiment, the proposed fuzzy system is evaluated separately 30 times for each value of α_1 over each test function. The results are recorded in Table 6.2. These values describe the average best fitness values ($\bar{f}$) and the standard deviations (σ_f) obtained in terms of a certain parameter combination of α_1, α_2 and γ. From Table 6.2, we can assume that the proposed algorithm with $\alpha_1 = 1.4$ keeps the best performance on the functions f_1–f_9, and f_{11}. Under this arrangement, the algorithm obtains the best results in 9 out of 14 functions. On the other hand, when the parameter α_1 is established to any other value, the performance of the algorithm is irregular, producing unsatisfactory results.

In the second part, the performance of the algorithm is measured considering different values for α_2. In the test, the values of α_2 are changed from 0.01 to 0.1 while the values of α_1 and γ prevail at 1.4 and 0.005. The statistical results collected by the fuzzy scheme considering different values of α_2 are exhibited in Table 6.3. From Table 6.3, it can be demonstrated that the fuzzy algorithm with $\alpha_2 = 0.05$ beats the other parameter configurations. Under this arrangement, the algorithm achieves the best results in 8 of the 14 functions.

Lastly, in the third step, the performance of the algorithm is judged considering different values for γ. In the experiment, the values of γ are modified from 0.001 to 0.01, while the values of α_1 and α_2 keep fixed at 1.4 and 0.05. Table 6.4 compiles the numerical results of this experiment. From the data provided by Table 6.4, it can be recognized that the proposed fuzzy method with $\gamma = 0.005$ achieves the best performance over the functions f_1, f_2, f_3, f_4, f_6, f_7, f_{10}, f_{12} and f_{13}. Nondetheless, when the parameter γ considers any other value, the fuzzy algorithm is irregular. Under this parameter arrangement, the algorithm presents the best possible performance index, since it gets the best performance criteria in 10 out of 14 functions.

In general, the numerical results presented in Tables 6.2, 6.3 and 6.4 suggest that a proper combination of the parameters can enhance the performance of the algorithm and the robustness of solutions. In this test, it can conclude that the best parameter configuration is formed by the subsequent values:$\alpha_1 = 1.4$, $\alpha_2 = 0.05$ and $\gamma = 0.005$.

Once the parameters α_1, α_2 and γ have been experimentally established, it is conceivable to study their influence in the optimization field. In the search procedure, united in the fuzzy system, α_1 changes the attraction that an individual experiment with respect to the best element in the current population. This action promotes the improvement of the solution quality of the individual, regarding that the unknown zone among the solution and the best element could include a better solution. On the other hand, α_2 regulates the repulsion to which a low-quality individual is supported. This procedure aims to improve the quality of the bad solution through a movement in opposite direction of the best current element. This repulsion is contemplated, since there is an indication that the unknown zone among the low quality solution

Table 6.2 Different values of α_1 obtained by the fuzzy method

α_1		0.6	0.7	0.8	0.9	1	1.2	1.3	1.4	1.5	1.6
f_1	$\bar{f}$	6.95E−55	7.74E−89	3.97E−167	1.01E−39	1.02E−193	0.00E+00	4.26E−29	**3.08E−281**	1.15E−28	6.85E−28
	σ_f	3.67E−54	4.24E−88	0.00E+00	5.54E−39	0.00E+00	0.00E+00	2.19E−28	**0.00E+00**	3.41E−28	1.08E−27
f_2	$\bar{f}$	6.10E−23	1.14E−53	1.12E−124	2.49E−139	1.08E−158	7.16E−22	1.31E−78	**2.66E−207**	2.03E−15	7.52E+00
	σ_f	3.34E−22	6.07E−53	4.81E−124	1.36E−138	0.00E+00	3.89E−21	7.18E−78	**0.00E+00**	4.31E−15	2.35E+01
f_3	$\bar{f}$	4.69E−10	1.80E−17	2.22E−22	3.23E−22	2.00E−27	4.96E−24	4.11E−27	**1.00E−27**	2.68E−18	1.93E−11
	σ_f	1.62E−09	6.81E−17	8.87E−22	1.64E−21	4.73E−27	2.66E−23	9.21E−27	**1.50E−27**	1.12E−17	5.52E−11
f_4	$\bar{f}$	1.55E−23	1.48E−30	1.51E−130	4.64E−180	2.84E−112	6.13E−19	2.00E−183	**3.85E−220**	9.09E−16	5.16E−15
	σ_f	7.09E−23	8.12E−30	8.25E−130	0.00E+00	1.56E−111	3.31E−18	0.00E+00	**0.00E+00**	2.76E−15	6.91E−15
f_5	$\bar{f}$	2.85E+01	2.85E+01	2.85E+01	2.55E+01	3.85E+01	1.75E+01	2.65E+01	**3.04E−03**	2.85E+01	2.99E+01
	σ_f	4.38E−02	3.86E−02	3.87E−02	4.37E−02	3.04E−02	3.72E−02	4.50E−02	**3.02E−02**	4.58E−02	4.72E−02
f_6	$\bar{f}$	2.15E−02	1.05E−02	1.19E−02	1.57E−02	1.59E−02	1.67E−02	1.69E−02	**7.94E−03**	1.98E−02	1.92E−02
	σ_f	1.90E−02	3.86E−03	1.13E−02	1.56E−02	1.49E−02	1.37E−02	1.49E−02	**1.89E−03**	1.80E−02	9.92E−03
f_7	$\bar{f}$	8.98E−03	3.22E−03	2.22E−03	1.88E−03	1.75E−03	2.07E−03	1.79E−03	**1.36E−03**	1.59E−03	1.64E−03
	σ_f	1.27E−02	2.65E−03	2.17E−03	1.48E−03	1.62E−03	2.26E−03	1.92E−03	**1.10E−03**	1.45E−03	1.50E−03
f_8	$\bar{f}$	−4.95E+03	−6.12E+03	−5.01E+04	−5.13E+04	−4.96E+03	−3.02E+03	−5.14E+04	**−5.58E+04**	−4.94E+03	−5.21E+03
	σ_f	4.36E+02	5.09E+02	4.15E+02	5.55E+02	5.08E+02	4.78E+02	4.24E+02	**4.10E+02**	4.85E+02	4.60E+02
f_9	$\bar{f}$	6.20E+01	2.91E+01	1.70E+01	1.57E+01	5.94E+00	5.21E+00	5.86E+00	**4.76E-01**	9.94E+00	2.02E+01
	σ_f	6.08E+01	5.32E+01	4.17E+01	4.27E+01	3.17E+01	2.77E+01	2.90E+01	**2.38E+00**	3.77E+01	5.22E+01
f_{10}	$\bar{f}$	8.70E−15	**7.16E−15**	7.99E-15	9.18E−15	9.41E-15	1.04E−14	1.19E−14	8.47E-15	1.07E−14	1.38E−14
	σ_f	5.39E−15	3.92E−15	3.61E−15	**2.53E−15**	2.57E−15	5.22E−15	6.00E−15	3.82E−15	6.64E−15	7.83E−15
f_{11}	$\bar{f}$	**0.00E+00**	**0.00E+00**	**0.00E+00**	**0.00E+00**	**0.00E+00**	**0.00E+00**	**0.00E+00**	**0.00E+00**	**0.00E+00**	3.46E−05

(continued)

Table 6.2 (continued)

α_1		0.6	0.7	0.8	0.9	1	1.2	1.3	1.4	1.5	1.6
	σ_f	**0.00E+00**	**0.00E+00**	**0.00E+00**	**0.00E+00**	**0.00E+00**	**0.00E+00**	**0.00E+00**	**0.00E+00**	**0.00E+00**	1.32E−04
f_{12}	$\bar{f}$	8.76E−02	8.32E−02	8.42E−02	7.82E−02	**7.70E−02**	8.73E−02	9.45E−02	9.59E−01	3.84E+00	6.49E+02
	σ_f	2.74E−02	3.35E−02	4.73E−02	2.30E−02	**1.80E−02**	2.78E−02	2.15E−02	3.26E+00	8.79E+00	2.65E+03
f_{13}	$\bar{f}$	**2.88E−01**	3.30E−01	3.69E−01	3.77E−01	3.99E−01	3.72E−01	4.27E−01	3.91E−01	2.44E+01	1.83E+06
	σ_f	**1.42E−01**	1.03E−01	1.63E−01	1.57E−01	2.20E−01	1.16E−01	3.78E−01	1.76E−01	9.67E+01	1.00E+07
f_{14}	$\bar{f}$	−8.43E+02	−8.33E+02	−8.31E+02	−8.29E+02	−8.43E+02	−8.97E+02	**−8.98E+02**	−8.90E+02	−8.86E+02	−8.84E+02
	σ_f	1.14E+01	9.37E+00	**1.06E+01**	1.12E+01	1.68E+01	2.54E+01	2.19E+01	1.80E+01	2.13E+01	2.48E+01

Bold data represent the best obtained values

Table 6.3 Different values of α_2 obtained by the fuzzy method

α_2		0.01	0.02	0.03	0.04	0.05	0.06	0.07	0.08	0.09	0.1
f_1	$\bar{f}$	1.01E−39	3.25E−28	3.39E−28	2.35E−28	**5.18E−49**	1.48E−28	1.36E−28	2.61E−28	2.15E−28	2.29E−28
	σ_f	5.54E−39	9.09E−28	1.02E−27	7.44E−28	**2.34E−48**	5.19E−28	4.95E−28	8.03E−28	6.36E−28	6.94E−28
f_2	$\bar{f}$	6.25E−16	3.61E−16	1.67E−22	2.66E−20	7.16E−22	1.15E−19	7.85E−16	3.29E−17	4.82E−16	**6.57E−30**
	σ_f	2.85E−15	1.97E−15	9.11E−22	1.45E−19	3.89E−21	5.41E−19	2.97E−15	1.53E−16	2.47E−15	**3.59E−29**
f_3	$\bar{f}$	4.96E−24	9.65E−27	9.29E−26	2.22E−25	**1.97E−27**	2.52E−23	2.81E−21	4.94E−22	7.99E−23	6.69E−21
	σ_f	2.66E−23	3.49E−26	4.33E−25	1.12E−24	**2.02E−27**	1.37E−22	1.53E−20	2.64E−21	2.25E−22	2.13E−20
f_4	$\bar{f}$	1.96E−15	1.20E−15	3.44E−17	5.99E−18	**3.08E−29**	1.92E−20	4.91E−26	6.13E−19	3.98E−16	1.42E−28
	σ_f	5.12E−15	4.45E−15	1.33E−16	3.28E−17	**1.69E−28**	9.01E−20	2.69E−25	3.31E−18	2.18E−15	7.76E−28
f_5	$\bar{f}$	3.85E+01	3.85E+01	2.55E+01	1.55E+01	**1.99E−04**	1.85E−01	4.85E−02	1.23E+01	1.35E+01	2.85E+01
	σ_f	4.45E−02	4.42E−02	4.38E−02	4.52E−02	**3.74E−02**	5.02E−02	3.45E−02	4.12E−02	5.02E−02	4.04E−02
f_6	$\bar{f}$	1.55E−02	1.47E−02	2.11E−02	2.15E−02	**1.07E−03**	1.73E−02	1.94E−02	1.78E−02	2.35E−01	2.04E−02
	σ_f	9.87E−03	5.13E−03	2.10E−02	4.72E−03	**1.67E−02**	6.85E−03	1.90E−02	7.59E−03	1.17E+00	2.21E−02
f_7	$\bar{f}$	**1.03E−03**	1.56E−03	1.09E−03	1.42E−03	1.36E−03	2.17E−03	1.88E−03	2.12E−03	2.53E−03	2.55E−03
	σ_f	8.62E−04	1.60E−03	7.41E−04	**1.07E−03**	1.10E−03	1.22E−03	1.56E−03	2.59E−03	3.04E−03	2.10E−03
f_8	$\bar{f}$	−3.10E+03	−5.24E+03	−2.17E+03	−5.00E+03	−5.13E+03	−5.01E+03	**−6.29E+03**	−5.11E+03	−5.24E+03	−5.18E+03
	σ_f	5.08E+02	4.49E+02	4.65E+02	3.83E+02	4.77E+02	**3.68E+02**	5.35E+02	4.36E+02	5.03E+02	4.41E+02
f_9	$\bar{f}$	8.98E+00	3.41E−02	5.91E+00	**9.16E−02**	4.76E−01	8.28E+00	7.75E−02	4.07E+00	1.64E+01	5.83E+00
	σ_f	3.42E+01	**1.87E−01**	3.14E+01	2.81E−01	2.38E+00	3.15E+01	2.61E−01	2.18E+01	5.02E+01	3.13E+01
f_{10}	$\bar{f}$	1.12E−14	1.01E−14	1.10E−14	1.17E−14	**8.47E−15**	9.30E−15	9.89E−15	1.21E−14	1.21E−14	1.20E−14
	σ_f	4.88E−15	3.82E−15	4.86E−15	7.26E−15	**3.44E−15**	4.71E−15	3.82E−15	6.19E−15	7.11E−15	6.03E−15
f_{11}	$\bar{f}$	**0.00E+00**	**0.00E+00**	**0.00E+00**	**0.00E+00**	**0.00E+00**	**0.00E+00**	**0.00E+00**	**0.00E+00**	**0.00E+00**	**0.00E+00**

(continued)

Table 6.3 (continued)

α_2		0.01	0.02	0.03	0.04	0.05	0.06	0.07	0.08	0.09	0.1
	σ_f	**0.00E+00**	**0.00E+00**	**0.00E+00**	**0.00E+00**	**0.00E+00**	**0.00E+00**	**0.00E+00**	**0.00E+00**	**0.00E+00**	**0.00E+00**
f_{12}	$\bar{f}$	1.03E+00	7.77E−01	**9.55E−02**	1.93E+00	9.59E−01	4.14E−01	1.17E+00	1.26E+00	1.60E+00	6.41E+00
	σ_f	3.80E+00	2.63E+00	**3.50E−02**	4.30E+00	3.26E+00	1.75E+00	3.56E+00	4.41E+00	4.70E+00	2.28E+01
f_{13}	$\bar{f}$	3.54E−01	4.08E−01	1.69E+00	2.38E+00	**3.51E−01**	4.20E−01	1.17E+00	1.12E+00	8.94E−01	1.34E+00
	σ_f	1.91E−01	2.18E−01	5.25E+00	7.65E+00	**1.76E−01**	1.69E−01	4.03E+00	3.38E+00	2.53E+00	4.99E+00
f_{14}	$\bar{f}$	−8.85E+02	−8.84E+02	−8.91E+02	−8.88E+02	−8.90E+02	−8.87E+02	−8.82E+02	−8.86E+02	**−8.94E+02**	−8.82E+02
	σ_f	2.41E+01	1.81E+01	2.25E+01	2.39E+01	1.80E+01	**1.47E+01**	2.37E+01	1.63E+01	2.07E+01	1.54E+01

Bold data represent the best obtained values

Table 6.4 Different values of γ obtained by the fuzzy method

γ		0.001	0.002	0.003	0.004	0.005	0.006	0.007	0.008	0.009	0.01
f_1	$\bar{f}$	3.73E−28	5.52E−29	4.79E−29	1.14E−28	**1.01E−39**	2.04E−28	1.92E−28	1.20E−28	8.28E−29	1.27E−28
	σ_f	8.01E−28	2.74E−28	2.23E−28	5.10E−28	**5.54E−39**	7.66E−28	7.31E−28	4.91E−28	4.46E−28	6.93E−28
f_2	$\bar{f}$	1.78E−16	5.62E−16	6.03E−16	9.19E−17	**5.26E−35**	4.20E−16	2.44E−22	7.16E−22	3.97E−17	2.41E−18
	σ_f	8.72E−16	2.14E−15	2.30E−15	4.21E−16	**2.88E−34**	1.63E−15	1.34E−21	3.89E−21	2.18E−16	1.30E−17
f_3	$\bar{f}$	1.19E−23	5.91E−25	1.31E−23	1.74E−24	**2.35E−25**	4.96E−24	6.38E−22	1.91E−23	3.41E−21	7.75E−13
	σ_f	6.49E−23	2.61E−24	6.52E−23	5.07E−24	**8.29E−25**	2.66E−23	2.64E−21	1.04E−22	1.85E−20	4.24E−12
f_4	$\bar{f}$	3.34E−26	9.36E−16	5.20E−16	5.23E−16	**6.13E−19**	7.80E−18	5.46E−16	6.14E−16	7.75E−21	4.90E−16
	σ_f	1.52E−25	3.79E−15	2.85E−15	2.86E−15	**3.31E−18**	4.27E−17	2.99E−15	2.37E−15	4.24E−20	2.56E−15
f_5	$\bar{f}$	2.75E+01	2.85E+01	2.97E+01	**3.85E−04**	1.45E−01	3.55E−01	8.35E−01	1.23E+00	2.78E+01	2.85E+01
	σ_f	4.67E−02	4.29E−02	4.47E−02	**2.85E−02**	4.38E−02	4.52E−02	4.31E−02	3.96E−02	4.18E−02	4.49E−02
f_6	$\bar{f}$	1.89E−02	1.93E−02	1.60E−02	1.85E−02	**1.54E−02**	1.57E−02	2.17E−02	2.06E−02	2.15E−02	1.92E−02
	σ_f	1.27E−02	1.50E−02	7.99E−03	1.14E−02	**5.55E−03**	8.61E−03	1.45E−02	1.91E−02	1.90E−02	1.34E−02
f_7	$\bar{f}$	1.98E−03	1.76E−03	1.49E−03	1.58E−03	**1.32E−03**	1.71E−03	1.61E−03	1.95E−03	2.30E−03	1.36E−03
	σ_f	2.02E−03	1.62E−03	2.01E−03	1.56E−03	**1.03E−03**	1.38E−03	1.92E−03	2.03E−03	2.79E−03	1.10E−03
f_8	$\bar{f}$	−5.11E+02	−5.05E+03	**−5.27E+04**	−5.19E+03	−5.13E+04	−4.98E+03	−5.05E+03	−4.12E+02	−5.11E+02	−4.98E+03
	σ_f	5.20E+02	4.42E+02	5.69E+02	4.20E+02	4.77E+02	4.66E+02	4.54E+02	5.31E+02	**3.24E+02**	5.47E+02
f_9	$\bar{f}$	**1.14E−14**	6.94E+00	4.72E+00	1.07E+01	4.76E−01	6.13E+00	8.69E+00	2.16E+01	7.27E−02	1.41E+01
	σ_f	**2.75E−14**	2.55E+01	2.56E+01	4.05E+01	2.38E+00	3.18E+01	3.15E+01	5.64E+01	2.81E−01	4.31E+01
f_{10}	$\bar{f}$	1.23E−14	8.70E−15	9.06E−15	1.13E−14	**1.04E−14**	1.05E−14	9.06E−15	1.26E−14	8.47E−15	8.82E−15
	σ_f	6.29E−15	3.29E−15	2.97E−15	6.79E−15	**2.97E−15**	6.06E−15	3.82E−15	6.47E−15	5.55E−15	3.58E−15
f_{11}	$\bar{f}$	**0.00E+00**	**0.00E+00**	**0.00E+00**	**0.00E+00**	**0.00E+00**	4.78E−05	**0.00E+00**	**0.00E+00**	**0.00E+00**	**0.00E+00**

(continued)

Table 6.4 (continued)

γ		0.001	0.002	0.003	0.004	0.005	0.006	0.007	0.008	0.009	0.01
	σ_f	**0.00E+00**	**0.00E+00**	**0.00E+00**	**0.00E+00**	**0.00E+00**	1.85E−04	**0.00E+00**	**0.00E+00**	**0.00E+00**	**0.00E+00**
f_{12}	$\bar{f}$	2.11E+00	9.59E−01	9.95E−01	7.95E−01	**9.38E−02**	8.96E−01	5.60E−01	1.08E+00	1.32E−01	6.74E−01
	σ_f	7.38E+00	3.26E+00	3.37E+00	3.83E+00	**1.99E−02**	3.17E+00	2.48E+00	3.73E+00	1.76E−01	3.12E+00
f_{13}	$\bar{f}$	3.98E−01	4.28E−01	4.12E−01	3.90E−01	**3.85E−01**	1.06E+00	3.91E−01	4.08E−01	1.14E+00	3.88E−01
	σ_f	2.68E−01	2.24E−01	3.49E−01	1.90E−01	**1.51E−01**	3.82E+00	1.76E−01	2.24E−01	4.12E+00	2.08E−01
f_{14}	$\bar{f}$	−8.86E+02	−8.82E+02	−8.91E+02	−8.91E+02	−8.90E+02	−8.93E+02	−8.93E+02	**−8.97E+02**	−8.89E+02	−8.89E+02
	σ_f	2.29E+01	2.03E+01	2.22E+01	2.03E+01	**1.80E+01**	2.72E+01	1.91E+01	2.24E+01	1.97E+01	1.98E+01

Bold data represent the best obtained values

and the best element does not embed promising solutions. Lastly, γ represents the area around a promising solution, from which a local search procedure is carried. The goal of this procedure is to improve the quality of the solutions that a satisfactory fitness value.

Regarding their magnitude, the values of $\alpha_1 = 1.4$, $\alpha_2 = 0.05$ and $\gamma = 0.005$ show that the attraction method is the most important operation in the optimization strategy. This event confirms that the attraction procedure describes the most productive operation in the fuzzy strategy, since it explores new solutions in the direction where high fitness values are expected. Regarding to its meaning, the repulsion procedure holds the second place. Repulsion provides meaningful small adjustments of solutions in comparison to the attraction method. This result manifests that the repulsion process requires an exploration with a higher uncertainty compared with the attraction process. This uncertainty is considered because of the lack of information, if the opposite movement may lead a position with a better fitness value. The only possible evidence is that in direction of the attraction movement, it is not conceivable to obtain promising solutions. Lastly, the small value of γ causes a minor fluctuation for each satisfactory solution, to refine its fitness value.

6.6.2 Comparison with Another Optimization Techniques

In this section, the proposed algorithm is judged in comparison with other traditional optimization algorithms based on natural principles. In the experiments, it is evaluated the 19 functions from Appendix B, and the results are compared to those provided by the Harmony Search (HS) method [14], the Bat (BAT) algorithm [15], the Differential Evolution (DE) [19], the Particle Swarm Optimization (PSO) method [12], the Artificial Bee Colony (ABC) algorithm [13] and the Co-variance Matrix Adaptation Evolution Strategies (CMA-ES) [49]. These are considered as the most popular evolutionary algorithms [50]. In the experiments, the population has been set to 50 individuals. The execution of the test functions is carried in 50 and 100 dimensions. To eliminate the stochastic effect, each function is evaluated for 30 independent executions. In the comparison, a fixed number *FN* of function evaluations has been implemented as the stop condition. Hence, every execution of a benchmark function consists of $FN = 10^4 \cdot n$ evaluations. This condition has been considered to maintain compatibility with published literature [51–54].

For the experimental comparison, all the optimization methods have been configured with their corresponding parameter values, which according to their related references, they lead to the best performance. Such arrangements are defined as follows:

1. **HS** [14]: The rate *HCMR* has been set to 0.7 and the *PArate* to 0.3.
2. **BAT** [15]: The Loudness value has been set to 2, the Pulse Rate value to 0.9, The minimum Frequency to 0 and the maximum Frequency to 1.

3. **DE** [19]: The crossover ratio CR has been set to 0.5 and the F proportional factor to 0.2.
4. **PSO** [12]: The weight factor decreases linearly from 0.9 to 0.2. The constants $c_1 = 2$ and $c_2 = 2$.
5. **ABC** [13]: The population limit has been set to 50 individuals.
6. **CMA-ES** [47]: The algorithm has been implemented according to the source code [55].
7. **FUZZY**: $\alpha_1 = 1.4$, $\alpha_2 = 0.05$ and $\gamma = 0.005$.

6.6.2.1 Unimodal Test Functions

In this experiment, the performance of the proposed algorithm is compared with HS, BAT, DE, PSO, CMA-ES and ABC, examining functions which contains only one optimum. Such functions are described by the functions f_1 to f_7 in Table B1. In the experiment, all the benchmark functions have been evaluated in 50 dimensions. The experimental results obtained from 30 independent executions are displayed in Table 6.5. They describe the averaged best fitness values ($\bar{f}$) and the standard deviations (σ_f) taken among the executions. Also it is included the best (f_{Best}) and the worst (f_{Worst}) fitness values collected through the total amount of runs. The best entries in Table 6.5 are highlighted. From Table 6.5, regarding to the $\bar{f}$ value, it can be concluded that the proposed method performs better than the other algorithms in functions f_1, f_3, f_4 and f_7. For functions f_2, f_5 and f_6, the CMA-ES algorithm achieves the best outcomes. As opposite, the rest of the algorithms sponsor different levels of precision, with ABC being the most consistent. These results designate that the proposed method provides better performance than HS, BAT, DE, PSO and ABC for all functions but for the CMA-ES which produces related results to those provided by the proposed method. By examining the standard deviation (σ_f) value in Table 6.5, it shifts clear that the proposed method presents the best results with the smallest deviations.

To statistically examine the numerical results of Table 6.5, a non-parametric test identified as the Wilcoxon test [56, 57] has been carried. It enables us to indicate the differences between the two related methods. The test is performed by considering the 5% (0.05) of the significance level over the average fitness values. Table 6.6 summarizes the p-values produced by Wilcoxon analysis for the pairwise comparison of the algorithms. For the study, five groups are produced: FUZZY versus HS, FUZZY versus BAT, FUZZY versus DE, FUZZY versus PSO, FUZZY versus CMA-ES and FUZZY versus ABC. In the statistical study, the null hypothesis has been considered as there is no difference among the two approaches. Also, it is accepted as an alternative hypothesis that there is an essential difference between the two approaches. To facilitate the statistical analysis of Table 6.6, the following symbols have been adopted: ▲, ▼, and ►. ▲ indicates that the proposed method achieves significantly better than the tested algorithm for the specified function. ▼ expresses that the proposed algorithm performs worse than the tested algorithm, and ► means that the

Table 6.5 Results of the minimization of Table B1 with $n = 50$

		HS	BAT	DE	PSO	CMA−ES	ABC	FUZZY
f_1	$\bar{f}$	87035.2235	121388.0212	61.1848761	4.39E+03	1.34E−11	3.09E−06	**2.30E−29**
	σ_f	5262.26532	6933.129294	163.555175	1261.19173	5.1938E−12	3.4433E−06	**1.17237E−28**
	f_{Best}	76937.413	108807.878	0.03702664	1.65E+03	5.69E−12	2.47E−07	**5.17E−114**
	f_{Worst}	95804.9747	138224.1125	878.436103	7.37E+03	2.55E−11	1.72E−05	**6.42E−28**
f_2	$\bar{f}$	1.3739E+14	4.31636E+17	0.04057031	4.54E+01	**9.92E−06**	1.39E−03	4.15E−04
	σ_f	3.188E+14	1.53734E+18	0.09738928	16.386199	**2.5473E−06**	0.00071159	0.00227186
	f_{Best}	1.0389E+10	1633259021	4.03E−12	2.61E+01	5.87E−06	5.62E−04	**7.20E−59**
	f_{Worst}	1.64E+15	7.60E+18	0.45348954	9.75E+01	**1.44E−05**	2.98E−03	0.01244379
f_3	$\bar{f}$	130472.801	297342.4211	55982.8182	1.57E+04	2.89E−03	4.14E+04	**1.93E−05**
	σ_f	11639.2864	99049.83213	9234.85975	9734.92204	0.00164804	4785.18216	**4.2843E−05**
	f_{Best}	104514.012	164628.01	36105.5799	4.23E+03	9.88E−04	2.85E+04	**1.66E−10**
	f_{Worst}	147659.604	563910.1737	70938.4205	4.96E+04	8.88E−03	4.84E+04	**0.00018991**
f_4	$\bar{f}$	80.1841708	90.17564768	25.8134455	2.32E+01	3.96E−04	7.35E+01	**3.37E−16**
	σ_f	2.55950002	1.862675447	6.30765469	3.51409694	8.2083E−05	3.60905231	**1.8484E−15**
	f_{Best}	73.2799506	86.11297617	15.7894785	1.73E+01	2.57E−04	6.55E+01	**7.52E−70**
	f_{Worst}	83.8375161	92.78058061	38.8210447	3.06E+01	5.65E−04	7.90E+01	**1.01E−14**
f_5	$\bar{f}$	1024.70257	276.2438329	52.5359064	6.04E+02	**3.51E−05**	4.53E+01	4.85E−04
	σ_f	100.932656	45.12095642	7.69858817	198.334321	0.49723274	1.13628434	**0.0389642**
	f_{Best}	783.653134	211.6001157	47.1421071	289.29993	**1.21E−09**	42.1783081	3.30E−09
	f_{Worst}	1211.08532	399.1608511	75.1362468	1126.38574	**3.0654249**	47.7422282	4.6323356
f_6	$\bar{f}$	88027.4244	119670.6412	43.5155273	4.51E+03	**1.42E−11**	4.15E−06	2.18E−07

(continued)

Table 6.5 (continued)

		HS	BAT	DE	PSO	CMA−ES	ABC	FUZZY
	σ_f	5783.21576	6818.723503	80.4217558	2036.72193	**5.5321E−12**	8.5588E−06	0.84607249
	f_{Best}	77394.5062	105958.6224	0.01832758	1705.47866	**5.88E−12**	6.00E−07	1.18513127
	f_{Worst}	97765.4819	130549.7364	306.098587	13230.6439	**2.85E−11**	4.79E−05	5.18913374
f_7	$\bar{f}$	197.476174	116.8196698	0.08164158	4.43E+01	2.82E−02	6.86E−01	**3.43E−04**
	σ_f	28.808573	16.46542385	0.12240289	17.8200508	0.00499868	0.14547266	**0.00447976**
	f_{Best}	116.483527	87.64501186	0.01387586	15.7697307	0.0201694	0.41798576	**0.00018152**
	f_{Worst}	263.233333	156.0245904	0.65353574	85.526355	0.03888318	0.8957427	**0.02057915**

Bold data represent the best obtained values

Table 6.6 *p*-values produced by Wilcoxon test over the average best fitness values from Table 6.5

FUZZY versus	HS	BAT	DE	PSO	CMA–ES	ABC
f_1	5.01E−07▲	9.49E−08▲	7.36E−07▲	7.89E−05▲	7.23E−03▲	5.14E−04▲
f_2	4.05E−07▲	2.46E−08▲	2.07E−03▲	2.01E−05▲	0.0937►	3.04E−03▲
f_3	2.01E−08▲	3.74E−08▲	1.04E−07▲	4.15E−05▲	0.0829►	2.76E−06▲
f_4	3.54E−07▲	2.14E−08▲	3.40E−06▲	2.01E−06▲	8.14E−03▲	4.16E−07▲
f_5	1.07E−08▲	4.04E−09▲	2.04E−07▲	8.13E−09▲	0.1264►	1.25E−07▲
f_6	6.17E−07▲	6.54E−08▲	4.59E−06▲	2.15E−07▲	0.0741►	2.15E−03▲
f_7	4.36E−07▲	1.92E−08▲	2.80E−04▲	5.48E−06▲	0.1031►	1.04E−03▲
▲	7	7	7	7	2	7
▼	0	0	0	0	0	0
►	0	0	0	0	5	0

Wilcoxon rank sum test cannot discriminate among the fitness values and the tested algorithm. The number of events that happen in these conditions are presented at the bottom of the table.

For the groups FUZZY versus HS, FUZZY versus BAT, FUZZY versus DE, FUZZY versus PSO and FUZZY versus ABC are less than 0.05 which is substantial proof against the null hypothesis and designates that the proposed approach performs better than the HS, BAT, DE, PSO and ABC techniques. This information is statistically meaningful and determines that it has not happened by coincidence. In the case of the FUZZY and CMA-ES techniques, the FUZZY algorithm keeps a better performance in functions f_1 and f_4. For functions f_2, f_3, f_5, f_6 and f_7 the CMA-ES presents quite similar performance than the FUZZY proposal. This event can be examined from the column FUZZY versus CMA-ES, where the *p*-values of functions f_2, f_3, f_5, f_6 and f_7 are higher than the significant level. These results exhibit that there is no statistical variation in terms of accuracy among FUZZY and CMA-ES. In general, the *p*-values of the Wilcoxon rank test prove that the proposed fuzzy method performs better its competitors.

Additionally to the simulations in 50 dimensions, the performance of the fuzzy method is tested on 100 dimensions to prove its scalability. For the comparison, the same criteria have been applied as the experiment on 50 dimensions. The numerical results are presented in Tables 6.7 and 6.8, which describe the outcomes generated through 30 runs and the Wilcoxon *p*-values. Regarding to $\bar{f}$, from Table 6.7, the proposed method performs better than the rest of tested algorithms in functions f_1, f_2, f_3, f_4 and f_7. For functions f_5 and f_6, the CMA-ES algorithm the proposed method. On the other hand, the remaining algorithms exhibit different levels of efficiency. From Table 6.7, it is quite clear that the proposed algorithm presents slightly better numerical results than CMA-ES in 100 dimensions. In Table 6.8, it is indisputable that the *p*-values for the groups FUZZY versus HS, FUZZY versus BAT, FUZZY versus DE, FUZZY versus PSO and FUZZY versus ABC are less than the significant

Table 6.7 Results of the minimization of Table B1 with $n = 100$

		HS	BAT	DE	PSO	CMA−ES	ABC	FUZZY
f_1	$\bar{f}$	2.19E+05	2.63E+05	3.89E+02	1.43E+04	1.32E−05	2.45E−02	**1.89E−16**
	σ_f	10311.7753	14201.563	323.607739	2920.78144	3.2095E−06	0.02575058	**1.0358E−15**
	f_{Best}	1.74E+05	2.30E+05	1.26E+01	9.64E+03	8.00E−06	4.46E−03	**6.43E−68**
	f_{Worst}	2.30E+05	2.89E+05	1.38E+03	2.09E+04	2.04E−05	1.26E−01	**5.67E−15**
f_2	$\bar{f}$	7.24E+37	1.31E+45	5.73E−01	1.51E+02	1.26E−02	9.28E−02	**1.89E−08**
	σ_f	2.587E+38	6.9056E+45	0.57775559	41.1235147	0.00287131	0.02545392	**1.03445E−07**
	f_{Best}	5.45E+31	3.36E+34	2.94E−02	9.05E+01	8.91E−03	5.40E−02	**4.84E−46**
	f_{Worst}	1.35E+39	3.79E+46	2.56E+00	2.52E+02	2.38E−02	0.18353771	**5.67E−07**
f_3	$\bar{f}$	4.98E+05	1.15E+06	2.84E+05	7.90E+04	8.76E−04	1.84E+05	**2.07E−08**
	σ_f	58467.2769	312595.437	27132.9955	34174.7164	0.7743521	20108.5821	**0.54899784**
	f_{Best}	3.37E+05	4.69E+05	2.31E+05	3.71E+04	8.76E−04	1.27E+05	**3.94E−09**
	f_{Worst}	616974.994	1942095.52	356515.579	160139.954	145362.58	219982.397	**2.25159901**
f_4	$\bar{f}$	9.01E+01	9.46E+01	4.05E+01	2.81E+01	2.16E−06	9.04E+01	**2.63E−15**
	σ_f	1.11508888	0.85959226	6.84190892	3.2581793	0.03982112	1.8297347	**6.0638E−15**
	f_{Best}	8.69E+01	9.29E+01	2.70E+01	2.15E+01	1.39E−08	8.37E+01	**1.24E−67**
	f_{Worst}	9.16E+01	9.62E+01	5.55E+01	3.40E+01	2.91E−01	9.30E+01	**2.02E−14**
f_5	$\bar{f}$	2.70E+03	1.15E+03	1.29E+02	3.95E+03	**9.09E−04**	1.04E+02	9.86E−04
	σ_f	540.831375	88.3793364	18.7799545	641.937017	1.45575683	6.27511831	**0.14584341**
	f_{Best}	**0**	980.500576	101.873611	2571.71519	**1.32E−05**	96.9508931	2.43E−06
	f_{Worst}	3247.18778	1308.54456	167.973469	5316.76433	**94.8276069**	121.37168	99.203712
f_6	$\bar{f}$	2.21E+05	2.64E+05	3.70E+02	1.45E+04	**1.51E−05**	1.92E−02	2.02E−05

(continued)

Table 6.7 (continued)

		HS	BAT	DE	PSO	CMA−ES	ABC	FUZZY
	σ_f	9381.48118	18216.585	280.553973	2798.89068	**2.2704E−06**	0.01816388	2.87413087
	f_{Best}	198863.662	226296.005	16.8023385	9243.44125	**1.10E−05**	0.0025079	0.15E−05
	f_{Worst}	235307.769	288557.483	1278.66865	20954.2719	**1.94E−05**	0.09120152	23.7436855
f_7	$\bar{f}$	1.28E+03	4.67E+02	7.35E−01	4.85E+02	7.03E−02	2.20E+00	**4.51E−03**
	σ_f	100.811769	54.7947564	0.58830651	125.201537	0.01043989	0.36193003	**0.00535871**
	f_{Best}	975.480173	351.38694	0.08400938	233.702775	0.04689927	1.43278953	**3.59E−05**
	f_{Worst}	1424.35137	609.120842	2.38315757	806.350617	0.08989015	2.98777544	**0.0213576**

Bold data represent the best obtained values

Table 6.8 *p*-values produced by Wilcoxon test over the average best fitness values from Table 6.7

FUZZY versus	HS	BAT	DE	PSO	CMA–ES	ABC
f_1	3.01E–08▲	8.07E–08▲	6.49E–06▲	2.01E–07▲	7.94E–04▲	4.01E–05▲
f_2	5.34E–11▲	1.13E–11▲	7.49E–05▲	4.70E–06▲	2.49E–04▲	8.29E–04▲
f_3	6.01E–07▲	7.06E–08▲	4.67E–07▲	8.46E–06▲	0.0743►	2.30E–07▲
f_4	1.49E–07▲	3.79E–07▲	2.01E–06▲	1.49E–06▲	7.46E–04▲	2.19E–07▲
f_5	8.79E–06▲	5.49E–06▲	9.46E–05▲	2.15E–07▲	0.1851►	4.72E–05▲
f_6	2.73E–07▲	4.79E–07▲	8.04E–05▲	5.79E–06▲	0.2451►	1.41E–04▲
f_7	1.04E–07▲	5.42E–06▲	2.00E–04▲	7.61E–06▲	0.0851►	4.6E–05▲
▲	7	7	7	7	3	7
▼	0	0	0	0	0	0
►	0	0	0	0	4	0

level, which means that the FUZZY performs better than the HS, BAT, DE, PSO and ABC methods.

6.6.2.2 Multimodal Test Functions

Multimodal functions involve the presence of many local optima. For this reason, they are more difficult to obtain an optimal solution. In this experiment, the performance of the fuzzy method is compared with HS, BAT, DE, PSO, CMA-ES and ABC considering multimodal surfaces. Multimodal functions ($f_8 - f_{14}$) are described in Table B2, where the number of local optima rises as the dimension of the optimization problem also increases. Under such circumstances, the test exhibits the capacity of each algorithm to find the global solution in presence of multiple local optima. The experiment evaluates each function in 50 dimensions. The numerical results are reported in Table 6.9. The best results are also highlighted. Furthermore, the *p*-values of the Wilcoxon rank test presented in Table 6.10. In the case of functions f_8, f_{10}, f_{11} and f_{14}, the proposed algorithm presents an outstanding performance than HS, BAT, DE, PSO, CMA-ES and ABC. For functions f_{12} and f_{13}, the fuzzy proposal presents worse performance than CMA-ES. In the case of the function f_9, the fuzzy algorithm and ABC keep the best performance compared to HS, BAT, DE, PSO and CMA-ES.

The statistical results obtained by the Wilcoxon test are presented in Table 6.10, from the results, it cvan be seen that the proposed method performs better than HS, DE, BAT, DE and PSO in all the test functions. Considering the group among FUZZY and CMA-ES, the FUZZY method keeps a better (▲) performance in functions f_8, f_9, f_{10}, f_{11} and f_{14}. Meanwhile, in functions f_{12} and f_{13} the FUZZY method presents worse results (▼) than CMA-ES. Nonetheless, from Table 6.10, the proposed FUZZY

Table 6.9 Results of the minimization of Table B2 with $n = 50$

		HS	BAT	DE	PSO	CMA−ES	ABC	FUZZY
f_8	$\bar{f}$	−5415.83905	−3270.254967	−20232.0393	−1.00E+04	−6.22E+03	−1.94E+04	**−2.69E+05**
	σ_f	**318.326084**	474.2974644	799.670519	1139.72504	577.603827	328.074723	338131.298
	f_{Best}	−6183.25306	−4253.007751	−20830.6009	−12207.27	−7910.14987	−20460.0202	**−1602802.18**
	f_{Worst}	−4937.01673	−2560.016959	−16849.3789	−7417.20923	−5277.36121	−18598.5718	**−51860.8124**
f_9	$\bar{f}$	637.314967	370.181231	94.9321639	2.78E+02	9.12E−03	**6.43E−06**	1.03E−06
	σ_f	25.9077403	31.0789956	23.8913991	38.1965572	8.314267	**2.34825349**	1.6781696
	f_{Best}	581.055495	321.398632	49.5787092	2.09E+02	6.91E−04	1.99899276	**0**
	f_{Worst}	681.505155	450.919651	143.664199	375.979507	342.621828	**11.2277505**	310.43912
f_{10}	$\bar{f}$	20.2950743	19.2995398	1.04072229	1.21E+01	8.74E−07	1.69E−02	**1.40E−14**
	σ_f	0.09866263	0.11146929	0.69779278	1.03376556	1.4921E−07	0.01136192	**4.489E−15**
	f_{Best}	19.9003135	18.9741996	0.0040944	9.35E+00	5.62E−07	2.92E−03	**7.99E−15**
	f_{Worst}	20.472367	19.576839	2.78934934	1.38E+01	1.18E−06	5.30E−02	**2.22E−14**
f_{11}	$\bar{f}$	786.564993	1072.40695	0.98725915	4.05E+01	9.87E−10	6.23E−03	**0.00E+00**
	σ_f	49.0195978	70.0220465	0.62998733	12.8397453	4.8278E−10	0.01154936	**0**
	f_{Best}	658.158623	926.062051	0.00107768	14.1594978	2.92E−10	0.01503879	**0**
	f_{Worst}	860.983823	1186.93137	2.4153938	64.7187195	2.32E−09	0.053391	**0**
f_{12}	$\bar{f}$	557399404	1029876322	1309.87126	4.08E+01	**2.58E−12**	3.67E−07	1.91E−08
	σ_f	68320767.6	150067294	4319.40539	27.0146375	**1.0706E−12**	4.1807E−07	0.63138873
	f_{Best}	444164964	763229039	0.20366113	16.607083	**9.05E−13**	2.14E−08	1.95E−08
	f_{Worst}	700961313	1277767934	17508.9826	136.891908	**5.63E−12**	1.84E−06	19.8206283
f_{13}	$\bar{f}$	1163989772	1982187734	29551.4297	1.41E+05	**5.03E−11**	5.98E−06	9.06E−09

(continued)

Table 6.9 (continued)

		HS	BAT	DE	PSO	CMA−ES	ABC	FUZZY
	σ_f	123421334	291495991	137415.981	186740.994	**2.7928E−11**	7.5042E−06	0.0439066
	f_{Best}	898858903	1250159582	4.37613356	1594.63864	**1.37E−11**	5.06E−07	6.91E−10
	f_{Worst}	1453640537	2558128052	754699.3	735426.524	**1.38E−10**	3.48E−05	60.2604938
f_{14}	$\bar{f}$	−958.679663	−1066.80779	−1937.10075	−1.40E+03	−1.84E+03	−1.32E+03	**−1.96E+03**
	σ_f	40.8885038	61.5793102	13.199329	77.8992722	41.6063335	29.4904745	**0.06633944**
	f_{Best}	−1060.29598	−1213.46466	−1958.29881	−1527.69665	−1915.89813	−1395.05118	**−1958.30818**
	f_{Worst}	−893.619964	−988.87492	−1915.09727	−1275.90292	−1732.12078	−1262.8218	**−1958.04305**

Bold data represent the best obtained values

Table 6.10 *p*-values produced by Wilcoxon test over the average best fitness values from Table 6.9

FUZZY versus	HS	BAT	DE	PSO	CMA−ES	ABC
f_8	7.10E−05▲	5.40E−05▲	3.47E−05▲	1.14E−05▲	4.72E−06▲	2.49E−05▲
f_9	4.16E−05▲	2.48E−05▲	6.14E−04▲	2.12E−05▲	4.29E−04▲	0.0783▶
f_{10}	3.15E−05▲	3.01E−05▲	7.49E−04▲	1.49E−05▲	3.11E−04▲	9.48E−04▲
f_{11}	5.49E−08▲	9.33E−09▲	7.13E−04▲	5.37E−06▲	8.49E−03▲	6.15E−03▲
f_{12}	6.48E−11▲	8.40E−11▲	4.68E−08▲	5.29E−06▲	7.23E−04▼	4.03E−03▲
f_{13}	7.98E−11▲	9.79E−11▲	4.16E−10▲	7.46E−11▲	9.40E−04▼	4.55E−05▲
f_{14}	7.13E−06▲	4.58E−07▲	5.79E−04▲	8.16E−05▲	9.64E−04▲	5.68E−05▲
▲	7	7	7	7	5	6
▼	0	0	0	0	2	0
▶	0	0	0	0	0	1

algorithm achieves better numerical results than ABC in most of the cases except for function, where there is no significant difference in the outcomes among the two.

Additionally to the 50 dimensional multimodal experiment, the performance of the proposal is also compared considering 100 dimensions by using the benchmark functions in Table B2. The numerical results are presented in Tables 6.11 and 6.12, which summarize the outcomes produced through the 30 runs, as well as the Wilcoxon test. In Table 6.11, it can be observed that the proposed methodology achieves better outcomes than HS, BAT, DE, PSO, CMA-ES and ABC for functions: f_8, f_9, f_{10}, f_{11} and f_{13}. On the other hand, the CMA-ES keeps better results than HS, BAT, DE, PSO, ABC and the proposed method for function f_{12}. Furthermore, the DE method reaches better performance index than the rest of testes algorithms for function f_{14}. From the Wilcoxon test, shown in Table 6.12, the *p*-values indicate that the proposed algorithm performs better than the HS, BAT, DE, PSO and ABC algorithms. Considering the group FUZZY versus CMA-ES, the FUZZY method keeps a better performance in the majority of test functions where the CMA-ES delivers better outcomes than the proposed approach.

6.6.2.3 Hybrid Test Functions

In this experiment, hybrid functions are operated to analyze the performance of the proposed method in hybrid environments. Hybrid functions ($f_{15} - f_{19}$), shown in Table B3, are considered as multimodal functions with complex behaviors since they are created from diverse multimodal single functions. A comprehensive implementation of the hybrid benchmark functions can be found in [46]. In the tests, the performance of the proposed fuzzy methodology is compared with HS, BAT, DE, PSO, CMA-ES and ABC.

Table 6.11 Results of the minimization of Table B2 with $n = 100$

		HS	BAT	DE	PSO	CMA−ES	ABC	FUZZY
f_8	$\bar{f}$	−7.87E+03	−4.47E+03	−2.34E+04	−1.48E+04	−8.66E+03	−3.51E+04	**−1.62E+05**
	σ_f	584.340059	799.31231	3914.77561	1729.15916	831.547363	**554.101741**	47514.2941
	f_{Best}	−9141.81789	−6497.01004	−32939.624	−19362.7097	−10177.5382	−36655.8271	**−272938.668**
	f_{Worst}	−6669.02042	−2858.38997	−18358.9463	−12408.7377	−7375.40434	−33939.9476	**−87985.343**
f_9	$\bar{f}$	1.44E+03	9.12E+02	3.98E+02	7.49E+02	2.37E+02	6.49E+01	**4.00E−05**
	σ_f	41.464352	60.4785656	67.0536747	67.5932069	209.403173	7.90879809	**0.00015207**
	f_{Best}	1293.19918	772.978721	205.209989	635.167862	74.8466353	49.9922666	**0**
	f_{Worst}	1503.0147	1033.80416	522.504376	865.893944	806.576683	79.9136576	**0.0005994**
f_{10}	$\bar{f}$	2.06E+01	1.98E+01	2.42E+00	1.33E+01	6.14E−04	3.01E+00	**3.96E−12**
	σ_f	0.06052521	0.08670819	0.82882063	0.87767923	9.6205E−05	0.31154878	**2.1504E−11**
	f_{Best}	2.05E+01	1.96E+01	1.12E+00	1.16E+01	4.14E−04	2.26E+00	**1.51E−14**
	f_{Worst}	2.08E+01	2.00E+01	4.63E+00	1.51E+01	8.49E−04	3.59E+00	**1.18E−10**
f_{11}	$\bar{f}$	1.96E+03	2.38E+03	4.46E+00	1.18E+02	1.20E−03	1.61E−01	**0.00E+00**
	σ_f	81.7177655	134.986806	2.67096611	22.637086	0.00028091	0.14553457	**0**
	f_{Best}	1773.68789	2035.67623	1.02086153	86.1907763	0.00066065	0.01309143	**0**
	f_{Worst}	2083.84569	2557.11279	12.932624	170.763675	2.25E−03	0.64617359	**0**
f_{12}	$\bar{f}$	1.88E+09	2.55E+09	4.85E+04	2.06E+04	**2.20E−06**	5.96E−03	2.67E+02
	σ_f	110307213	242921094	140231.648	50859.5682	**7.1926E−07**	0.01533206	1423.14186
	f_{Best}	1.64E+09	1.87E+09	1.86E+00	2.93E+01	**1.27E−06**	2.91E−05	4.21E−01
	f_{Worst}	2090764763	2966731722	769582.376	251450.989	**4.07E−06**	0.06229125	7801.38816
f_{13}	$\bar{f}$	3.54E+09	4.83E+09	6.63E+05	1.77E+06	2.20E+01	4.74E−03	**4.71E−05**

(continued)

Table 6.11 (continued)

		HS	BAT	DE	PSO	CMA−ES	ABC	FUZZY
	σ_f	255789666	477261470	1076152.4	1373210.26	33.1622741	0.00582626	**1.2473E−05**
	f_{Best}	2936347707	3998731504	3147.16418	417561.443	9.98142144	0.00075547	**2.65E−05**
	f_{Worst}	3928279153	5901118241	4651437.83	7741055.37	176.314279	0.02480097	**7.86E−05**
f_{14}	$\bar{f}$	−1.57E+03	−1.82E+03	**−3.83E+03**	−2.46E+03	−3.57E+03	−3.78E+03	−2.30E+03
	σ_f	79.1693639	76.7852589	36.6930452	97.0301039	59.5947211	**22.8558428**	65.2812493
	f_{Best}	−1722.36681	−1945.12523	**−3900.49741**	−2683.50044	−3676.29234	−3823.95029	−2455.01558
	f_{Worst}	−1432.17812	−1695.67279	**−3740.63055**	−2280.67655	−3393.55796	−3724.2658	−2156.74064

Bold data represent the best obtained values

Table 6.12 *p*-values produced by Wilcoxon test over the average best fitness values from Table 6.11

FUZZY versus	HS	BAT	DE	PSO	CMA–ES	ABC
f_8	8.12E−05▲	6.47E−05▲	4.69E−05▲	3.16E−05▲	7.11E−05▲	3.79E−05▲
f_9	1.36E−07▲	9.49E−06▲	7.60E−06▲	8.66E−06▲	5.49E−06▲	3.13E−06▲
f_{10}	6.94E−07▲	6.34E−07▲	3.56E−06▲	6.13E−06▲	3.16E−04▲	3.93E−06▲
f_{11}	4.98E−07▲	8.16E−07▲	7.13E−04▲	5.31E−06▲	2.02E−03▲	9.48E−03▲
f_{12}	4.26E−08▲	7.68E−08▲	8.46E−07▲	7.46E−07▲	3.05E−06▼	1.64E−05▲
f_{13}	4.59E−09▲	6.47E−09▲	6.16E−07▲	7.46E−08▲	9.17E−06▲	7.46E−04▲
f_{14}	8.15E−05▲	8.96E−05▲	6.49E−04▼	9.42E−03▲	5.46E−04▲	6.01E−04▲
▲	7	7	6	7	6	7
▼	0	0	1	0	1	0
►	0	0	0	0	0	0

In the first examination, the hybrid functions have been evaluated in 50 dimensions. The numerical results gathered from 30 independent runs are presented in Tables 6.13 and 6.14. In Table 6.13, the $\bar{f}, \sigma_f, f_{Best}$ and f_{Worst}, gathered through the total number of executions, are listed. Moreover, Table 6.14 displays the statistical Wilcoxon judgment of the fitness values $\bar{f}$ from Table 6.13.

Regarding to Table 6.13, the proposed method preserves a higher performance than the remaining optimization methods. In the case of f_{15}, f_{16} and f_{18}, the proposal outperforms HS, BAT, DE, PSO, CMA-ES and ABC. For function f_{19}, the fuzzy approach presents a worst performance than CMA-ES or ABC. Nonetheless, in functions f_{16} and f_{18}, the proposal as well as ABC keep a better performance than HS, BAT, DE, PSO and CMA-ES. For function f_{17} the FUZZY and CMA-ES techniques perform better than the remaining methods. Consequently, the proposed methodology gives better fitness values in 4 hybrid functions. This case proves that the fuzzy design is capable of producing more reliable outcomes than its competitors. From Table 6.13, it is explicit that the fuzzy proposal achieves better consistency than the other algorithms, since its assembled solutions present sparsity. As it can be assumed, the single exception is function, where the fuzzy algorithm does not reach the best performance In addition to such results, Table 6.13 determines that the fuzzy method achieves the best composed solution through the 30 independent runs than the rest algorithms, but for function.

Table 6.14 exposes the statistical results of the Wilcoxon test over the fitness values from Table 6.13. They suggest that the proposed approach performs better than HS, BAT, DE and PSO. In the group, FUZZY versus CMA-ES, the FUZZY method keeps a better performance in most of the test functions but in problem f_{19}, where the CMA-ES provides better outcomes than the FUZZY methodology. Nevertheless, in the comparison among FUZZY and ABC, FUZZY reaches the best results in all the functions except for functions f_{16} and f_{18} where there is no statistical difference among these methods.

Table 6.13 Results of the minimization of Table B3 with $n = 50$

		HS	BAT	DE	PSO	CMA−ES	ABC	FUZZY
f_{15}	$\bar{f}$	7.9969E+13	5.1022E+21	12.3509776	9.64E+03	6.36E−06	5.23E−04	**2.49E−15**
	σ_f	1.3868E+14	2.7945E+22	16.9157032	5195.1729	2.0915E−06	0.00018266	**6.6512E−15**
	f_{Best}	1.5309E+10	8.1918E+12	0.01189652	4.19E+03	4.16E−06	2.68E−04	**3.17E−58**
	f_{Worst}	4.7627E+14	1.53E+23	65.8109321	2.50E+04	1.48E−05	1.02E−03	**2.53E−14**
f_{16}	$\bar{f}$	2706.73644	3508.41961	73.699317	5.99E+02	5.88E+02	4.91E+01	**4.90E+01**
	σ_f	103.253645	252.48393	13.915609	114.611979	9.03874746	0.19971751	**0.00025806**
	f_{Best}	2491.6759	2883.46006	49.0005972	393.824001	**48.9964485**	48.998348	48.9973691
	f_{Worst}	2876.20757	3869.21201	103.690098	878.998232	81.0717712	49.6854294	**48.9981462**
f_{17}	$\bar{f}$	1151151029	2105822689	67405.5759	3.13E+05	**5.40E+01**	8.96E+02	**5.40E+01**
	σ_f	113601255	190215425	193321.347	421404.743	0.00020018	132.212771	**9.8857E−05**
	f_{Best}	909668213	1637035871	413.088353	15529.3525	53.9999308	546.710586	**53.9998073**
	f_{Worst}	1428940501	2452560936	936409.163	1947952.34	54.0007602	1155.9351	**54.000177**
f_{18}	$\bar{f}$	2.0155E+14	3.3427E+19	54.2557544	9.06E+02	5.99E+01	**4.90E+01**	**4.90E+01**
	σ_f	3.9073E+14	1.83E+20	10.4970565	291.082635	12.4089453	0.01211373	**0**
	f_{Best}	911061731	3.2735E+13	49.0002421	530.607446	49.0000116	49.0021079	**49**
	f_{Worst}	1.85E+15	1.00E+21	97.123625	1597.0748	86.0099556	49.0607764	**49**
f_{19}	$\bar{f}$	7.7416E+14	7.7174E+18	−19.5833354	1.17E+06	**−1.44E+02**	−1.43E+02	2.18E+01
	σ_f	1.6757E+15	4.2201E+19	117.854019	5835860.88	0.39733093	**0.29998167**	472.608012
	f_{Best}	1343488881	1.939E+10	−143.748394	4511.91824	**−144.056723**	−143.608756	−83.2609165
	f_{Worst}	7.53E+15	2.31E+20	334.753741	32053489.3	**−142.208256**	−143.0037	2523.59236

Bold data represent the best obtained values

Table 6.14 p-values produced by Wilcoxon test over the average best fitness values from Table 6.13

FUZZY versus	HS	BAT	DE	PSO	CMA−ES	ABC
f_{15}	4.61E−10▲	7.68E−11▲	8.12E−07▲	3.16E−09▲	1.35E−04▲	5.69E−05▲
f_{16}	5.69E−08▲	6.49E−09▲	6.31E−04▲	8.40E−05▲	6.49E−05▲	0.1560►
f_{17}	8.65E−11▲	9.46E−11▲	6.33E−09▲	7.34E−10▲	0.0956►	4.65−04▲
f_{18}	3.49E−11▲	7.68E−12▲	4.68E−04▲	5.31E−06▲	4.82E−04▲	0.1986►
f_{19}	7.63E−10▲	9.31E−11▲	4.33E−07▲	6.00E−09▲	6.33E−07▼	5.89E−07▼
▲	5	5	5	5	4	3
▼	0	0	0	0	1	1
►	0	0	0	0	0	1

The second part of the experiments have been considered within a 100 dimensional search space, Tables 6.15 and 6.16 introduce the numerical results of the examination in 100 dimensions. In Table 6.15, the indexes $\bar{f}$, σ_f,f_{Best} and f_{Worst}, are collected through the total amount of runs, are reported. On the other hand, Table 6.14 exhibits the statistical Wilcoxon test from Table 6.15.

Table 6.15 verifies the benefit of the proposed approach over HS, BAT, DE, PSO, CMA-ES and ABC techniques. From the results, it is evident that the fuzzy method provides better results than HS, BAT, DE, PSO, CMA-ES and ABC for the functions f_{15}–f_{18}. Nevertheless, it can be observed that the proposed methodology performs worse than CMA-ES and ABC in function f_{19}. Comparable to the case of 50 dimensions, in the case of 100 dimensions, the fuzzy proposal gets solutions with the smallest sparcity. This consistency is valid for most of the benchmark functions. Considering the results in 100 dimension, it is further remarked that the fuzzy method exceeds all algorithms.

The produces outcomes obtained from the Wilcoxon test, demonstrates that the fuzzy algorithm performs better than the rest of metaheuristics in most of the benchmark functions. In Table 6.16, it is also compiled the results of the study within the symbols ▲, ▼, and ►. The conclusions of the Wilcoxon rank-sum statistically validate the results from Table 6.15. They designate that the outstanding performance of the proposal is as a consequence of a better search mechanism and not for stochastic effects.

6.6.3 Convergence Analysis

In this part of the experimental study, a convergence analysis is performed. The objective of this test is to evaluate the speed at which each evolutionary method leads the optimum value. In the analysis, the performance of each algorithm is analyzed over all the benchmark functions from Appendix B, evaluated within 50-dimensional

Table 6.15 Results of the minimization of Table B3 with $n = 100$

		HS	BAT	DE	PSO	CMA−ES	ABC	FUZZY
f_{15}	$\bar{f}$	1.07E+38	1.45E+46	1.67E+02	3.58E+04	8.48E−03	8.13E−02	**1.03E−05**
	σ_f	2.6648E+38	7.9319E+46	212.77092	19777.8798	0.00248591	0.04144673	**5.64261E−05**
	f_{Best}	1.88E+28	1.31E+37	5.17E+00	1.58E+04	5.99E−03	4.35E−02	**4.50E−44**
	f_{Worst}	1.21E+39	4.34E+47	1.15E+03	9.15E+04	1.75E−02	2.66E−01	**3.09E−04**
f_{16}	$\bar{f}$	6.46E+03	8.00E+03	1.90E+02	1.36E+03	1.55E+02	1.81E+02	**9.90E+01**
	σ_f	316.896114	426.761823	28.0273818	129.05773	18.161094	18.5185506	**0.00187142**
	f_{Best}	5753.69747	7125.84623	148.521771	1116.8234	122.010095	129.807203	**98.9958572**
	f_{Worst}	6997.63377	8965.08163	260.542179	1681.54333	187.628218	206.634292	**99.0057217**
f_{17}	$\bar{f}$	3.57E+09	5.03E+09	6.85E+05	2.10E+06	5.95E+02	3.72E+03	**1.09E+02**
	σ_f	251070990	401513619	1158365.05	1426951.79	84.33472	459.761456	**0.0026802**
	f_{Best}	2908492728	4028081811	4767.47553	428429.323	428.761773	2973.30401	**108.99997**
	f_{Worst}	3953742523	5605713616	4789383.94	6274361.22	784.972289	4756.54052	**109.01469**
f_{18}	$\bar{f}$	3.30E+38	5.29E+43	1.33E+02	2.09E+03	1.49E+02	3.34E+02	**1.08E+02**
	σ_f	1.2038E+39	2.669E+44	20.7615151	481.839456	23.6851947	908.915277	**8.76059629**
	f_{Best}	5.02E+29	4.38E+34	1.01E+02	1.24E+03	1.12E+02	**9.90E+01**	99.6507148
	f_{Worst}	5.86E+39	1.47E+45	1.92E+02	3.51E+03	2.10E+02	4.24E+03	**134.545048**
f_{19}	$\bar{f}$	1.01E+38	1.49E+44	9.38E+03	6.45E+07	**−2.94E+02**	−2.01E+02	4.29E+07
	σ_f	3.9907E+38	6.2388E+44	40545.7303	217095412	**0.55047399**	1.27097202	200132558
	f_{Best}	4.34E+29	1.43E+36	−1.10E+02	4.71E+04	−2.95E+02	**−295.68221**	−9.45E+01
	f_{Worst}	2.18E+39	3.36E+45	2.23E+05	1.17E+09	**−2.92E+02**	−288.145624	1.08E+09

Bold data represent the best obtained values

Table 6.16 *p*-values produced by Wilcoxon test over the average best fitness values from Table 6.15

FUZZY versus	HS	BAT	DE	PSO	CMA-ES	ABC
f_{15}	8.46E−12▲	9.76E−12▲	6.49E−07▲	7.00E−08▲	3.12E−04▲	7.68E−04▲
f_{16}	7.63E−05▲	8.42E−05▲	6.50E−04▲	2.05E−04▲	3.96E−04▲	6.00E−11▲
f_{17}	5.34E−08▲	6.88E−08▲	3.30E−07▲	9.03E−07▲	4.29E−04▲	8.63E−05▲
f_{18}	4.93E−12▲	8.36E−12▲	5.63E−04▲	3.46E−05▲	6.03E−04▲	1.30E−05▲
f_{19}	6.92E−11▲	2.49E−12▲	6.33E−06▲	2.01E−04▲	6.30E−07▼	4.13E−07▼
▲	5	5	5	5	4	4
▼	0	0	0	0	1	1
▶	0	0	0	0	0	0

search space. The construction of the convergence graphs, considered the employment of the fitness information generated in Sects. 6.2.1, 6.2.2 and 6.2.3. Each test function is executed 30 runs by each algorithm, then, the selection of the convergence information is represented with the mean result. Figures 6.5, 6.6 and 6.7 depict the convergence information of the proposed method and its competitors. Figure 6.5 displays the convergence outcomes for functions f_1-f_6, Fig. 6.6 the convergence for functions f_7–f_{12}. Finally, Fig. 6.7 exposes the convergence for functions f_{13}-f_{19}.

Figure 6.5, indicates that the proposed fuzzy presents a better con-vergence degree than the rest of tested algorithms for functions f_1, f_2, f_4 and f_5. Nevertheless, for the benchmark functions f_3 and f_6 the CMA-ES achieves faster the optimal solution. In Fig. 6.6, the convergence plots indicate that the proposed method achieves the best replies for functions f_9, f_{10} and f_{11}. In function f_7, even if the fuzzy methodology finds the optimal solution, the DE method produces the best convergence result. An interesting case of study is function f_9, where many methods achieve an acceptable convergence rate. In case of function f_8, the DE and ABC schemes obtain the best convergence properties. Finally, in function f_{12}, the CMA-ES accomplishes the fastest reaction. Finally, in Fig. 6.7, the convergence rates for functions f_{13}–f_{19} are exhibited. In Fig. 6.7, the algorithms CMA-ES and ABC get the best replies. In case of function f_{14}, the DE and ABC methods obtain an optimal value in a punctual manner than the remaining techniques. Although for functions f_{15}-f_{18} the fuzzy algorithm gives the fastest convergence response, the CMA-ES method keeps a related reply.

Consequently, the convergence rate of the fuzzy for solving unimodal problems is faster than HS, BAT, DE, PSO, CMA-ES, and ABC. On the other hand, when solving multimodal optimization problems, the fuzzy proposal generally converges as quick as the compared algorithms. This aspect can be distinctly recognized in Figs. 6.6 and 6.7, where the proposed fuzzy system produces related convergence curve to the rest of the algorithms. Finally, after examining the performance of all the algorithms on hybrid functions, it is clear that the convergence response of the proposed method is not as fast as the one presented by CMA-ES. In particular, the proposed fuzzy and CMA-ES perform the best convergence features when they handle hybrid functions.

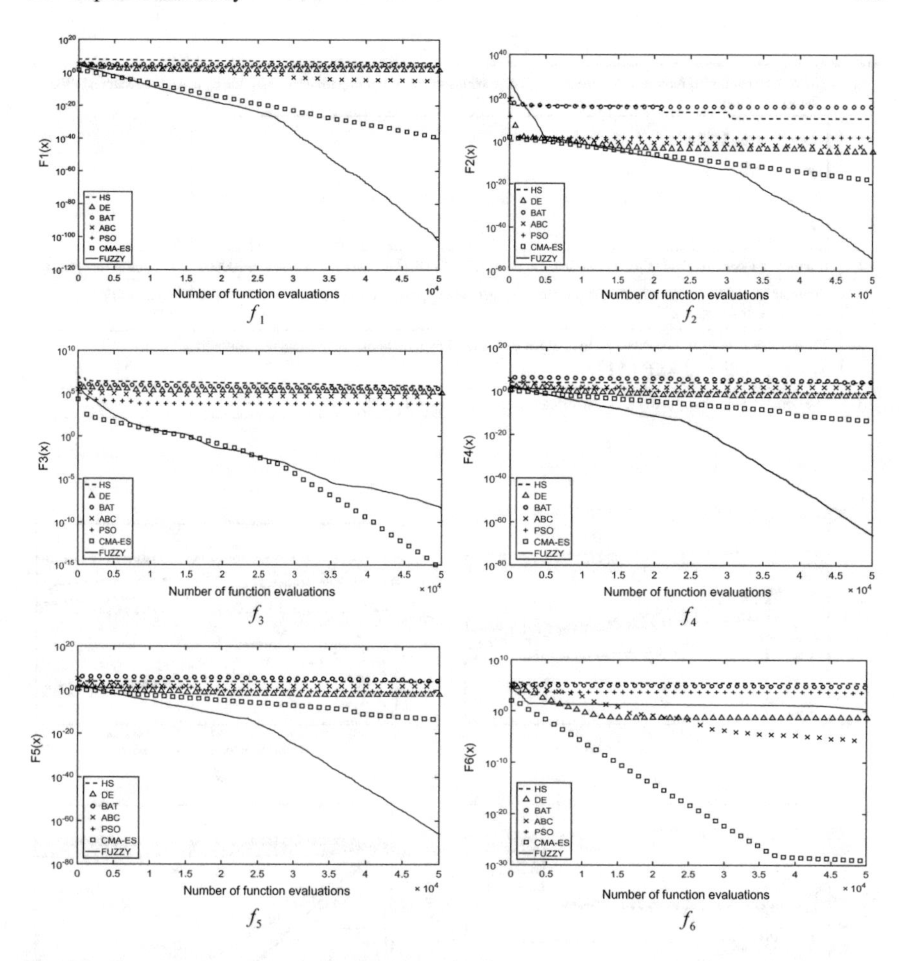

Fig. 6.5 Convergence test results for functions f_1–f_6

6.6.4 Computational Complexity Analysis

The computational complexity of the proposed method as well as its competitors are analyzed in this section. Commonly, optimization techniques are generally, complex processes with many stochastic procedures. Under such circumstances, the computational complexity (C) is employed to estimate the computational effort of each algorithm. C presents the averaged CPU time spent by an algorithm with respect to a normal time reference. In order to estimate the computational complexity, the procedure exhibited in [48] has been carried. Under this method, C is collected by the following method:

1.	The considered time reference v is computed T_0 corresponds to the computing time consumption by a single execution of the subsequent code:
	`for j = 1:1000000`
	`v = 0.55 + j`
	`v = v + v; v = v/2; v = v*v; v = sqrt(v); v = exp(v); v = v/(v + 2);`
	`end`

2.	Compute the computational time T_1 for function procedure. T_0 Displays the time involved in 200000 executions of function f_9
3.	Compute the execution time T_2 for the optimization method. T_2 presents the elapsed time involved in 200000 function evaluations of function f_9
4.	The average time $\bar{T}_2$ is calculated. At first, execute the Step 3 by five times. Later, compute their average value $\bar{T}_2 = \left(T_2^1 + T_2^2 + T_2^3 + T_2^4 + T_2^5\right)/5$
5.	The computational complexity C is then calculated as: $C = \left(\bar{T}_2 - T_1\right)/T_0$

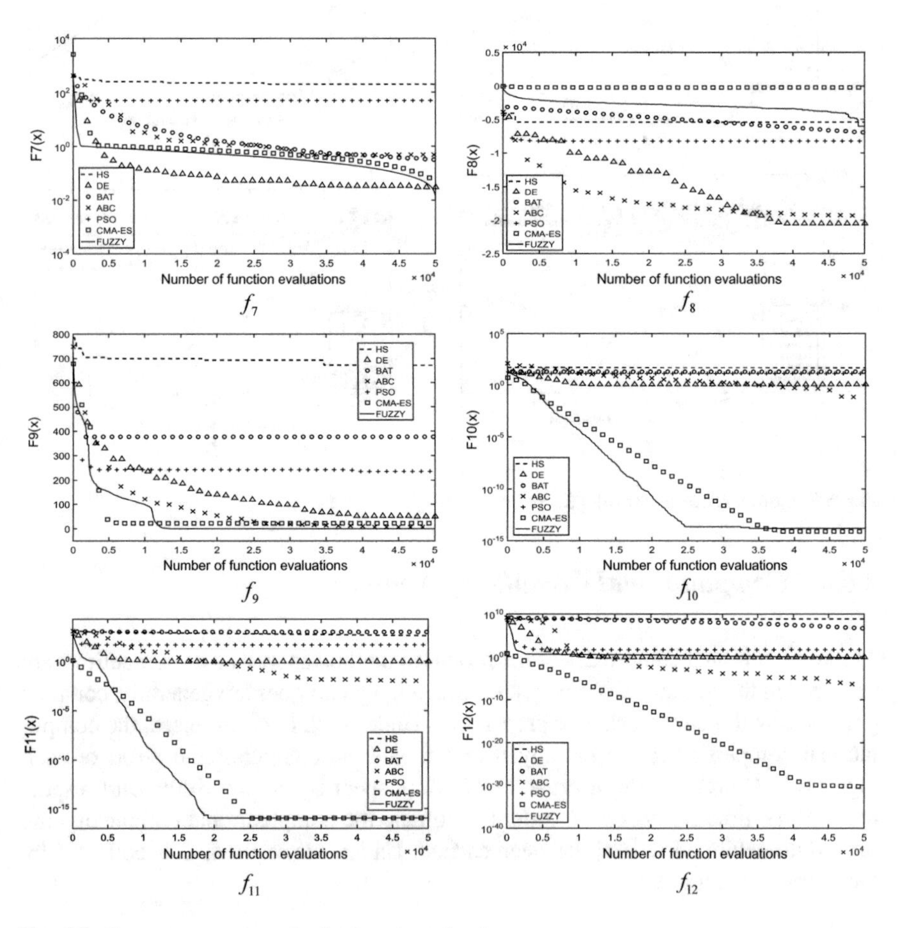

Fig. 6.6 Convergence test results for functions f_7–f_{12}

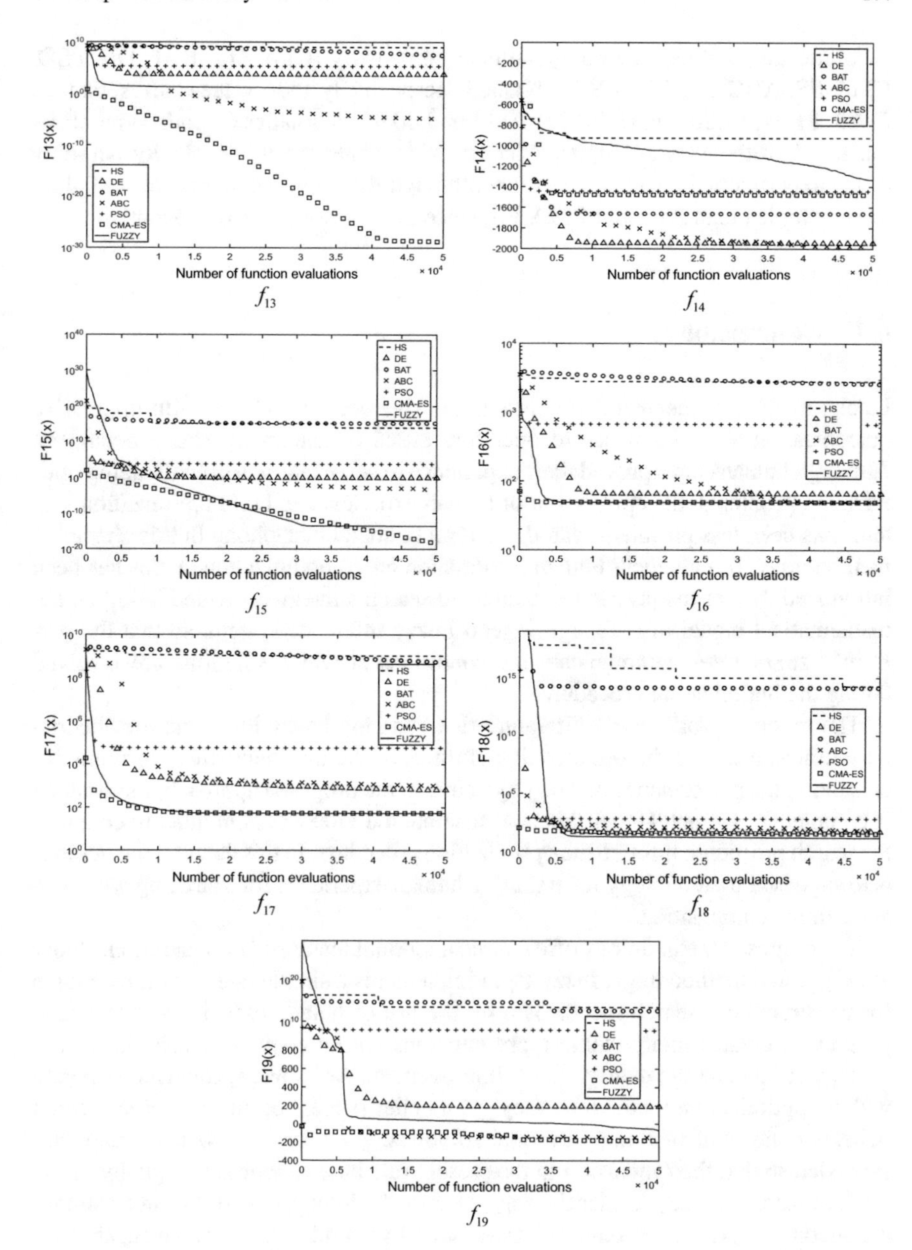

Fig. 6.7 Convergence test results for functions f_{13}–f_{129}

Under this method, the computational complexity (C) of HS, BAT, DE, PSO, CMA-ES, ABC, and FUZZY is obtained. Respectively, their values corresponds to 77.23, 81.51, 51.20, 36.87, 40.77, 70.17 and 40.91. A smaller C grade symbolizes that the algorithm is less complex, which provides faster execution velocity. An study of the experimental results reveals that although the proposed algorithm is slightly more complex than PSO and CMA-ES, since their C values are comparable.

6.7 Conclusions

Lately, many new metaheuristic algorithms have been introduced with impressive outcomes. Most of them utilize operators based on natural or social behaviors. Although humans have proved their experience to solve real-world challenging optimization problems, the application of human experience to build optimization systems has been less prevalent than the natural or social metaphors. In this chapter, a methodology to implement human information based optimization tactics has been introduced. Under the procedure, a directed search strategy is created based on the mathematical model of a Takagi-Sugeno Fuzzy inference system, so that the performed fuzzy rules communicate the conditions at which solutions are involved during the optimization procedure.

The reported works in the literature that combine Fuzzy logic and metaheuristic routines, examine the optimization abilities of the metaheuristic algorithms for enhancing the performance of fuzzy systems. In the proposed approach, the strategy is completely distinct. Under this new schema, the Fuzzy system quickly conducts the search procedure through the optimization procedure. In this chapter, the purpose is to suggest a methodology for imitating human experience for searching within an algorithmic composition.

The proposed methodology offers three important features: (1) Generation. Under the suggested methodology, fuzzy logic implements a simple and known technique for constructing a search mechanism by the use of human experience. (2) Transparency. It creates interpretable representations whose content reveals the search strategy as humans can manage it. (3) Improvement. As human specialists cooperate with an optimization method, so they get a better perception of prosperous search policies competent of finding optimal solutions. As a result, new commands are appended so that their inclusion in the present rule base enhances the quality of the original search strategy. Under the suggested methodology, new rules can be simply included to an existent system. The extension of such rules provides the capabilities of the original system to be robust.

In order to demonstrate the capability and robustness of the fuzzy approach, the proposed algorithm has been experimentally evaluated considering a benchmark set of mathematical functions. To judge the performance of the fuzzy algorithm, it has been compared to different popular optimization strategies based on evolutionary strategies. The numerical results are statistically validated. The results confirmed that the proposed scheme beats its competitors for most of the benchmark functions in terms of accuracy and convergence.

References

1. L.A. Zadeh, Fuzzy sets. Inf. control **8**, 338–353 (1965)
2. Yingdong He, Huayou Chen, Zhen He, Ligang Zhou, Multi-attribute decision making based on neutral averaging operators for intuitionistic fuzzy information. Appl. Soft Comput. **27**, 64–76 (2015)
3. J. Taur, C.W. Tao, Design and analysis of region-wise linear fuzzy controllers. Systems, Man, Cybern. Part B: Cybern. IEEE Trans. **27**(3), 526–532 (1997)
4. M.I. Ali, M. Shabir, Logic connectives for soft sets and fuzzy soft sets. Fuzzy Syst. IEEE Trans. **22**(6), 1431–1442 (2014)
5. V. Novák, P. Hurtík, H. Habiballa, M. Štepnička, Recognition of damaged letters based on mathematical fuzzy logic analysis. J. Appl. Logic **13**(2), Part A, 94–104 (2015)
6. G.A. Papakostas, A.G. Hatzimichailidis, V.G. Kaburlasos, Distance and similarity measures between intuitionistic fuzzy sets: a comparative analysis from a pattern recognition point of view. Pattern Recogn. Lett. **34**(14), 1609–1622 (2013)
7. Xinyu Wang, Fu Mengyin, Hongbin Ma, Yi Yang, Lateral control of autonomous vehicles based on fuzzy logic. Control Eng. Pract. **34**, 1–17 (2015)
8. O. Castillo, P. Melin, A review on interval type-2 fuzzy logic applications in intelligent control. Inf. Sci. **279**, 615–631 (2014)
9. G. Raju, M.S. Nair, A fast and efficient color image enhancement method based on fuzzy-logic and histogram. AEU Int. J. Electron. Commun. **68**(3), 237–243 (2014)
10. H. Zareiforoush, S. Minaei, M.R. Alizadeh, A. Banakar, A hybrid intelligent approach based on computer vision and fuzzy logic for quality measurement of milled rice. Measurement **66**, 26–34 (2015)
11. S.J. Nanda, G. Panda, A survey on nature inspired metaheuristic algorithms for partitional clustering. Swarm Evol. Comput. **16**, 1–18 (2014)
12. J. Kennedy, R. Eberhart, Particle swarm optimization, in *Proceedings of the 1995 IEEE International Conference on Neural Networks*, vol. 4, pp. 1942–1948, December 1995
13. Karaboga, D, An idea based on honey bee swarm for numerical optimization. TechnicalReport-TR06. Engineering Faculty, Computer Engineering Department, Erciyes University, 2005
14. Z.W. Geem, J.H. Kim, G.V. Loganathan, A new heuristic optimization algorithm: harmony search. Simulations **76**, 60–68 (2001)
15. X.S. Yang, A new metaheuristic bat-inspired algorithm, in *Nature Inspired Cooperative Strategies for Optimization (NISCO 2010), Studies in computational intelligence*, vol. 284, ed. by C. Cruz, J. González, G.T.N. Krasnogor, D.A. Pelta (Springer, Berlin, 2010), pp. 65–74
16. X.S. Yang, Firefly algorithms for multimodal optimization, in: *Stochastic Algorithms: Foundations and Applications*, SAGA 2009, Lecture notes in computer sciences, vol. 5792, 2009, pp. 169–178
17. Erik Cuevas, Miguel Cienfuegos, Daniel Zaldívar, Marco Pérez-Cisneros, A swarm optimization algorithm inspired in the behavior of the social-spider. Expert Syst. Appl. **40**(16), 6374–6384 (2013)
18. Cuevas, E., González, M., Zaldivar, D., Pérez-Cisneros, M., García, G, An algorithm for global optimization inspired by collective animal behaviour, *Discrete Dynamics in Nature and Society* 2012, art. no. 638275
19. R. Storn, K. Price, Differential evolution-a simple and efficient adaptive scheme for global optimisation over continuous spaces. TechnicalReportTR-95–012, ICSI, Berkeley, CA, 1995
20. D.E. Goldberg, genetic algorithm in search optimization and machine learning, Addison-Wesley, 1989

21. F. Herrera, Genetic fuzzy systems: taxonomy, current research trends and prospects. Evol. Intel. **1**, 27–46 (2008)
22. A. Fernández, V. López, M.J. del Jesus, F. Herrera, Revisiting evolutionary fuzzy systems: taxonomy, applications, new trends and challenges. Knowl.-Based Syst. **80**, 109–121 (2015)
23. C. Caraveo, F. Valdez, O. Castillo, Optimization of fuzzy controller design using a new bee colony algorithm with fuzzy dynamic parameter adaptation. Appl. Soft Comput. **43**, 131–142 (2016)
24. O. Castillo, H. Neyoy, José Soria, P. Melin, F. Valdez, A new approach for dynamic fuzzy logic parameter tuning in Ant Colony optimization and its application in fuzzy control of a mobile robot. Appl. Soft Comput. **28**, 150–159 (2015)
25. F. Olivas, F. Valdez, O. Castillo, P. Melin, Dynamic parameter adaptation in particle swarm optimization using interval type-2 fuzzy logic. Soft. Comput. **20**(3), 1057–1070 (2016)
26. O. Castillo, P. Ochoa, J. Soria, Differential evolution with fuzzy logic for dynamic adaptation of parameters in mathematical function optimization. in *Imprecision and Uncertainty in Information Representation and Processing*, pp. 361–374, 2016
27. M. Guerrero, O. Castillo, M. García Valdez, Fuzzy dynamic parameters adaptation in the cuckoo search algorithm using fuzzy logic. in *CEC 2015*, pp. 441–448, 2015
28. R. Alcala, M.J. Gacto, F. Herrera, A fast and scalable multiobjective genetic fuzzy system for linguistic fuzzy modeling in high-dimensional regression problems. IEEE Trans. Fuzzy Syst. **19**(4), 666–681 (2011)
29. J. Alcala-Fdez, R. Alcala, M.J. Gacto, F. Herrera, Learning the membership function contexts for mining fuzzy association rules by using genetic algorithms. Fuzzy Sets Syst. **160**(7), 905–921 (2009)
30. R. Alcala, J. Alcala-Fdez, F. Herrera, A proposal for the genetic lateral tuning of linguistic fuzzy systems and its interaction with rule selection. IEEE Trans. Fuzzy Syst. **15**(4), 616–635 (2007)
31. J. Alcala-Fdez, R. Alcala, F. Herrera, A fuzzy association rule-based classification model for high-dimensional problems with genetic rule selection and lateral tuning. IEEE Trans. Fuzzy Syst. **19**(5), 857–872 (2011)
32. C.J. Carmona, P. Gonzalez, M.J. del Jesus, M. Navio-Acosta, L. Jimenez-Trevino, Evolutionary fuzzy rule extraction for subgroup discovery in a psychiatric emergency department. Soft. Comput. **15**(12), 2435–2448 (2011)
33. O. Cordon, A historical review of evolutionary learning methods for Mamdani-type fuzzy rule-based systems: designing interpretable genetic fuzzy systems. Int. J. Approx. Reason. **52**(6), 894–913 (2011)
34. M. Cruz-Ramirez, C. Hervas-Martinez, J. Sanchez-Monedero, P.A. Gutierrez, Metrics to guide a multi-objective evolutionary algorithm for ordinal classification. Neurocomputing **135**, 21–31 (2014)
35. Stefan Lessmann, Marco Caserta, Idel Montalvo Arango, Tuning metaheuristics: A data mining based approach for particle swarm optimization. Expert Syst. Appl. **38**(10), 12826–12838 (2011)
36. Kenneth Sörensen, Metaheuristics—the metaphor exposed. Int. Trans. Oper. Res. **22**(1), 3–18 (2015)
37. M. Omid, M. Lashgari, H. Mobli, R. Alimardani, S. Mohtasebi, R. Hesamifard, Design of fuzzy logic control system incorporating human expert knowledge for combine harvester. Expert Syst. Appl. **37**(10), 7080–7085 (2010)
38. R. Fullér, L. Canós Darós, M.J. Canós Darós, Transparent fuzzy logic based methods for some human resource problems. Revista Electrónica de Comunicaciones y Trabajos de ASEPUMA **13**, 27–41 (2012)
39. O. Cordón, F. Herrera, A three-stage evolutionary process for learning descriptive and approximate fuzzy-logic-controller knowledge bases from examples. Int. J. Approximate Reasoning **17**(4), 369–407 (1997)
40. T. Takagi, M. Sugeno, Fuzzy identification of systems and its applications to modeling and control, *IEEE Trans. Syst. Man Cybern. SMC-15*, 116–132 (1985)

41. E. Mamdani, S. Assilian, An experiment in linguistic synthesis with a fuzzy logic controller. Int. J. Man Mach. Stud. **7**, 1–13 (1975)
42. Aytekin Bagis, Mehmet Konar, Comparison of Sugeno and Mamdani fuzzy models optimized by artificial bee colony algorithm for nonlinear system modelling. Trans. Inst. Measurement Control **38**(5), 579–592 (2016)
43. K. Guney, N. Sarikaya, Comparison of Mamdani and Sugeno fuzzy inference system models for resonant frequency calculation of rectangular microstrip antennas. Progr Electromagn. Res. B **12**, 81–104 (2009)
44. R. Baldick, *Applied Optimization* (Cambridge University Press, 2006)
45. D. Simon, Evolutionary Algorithms -Biologically Inspired and Population Based Approaches To Computer Intelligence (John Wiley & Sons, Inc, 2013)
46. S.Y. Wong, K.S. Yap, H.J. Yap, S.C. Tan, S.W. Chang, On equivalence of FIS and ELM for interpretable rule-based knowledge representation. IEEE Trans. Neural Netw. Learning Syst. **27**(7), 1417–1430 (2015)
47. K.S. Yap, S.Y. Wong, S.K. Tiong, Compressing and improving fuzzy rules using genetic algorithm and its application to fault detection. in *IEEE 18th Conference on Emerging Technologies & Factory Automation (ETFA)*, vol. 1 (2013), pp. 1–4
48. J.J. Liang, B.-Y. Qu, P.N. Suganthan, *Problem Definitions and Evaluation Criteria for the CEC 2015 Special Session and Competition On Single Objective Realparameter Numerical Optimization*, Technical Report 201311, Computational Intelligence Laboratory, Zhengzhou University, Zhengzhou China and Nanyang Technological University, Singapore (2015)
49. N. Hansen, A. Ostermeier, A. Gawelczyk, On the adaptation of arbitrary normal mutation distributions in evolution strategies: the generating set adaptation. in *Proceedings of the 6th International Conference on Genetic Algorithms* (1995), pp. 57–64
50. I. Boussaïda, J. Lepagnot, P. Siarry, A survey on optimization metaheuristics. Inf. Sci. **237**, 82–117 (2013)
51. J.Q.Y. James, V.O.K. Li, A social spider algorithm for global optimization, Appl. Soft Comput. **30**, 614–627 (2015)
52. M.D. Li, H. Zhao, X.W. Weng, T. Han, A novel nature-inspired algorithm for optimization: virus colony search. Adv. Eng. Softw. **92**, 65–88 (2016)
53. M. Han, C. Liu, J. Xing, An evolutionary membrane algorithm for global numerical optimization problems. Inf. Sci. **276**, 219–241 (2014)
54. Z. Meng, J.S. Pan, Monkey king evolution: a new memetic evolutionary algorithm and its application in vehicle fuel consumption optimization. Knowl.-Based Syst. **97**, 144–157 (2016)
55. https://www.lri.fr/~hansen/cmaesintro.html
56. F. Wilcoxon, Individual comparisons by ranking methods. Biometrics **1**, 80–83 (1945)
57. S. Garcia, D. Molina, M. Lozano, F. Herrera, A study on the use of non-parametric tests for analyzing the evolutionary algorithms' behavior: a case study on the CEC'2005 Special session on real parameter optimization. J. Heurist. (2008), https://doi.org/10.1007/s10732-008-9080-4

Chapter 7
Neighborhood Based Optimization Algorithm

7.1 Introduction

Optimization field comprises an optimal solution, to be obtained from a probable set of solutions, that expresses complex optimization problems [1]. Complex optimization problems are encountered in several science disciplines such as engineering, medicine, biology, and others where mathematical constructions require to be formed, to reach its inherent features [2]. Over the last decade, several optimization methods have been introduced. Such search strategies can be split into derivative-based techniques and Evolutionary Computation (EC) methods. Derivative schemes can adequately determine the optimal solution over unimodal functions. However, most of the optimization problems operate multimodal functions [3]. Under such situations, derivative-based mechanisms tend to be caught by local minima, achieving suboptimal solutions. On the other hand, EC algorithms are random search mechanisms, which can work through multimodal functions, where derivative-based techniques are useless [4].

In the last decade, some EC have been introduced to sponsoring competitive performance results. Such methods are motivated principally by our knowledge of physical, natural, and even cultural phenomena. Some of these methods are the Particle Swarm Optimization algorithm (PSO) [5], the Genetic Algorithm (GA) [6, 7], the Gravitational Search Algorithm (GSA) [8], and the Electromagnetism-like Optimization (EMO) [9].

The diversity of beliefs of natural methods to produce EC algorithms is supported by the so-called No-Free-Lunch theorem [10], formulated by Wolpert. Such theorem affirms that there is no single optimization method, that will perform adequately to solve an optimization problem. Therefore, the evaluation of an EC technique over a given problem, might not exhibit precisely the performance of such a methodology, analyzing different problems. Hence, to assure the versatility of an EC method to solve most of the optimization problems, the prototype phase of EC techniques depends essentially on its capacity to obtain, and sustain a correct balance between the exploration, and the exploitation [11]. The exploration process consists in the

E. Cuevas et al., *Recent Metaheuristics Algorithms for Parameter Identification*, Studies in Computational Intelligence 854, https://doi.org/10.1007/978-3-030-28917-1_7

search of new promissory positions; while the exploitation stage, expresses the local search within earlier recognized positions, with the intention of increasing their quality. Using principally exploration operators diminishes the accuracy of the optimal solution, but improves the abilities to discover new potential solutions [12]. Contrary to that, if only the exploitation operators are employed, suboptimal solutions are achieved but enhances the polished version of earlier visited positions [13]. Under such situations, each optimization problem needs a different trade-off among both evolutionary stages [14]. For that, each EC algorithm manifests a combination of stochastic, and deterministic operators, to adequately identify candidate solutions.

The majority of the design of exploration, and exploitation elements for EC methods, fall on complex mathematical formulations, to mimic the natural abstraction of the considered metaphor which promotes the increase of the computational overload, decreasing the versatility, and scalability of the EC techniques, operating on various objective functions generating an unbalance among evolutionary steps. On the other hand, the Swarm Robotics Intelligence (SRI) paradigm [15], proffers attractive features, allowing complex interactions among individuals to be modeled by simple behaviors, improving sensing abilities, self-organization, and decentralized control strategies [16]. SRI models use the collaboration, and coordination of autonomous agents, which can be employed for multiple attractive purposes including environmental monitoring [17], target location in unknown environments [18], and search and rescue [19].

Contrary to natural inspirations to construct EC methods, SRI approaches are conceived by the synchronization actions of individuals, multi-agent systems analyze autonomous action judgments based on sensed environments, to achieve a collection of intentions using simplistic rules to mimic complex actions. The autonomous decision-based knowledge of SRI, is classified as a non-centralized coordinated tool, which provides asynchronized functionality [20]. These asynchronous abilities guarantee that each individual can continuously communicate by itself, on its surroundings, with or without analyzing the actual state of the remaining individuals within the multi-agent system, enhancing versatility, and scalability of the whole system. Under such conditions, the combination of SRI and EC concepts presents an attractive choice to prototype complex relations among individuals.

In the integration situation, the principal feature of an autonomous decision-based mechanism proposed by SRI, is formulated by applying a combination of manageable steering behaviors [21], which will provide complex interactions between agents, over the whole multi-agent system. The steering behaviors are viewed as movement forces, that will adaptively respond regarding the information of the neighborhood of each autonomous agent. These reactive force features capture local, and global knowledge of the current state of certain agent that will labor, as input data vector to adjust the actual state of the agent into a future position. Opposite to classical EC methods, this autonomous policy can be used to express simple, but persuasive evolutionary components, promoting interaction mechanisms within a certain neighborhood, using leaderless movement judgments, regarding the spatial connections between agents, to efficiently explore, and exploit likely search zones. As a result, the flexibility of reactive responses, with the adaptability of neighborhood agreement, can perform simplistic mathematical procedures for EC operators improving the population diversity of traditional EC approaches.

In this chapter, a reactive response approach for solving optimization problems is proposed. The strategy named Neighborhood-based Consensus for Continuous Optimization (NCCO), links the autonomous local consensus concepts, with neighborhood-based decisions to lead the search. The NCCO algorithm has been considered as an optimization method since it encourages synergy between evolutionary elements for its procedures and non-centralized movement judgments. The combination of evolutionary operators reflects a double pair of operators controlling a complex search behavior, based on simple area negotiations to direct the search into encouraging search zones. Opposed to classical EC, NCCO does not count on leader individual to lead the search. This scheme enables each individual to explore broadly, and exploit more efficiently search areas.

In the optimization scheme, NCCO identifies each individual of the community as an autonomous agent, which controls neighborhood consensus problems, based on simple reactive rules to conduct the search. To construct and keep a balance between evolutionary stages, NCCO analyzes two pairs of evolutionary operations to adjust the position, velocity, and acceleration vectors of all agents. The first pair; separation-alignment, supports the search strategy to be carried into broader zones until each agent is located far apart between the rest, while they align their heading vector towards encouraging search areas. The second pair of operators; cohesion-seek, nearby explores and exploits search areas collecting all agents into organizations. When all the groups are created, each group meets into a target location, which matches to an optimal value for the optimization problem. With such a strategy, the performance results of the proposed algorithm are judged, and compared, against the performance index achieved by some state-of-art evolutionary approaches, using test functions, and design problems. Regarding the performance results, NCCO outperforms the rest of the EC techniques. The numerical results are also statistically validated.

The remainder of the chapter is arranged as follows: Sect. 7.2, sponsors a literature survey concerning multi-agent systems. In Sect. 7.3, some introductory ideas are presented. In Sect. 7.4, the proposed method is defined. In Sect. 7.5, the computational procedure is exhibited. In Sect. 7.6, the experimental study of the proposed method, versus some EC methodologies are presented. Lastly, in Sect. 7.7, some conclusions are detailed.

7.2 Literature Review

Multi-agent schemes comprise the interaction among agents through information sharing to achieve an assignment. Such concepts, interconnect some elements employing simple local rules. The combination of simple cooperation rules, yields into complex cooperation models, where the sensing and cooperation among agents, overwhelm the limitations of actions performed by a single agent. Under such conditions, multiple multi-agent systems have been employed to manage diverse real-life problems. In the published literature, De Meo et al. [22], generates a multi-agent system for solving e-commerce activities called ec-XAMAS, where a client agent allocates user profiles between different sessions. In [23], Ardissono sponsors an intelligent web purchasing personalization system, based on the interaction of agents to manage user curiosities. Ursino et al. [24], suggests a multi-agent approach for e-commerce sites to uniformly designate customer profiles. Macskassy in [25], proposes a global system for information access, adopting diverse user devices to describe customer services.

Lately, several published investigations in smart grid, choose a multi-agent based solution [26, 27] for handling transmission of electricity networks, to decrease the energy consumption, improving the fidelity of the energy supply chain. In [28], Kumar suggests an agent-based business model, to fulfill the demand in micro-grids, using 6 agents to reproduce the market conditions, between customers, and the distribution system. Anvari in [29], sponsors a novel multi-agent solution for observing smart grids, with controllable loads employing renewable energy. In [30], Loia applies simulated agents with fuzzy systems, to solve optimal power flow problem, enhancing the online stage by monitoring grids. On the other hand, in mobile robotics, Zhang et al. [31], introduced a distributed consensus control, for resolve the leader-follower problem considering dynamic parameters [32–34]. Alonso [35], has suggested drone collision-free trajectories, by employing three methods for local motion preparation in 3D scenes. In [36], Hönig presents a polynomial solution for path finding, analyzing kinematic constraints in non-holonomic arrangements. Shalev [37], uses reinforcement learning for multi-agent self-governing driving policies. In [38], Zhao introduced distributed feedback for second-order multi-agent systems. Lastly, Nikou et al. [39], exhibits a cooperative task planning framework, based on automated procedures.

7.3 Preliminary Concepts

The proposed method merges the flexibility of reactive models, with the adaptability of the neighborhood consensus formulations, to lead its search strategy. In this part, the principal features of the adopted flocking model, as well as the principle of neighborhood consensus are detailed.

7.3.1 Reactive Models

Reactive models [40], are popular models for coordinating intelligence in multi-agent systems. Reactive models provide a set of interactions among agents, based on simplistic sensing rules. Such intercommunications serve as constructing blocks to model complex behaviors, enabling agents to achieve specific tasks, within their environment. The aptitude to model complex behaviors in multi-agent systems presents important information about self-organization, and collaboration of agents in multiple fields, including artificial intelligence [41], robotics [42], and complex systems [43].

In the proposed NCCO, the reactive model introduced by Craig Reynolds [44], has been regarded to create reactive intercommunications among agents. In this representation, each agent performs a movement decision according to steering behaviors [21], which response to the area in it can sense. The simplistic rules, allow global operations in the whole multi-agent system, without the requirement of a leader agent, which can prejudice the overall behavior.

The fundamental Reynolds' flock model studied in the proposed strategy, is composed of 4 kinds of simplistic rules; separation, alignment, cohesion, and seek. Separation keeps specific distance among agents, regard to its nearby agents within its sensed environment. This process bypasses crowding and, enables agents to operate inside a broader space. Alignment provides the capability to each agent to adjust its heading vector, analyzing all surrounding agents. This mechanism permits agents to adjust the flock over a given direction. Cohesion permits each agent to be part of a cluster. This procedure encourages the creation of clusters, where each agent shares connections with the remaining of nearby agents, leaving the navigation in a narrower space. Ultimately, the seek process consists of turning the heading vector for all agents into a target position.

7.4 Neighborhood-Based Consensus for Continuous Optimization

NCCO regards as the principal search step, the reactive responses employed by each agent using consensus agreements. With the responses, the aim is to produce movement judgments to lead the search into attractive regions through the generations. This strategy proposes an alternative methodology for constructing evolutionary operators, where each agent determines its following position based on the sensed information. To achieve such a decision-making process, the NCCO performs two exploration operators; and two exploitation operators. For the exploration step, separation and alignment methods keep a certain amount of distance respect all agents belonging to its neighborhood agents, meanwhile they turn their heading vectors into promissory search zones within a broader space. On the other hand, cohesion, and seek schemes organize agents into small groups, while they approach to the target (best). This synergy of double evolutionary operators, improves the search capabilities increasing the population diversity.

The majority of previously published EC methods employ a single pair of operators for exploration and exploitation. This popular evolutionary methodology does not ensure stability among evolutionary stages since it does not consider flexibility in the search strategy. As a consequence, the performance of such a paradigm, poor exploration, and inconsistent exploitation are presented.

Another essential feature of NCCO is the leaderless decision-making process. This characteristic avoids NCCO to catch false-positive positions; reducing the likelihood of being stuck inside suboptimal solutions. Also, it enhances the search abilities to explore into a wider space.

NCCO employs a set of individuals $\mathbf{Z}^k = \left(\left\{\mathbf{a}_1^k, \mathbf{a}_2^k, \ldots, \mathbf{a}_N^k\right\}\right)$ as search agents that are emerged from an initial iteration ($k = 0$), to a maximum number of generations ($k = gen$). Each solution (agent) $\mathbf{a}_{i\in[1,\ldots,N]}^k = \left(\left\{a_i^k \cdot \hat{\mathbf{p}}_i,\ a_i^k \cdot \hat{\mathbf{v}}_i,\ a_i^k \cdot \hat{\mathbf{g}}_i\right\} \cup \left\{\hat{\mathbf{s}}_{i,bias}, \hat{\mathbf{a}}_{i,bias}, \hat{\mathbf{c}}_{i,bias}\right\}\right)$, denotes a data set consisting of 6 vectors: position $\hat{\mathbf{p}}_i$, velocity $\hat{\mathbf{v}}_i$, acceleration $\hat{\mathbf{g}}_i$, separation bias $\hat{\mathbf{s}}_{i,bias}$, alignment bias $\hat{\mathbf{a}}_{i,bias}$, and cohesion bias $\hat{\mathbf{c}}_{i,bias}$. Each position vector $\hat{\mathbf{p}}_i = (\{p_1, p_2, \ldots, p_n\})$, matches to an n-dimensional vector, where each dimensional component matches to a decision variable of the optimization problem meant to be solved. Each velocity vector $\hat{\mathbf{v}}_i = (\{v_1, v_2, \ldots, v_n\})$, describes the heading direction for the ith agent. Ultimately, each acceleration vector $\hat{\mathbf{g}}_i = (\{g_1, g_2, \ldots, g_n\})$, describes the sampled acceleration force. The rest of the vectors are recognized as auxiliary vectors, which collect the accumulative force effects through each iteration. Separation bias vector $\hat{\mathbf{s}}_{i,bias}$, stores the repulsion effect between ith solution, and the nearby agents; the alignment bias vector $\hat{\mathbf{a}}_{i,bias}$, allocates the rotation of the heading vector for the ith agent, regard to its region, and the cohesion bias vector $\hat{\mathbf{c}}_{i,bias}$, enables each agent to set clusters among the remaining individuals.

To enhance the balance of the evolutionary stages, the hybridization of separation, and alignment procedures, and the hybridization among cohesion, and seek are divided into two phases. The first phase s = 1, resembles from 0 to 30% of the maximum number of iterations. The second phase s = 2, comprises from 31 to 100%. The purpose of such division was experimentally achieved, due to the exponential formulation of the separation, alignment, and cohesion procedures. The formulations provide fluid changes over the optimization process, beginning with exploration, and slowly changing into the exploitation stage. For the exploration stage, separation, and alignment procedures execute fewer times than the employment of cohesion and seek. This approach reflects the inherent nature of the consensus protocol. Separation and alignment keep agents far distant from each other, rushing up the exploration from the initial point. Under such conditions, there is no necessity to perform more exploration calculations. On the other hand, cohesion, and seek, do the reverse effect. Each agent begins remotely as a single search element then, each agent begins grouping into clusters between similar agents, while they spin towards the best position (target position).

7.4.1 Initialization

The suggested approach starts with the initialization of a population $\mathbf{Z}^k$ of N solutions (agents). Given the fact that every single agent contains a position, velocity, and acceleration vectors, and a separation, alignment, and cohesion auxiliary vectors, the initialization process also creates a task for each of such vectors. The position vector $a_i^k \cdot \hat{\mathbf{p}}_{i,j}$, is randomly initialized as:

$$a_i^k \cdot \hat{\mathbf{p}}_{i,j} = l_j + rand(0, 1) \cdot (u_j - l_j), \quad j = 1, 2, \ldots, n; \quad i = 1, 2, \ldots, N \tag{7.1}$$

where, $a_i^k \cdot \hat{\mathbf{p}}_{i,j}$, represents the jth component of the ith solution at the k generation. Additionally, the initialization process of the velocity vector $a_i^k \cdot \hat{\mathbf{v}}_{i,j}$ is as follows:

$$a_i^k \cdot \hat{\mathbf{v}}_{i,j} = rand(-1, 1), \quad j = 1, 2, \ldots, n; \quad i = 1, 2, \ldots, N \tag{7.2}$$

where the jth velocity element is a random value within $[-1, 1]$. Lastly, the acceleration vector $a_i^k \cdot \hat{\mathbf{g}}_{i,j}$, is denoted as the zero vector, since the initial point is considered causal:

$$a_i^k \cdot \hat{\mathbf{g}}_{i,j} = (0, 0, \ldots 0_n), \quad i = 1, 2, \ldots, N \tag{7.3}$$

The auxiliary vectors: $\hat{\mathbf{s}}_{i,bias}$, $\hat{\mathbf{a}}_{i,bias}$, and $\hat{\mathbf{c}}_{i,bias}$, for the ith agent are considered as zero vectors, in the initialization process:

$$\begin{aligned} \hat{\mathbf{s}}_{i,bias} &= (0, 0, \ldots 0_n) \\ \hat{\mathbf{a}}_{i,bias} &= (0, 0, \ldots 0_n), \quad i = 1, 2, \ldots, N \\ \hat{\mathbf{c}}_{i,bias} &= (0, 0, \ldots 0_n) \end{aligned} \tag{7.4}$$

7.4.2 Reactive Flocking Response

Reactive models' responses describe the fundamental process for autonomous decision-making behaviors, based on the sensed information. With such modeling, the aim is to create intelligent movement decisions, considering a neighborhood. This idea establishes an alternative communication channel, to improve the search over the search space. NCCO employs two pairs of operators to achieve strong optima finding. The separation and alignment schemes are employed to change the agent's position during the exploration meanwhile the cohesion and seek schemes modify the multi-agent structure in the exploitation. In this section, the formulations for each flocking model is detailed.

7.4.2.1 Separation

The separation system includes the calculation of a separation vector $\hat{\mathbf{s}}_f$, inside each sensed environment. Through the separation procedure, each agent $\mathbf{a}_i^k$, starts sensing its vicinity to reveal closely neighbors being part of its area. Then, based on the amount of discovered agents M, a separation vector is calculated by a consensus protocol between the nearby neighbors to produce a repulsion effect.

During the sensing stage, each agent sees a working radius. The working radius is specified as:

$$\rho = 10 \cdot n \tag{7.5}$$

where n denotes the number of dimensions and 10 is an experimentally defined constant. To explore nearby solutions inside a given area determined by a given agent $\mathbf{a}_i^k$; the Euclidean distances between the position vectors for the total number of agents, and the position vector for the agent $\mathbf{a}_i^k$ are examined with Eq. (7.5). The agents that present position distances, less than or equal to the working radius, will be cataloged as nearby neighbors of the agent $\mathbf{a}_i^k$. Once the nearby solutions have been identified, the separation vector $\hat{\mathbf{s}}_f$, is calculated as the arrangement defined as:

$$\hat{\mathbf{s}}_f = \frac{\sum_{m=1}^{M}\left(a_i^k \cdot \hat{\mathbf{p}}_i - a_m^k \cdot \hat{\mathbf{p}}_m\right)}{\left\|\sum_{m=1}^{M}\left(a_i^k \cdot \hat{\mathbf{p}}_i - a_m^k \cdot \hat{\mathbf{p}}_m\right)\right\|} - a_i^k \cdot \hat{\mathbf{p}}_i \tag{7.6}$$

where $a_i^k \cdot \hat{\mathbf{p}}_{i,j}$ corresponds to the position vector belonging to the agent $\mathbf{a}_i^k$, and $a_m^k \cdot \hat{\mathbf{p}}_m$ represents to the mth nearby neighbor. As Eq. (7.6) implies, the neighborhood arrangement for the separation operator is computed, generating the resultant repulsion vector. Such vector is employed to adjust the position vectors of each nearest agent, to ensure remoteness regard all agents inside the area. For that, a separation vector $\hat{\mathbf{s}}_{m,bias}$, for the mth nearby agent collect the accumulative repulsion effect over each iteration, this process is defined as:

$$\hat{\mathbf{s}}_{m,bias} = \left((\hat{\mathbf{s}}_{m,bias} + \hat{\mathbf{s}}_f) \cdot \frac{a_i^k \cdot \hat{\mathbf{p}}_i - a_m^k \cdot \hat{\mathbf{p}}_m}{\max\left(\left\{a_i^k \cdot \hat{\mathbf{p}}_i - a_m^k \cdot \hat{\mathbf{p}}_m, \forall m \in [1, 2, \ldots M]\right\}\right)} \right) \cdot \exp\left(\frac{-k}{gen - k}\right) \tag{7.7}$$

where the decaying function simulates powerful repulsion effect at the initial iteration ($k = 0$); and weak separation at the end of the exploration phase. Later, the location of each nearby neighbor is updated regarding the distance to the agent $\mathbf{a}_i^k$, the separation vector $\hat{\mathbf{s}}_f$, and the separation bias $\hat{\mathbf{s}}_{m,bias}$ as:

$$a_m^k \cdot \hat{\mathbf{p}}_m = a_m^k \cdot \hat{\mathbf{p}}_m + (a_m^k \cdot \hat{\mathbf{p}}_m \cdot \hat{\mathbf{s}}_{m,bias}) \tag{7.8}$$

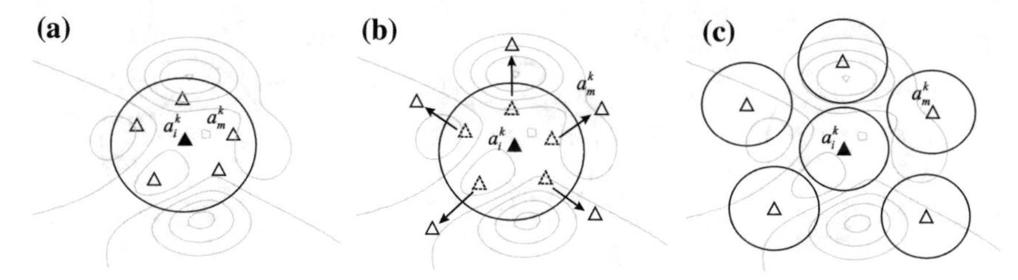

Fig. 7.1 Separation process

Lastly, the acceleration vector $a_i^k \cdot \hat{\mathbf{g}}_i$, for the agent $\mathbf{a}_i^k$ is changed by attaching the separation vector $\hat{\mathbf{s}}_f$ as:

$$a_i^k \cdot \hat{\mathbf{g}}_i = a_i^k \cdot \hat{\mathbf{g}}_i + \hat{\mathbf{s}}_f \tag{7.9}$$

To graphically demonstrate the separation process, Fig. 7.1a, describes the neighborhood of the $\mathbf{a}_i^k$ which includes its M neighbors. Figure 7.1b, shows the progressive movement for each agent based on Eqs. (7.6), (7.7); until all the neighbors are positioned outside the vicinity of agent $\mathbf{a}_i^k$, Fig. 7.1c.

7.4.2.2 Alignment

The alignment process establishes the second operator for the exploration phase. Alignment determines the force vector needed to twist the heading vector of an agent $\mathbf{a}_i^k$, into the mean direction the neighboring. In this procedure, each agent, senses its area. Then, based on the amount of neighbors N_m, an alignment vector is computed by an arrangement between neighbors. This vector provides finer modifications between position for every agent inside the neighborhood.

To apply the alignment scheme, every agent senses its surroundings by the working radius detailed in Eq. (7.5), to identify its neighbors. For that, the distances between the position vectors for all the agents are computed. Once the nearby solutions have been discovered, the alignment vector $\hat{\mathbf{a}}_f$, is computed as:

$$\hat{\mathbf{a}}_f = \frac{\sum_{m=1}^{M} a_m^k \cdot \hat{\mathbf{v}}_m}{M} - a_i^k \cdot \hat{\mathbf{v}}_i \tag{7.10}$$

where $a_m^k \cdot \hat{\mathbf{v}}_m$ matches to the velocity vector of the mth nearby neighbor, and $a_i^k \cdot \hat{\mathbf{v}}_i$ describes the velocity vector of the corresponding vector $\mathbf{a}_i^k$. Given that the agent $\mathbf{a}_i^k$, holds an alignment bias $\hat{\mathbf{a}}_{i,bias}$, this secondary vector will hold the accumulative rotation effect over each iteration, towards the mean direction in the neighborhood to promote finer search movements inside the neighborhood. The accumulative effect is specified as:

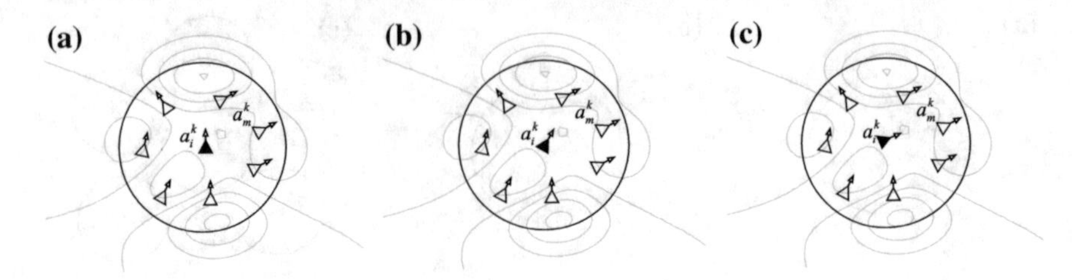

Fig. 7.2 Alignment process

$$\hat{\mathbf{a}}_{i,bias} = (\hat{\mathbf{a}}_{i,bias} + \hat{\mathbf{a}}_f) \cdot \left(\frac{1}{\exp\left(\frac{gen-k}{k}\right)} \right) \quad (7.11)$$

where the exponential term helps to mimic low alignment effect at the beginning, and powerful effect at the end. Lastly, the acceleration vector $a_i^k \cdot \hat{\mathbf{g}}_i$ is adjusted by the joining of the alignment bias as:

$$a_i^k \cdot \hat{\mathbf{g}}_i = a_i^k \cdot \hat{\mathbf{g}}_i + \hat{\mathbf{a}}_{i,bias} \quad (7.12)$$

To graphically demonstrate the alignment method, Fig. 7.2a, manifests the heading vectors for every agent inside the neighborhood. In Fig. 7.2b, the heading vector starts adjusting towards orientation of the neighborhood (Eq. 7.11). This toward movement is ended until the majority of neighbors are totally aligned Fig. 7.2c.

7.4.2.3 Cohesion

The cohesion process is the first process in the exploitation phase. This operation comprises the calculation of a cohesion vector. During this scheme, each agent begins forming into clusters. For that, each agent recognizes the belonging agents of its environment, and calculate a cohesion vector, enabling the exploitation of promissory search zones. The cohesion vector will then give an attraction force between agents intensifying the search strategy inside a narrow zone.

To correctly implement this strategy, each agent $\mathbf{a}_i^k$, recognizes its neighbors. Then, the cohesion force vector is computed by:

$$\hat{\mathbf{c}}_f = \frac{\sum_{m=1}^{M} \left(a_i^k \cdot \hat{\mathbf{p}}_i - a_m^k \cdot \hat{\mathbf{p}}_m \right)}{M} - a_i^k \cdot \hat{\mathbf{p}}_i - a_i^k \cdot \hat{\mathbf{v}}_i \quad (7.13)$$

where $a_i^k \cdot \hat{\mathbf{p}}_i$ and $a_i^k \cdot \hat{\mathbf{v}}_i$ describe the position, and velocity vectors for a certain agent $\mathbf{a}_i^k$, and $a_m^k \cdot \hat{\mathbf{p}}_m$, matches the position of the mth neighbor. As Eq. (7.13) indicates, the neighbors allow a consensus arrangement to designate the resultant attraction force. Given the fact that each agent holds an auxiliary cohesion bias $\hat{\mathbf{c}}_{i,bias}$, this bias vector

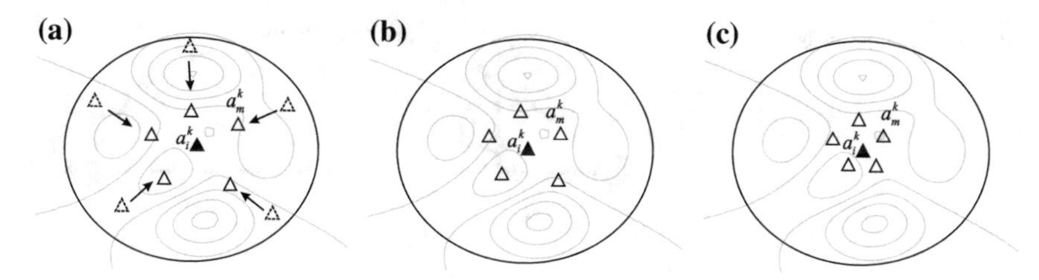

Fig. 7.3 Cohesion process

will store the accumulative attraction effect, during the optimization process. This accumulative effect is computed by:

$$\hat{\mathbf{c}}_{i,bias} = (\hat{\mathbf{c}}_{i,bias} + \hat{\mathbf{c}}_f) \cdot \left(\frac{1}{\exp\left(\frac{gen-k}{k}\right)} \right) \tag{7.14}$$

where the exponential term helps to mimic low alignment effect at the beginning, and powerful effect at the end. Lastly, the acceleration vector $a_i^k \cdot \hat{\mathbf{g}}_i$ is adjusted by the joining of the cohesion bias vector $\hat{\mathbf{c}}_{i,bias}$ as:

$$a_i^k \cdot \hat{\mathbf{g}}_i = a_i^k \cdot \hat{\mathbf{g}}_i + \hat{\mathbf{c}}_{i,bias} \tag{7.15}$$

To graphically depict the cohesion procedure, Fig. 7.3, shows the changing of this mechanism. At the initial point, each neighbor, which is positioned far away its corresponding agent Fig. 7.3a, starts moving closely to the agent (Eq. 7.13) based on the accumulative cohesion vector $\hat{\mathbf{c}}_{i,bias}$ Fig. 7.3b, until the neighbors are closer to the agent $\mathbf{a}_i^k$.

7.4.2.4 Seek

The Seek process steers the heading vector of each agent, towards a designated location inside the search space. The principal purpose of this process is the determination of a path where every agent should follow to pursuit and strike the target position. This mechanism enables the alignment of the velocity vector to be radially towards the target position. This action holds as target position the best position discovered until now $\mathbf{B}_G$.

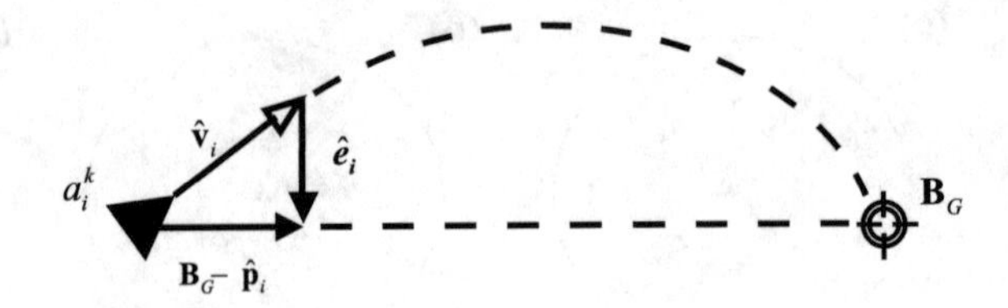

Fig. 7.4 Seek process

The steering vector $\hat{\mathbf{e}}_i$ is then computed as:

$$\hat{\mathbf{e}}_i = \left(\mathbf{B}_G - a_i^k \cdot \hat{\mathbf{p}}_i\right) - a_i^k \cdot \hat{\mathbf{v}}_i \tag{7.16}$$

This steering force will provide a radial alignment between all agents and the target solution during the optimization process. The acceleration vector $a_i^k \cdot \hat{\mathbf{g}}_i$ for the $\mathbf{a}_i^k$ agent is then modified as:

$$a_i^k \cdot \hat{\mathbf{g}}_i = a_i^k \cdot \hat{\mathbf{g}}_i + \left(\hat{\mathbf{e}}_i \cdot rand(0, 1)\right) \tag{7.17}$$

The seek mechanism can be compiled in Fig. 7.4. In the figure, the $\mathbf{a}_i^k$ agent should chase the best position ($\mathbf{B}_G$), the determined trajectory a given agent should follow is reached, by the steering vector which is calculated as Eq. (7.16). The radial trajectory adds flexibility to the model.

7.4.3 Update Mechanism

The steering responses specified in Sect. 7.4.2 adjust each agent's position, velocity, and acceleration vectors, according to an agreement with its sensed environment. Nevertheless, due to the mathematical definition of the consensus protocols, some position vectors will lead to overlap other position vectors of diverse agents, helping early convergence. Under such cases, a modification of the final position vectors must be employed. If the update mechanism is performed in the exploration phase, the velocity and position are updated following the Uniformly Accelerated Rectilinear Motion (UARM) scheme with stochastic terms as:

$$\begin{aligned} a_i^k \cdot \hat{\mathbf{v}}_{i,j} &= a_i^k \cdot \hat{\mathbf{v}}_{i,j} + a_i^k \cdot \hat{\mathbf{g}}_{i,j} + rand(-1, 1) \\ a_i^k \cdot \hat{\mathbf{p}}_i &= a_i^k \cdot \hat{\mathbf{p}}_i + a_i^k \cdot \hat{\mathbf{v}}_i \end{aligned} \tag{7.18}$$

Ultimately, the update procedure considered in the exploitation phase will adjust the velocity vector following:

$$a_i^k \cdot \hat{\mathbf{v}}_{i,j} = a_i^k \cdot \hat{\mathbf{v}}_{i,j} + a_i^k \cdot \hat{\mathbf{g}}_{i,j} \tag{7.19}$$

Next, the position vector for a certain agent will be adjusted by the next rule. If the maximum distance among the position vector of certain agent, with respect to the rest of the position vectors of the other agents is less or equal than eps, the updated position vector for the agent is adjusted by a binary random from the range $[-1, 1]$ to invert some position values. This rule can be defined by:

$$a_i^k \cdot \hat{\mathbf{p}}_{i,j} = \left(a_i^k \cdot \hat{\mathbf{p}}_{i,j} + a_i^k \cdot \hat{\mathbf{v}}_{i,j}\right) \cdot randi[-1, 1] \tag{7.20}$$

Oppositely, the position vector will be refreshed by the UARM model:

$$a_i^k \cdot \hat{\mathbf{p}}_i = a_i^k \cdot \hat{\mathbf{p}}_i + a_i^k \cdot \hat{\mathbf{v}}_i \tag{7.21}$$

7.5 Computational Procedure

The NCCO is conceived as an iterative process, in which various operations are accomplished. Such operations match the reactive models previously detailed in Sect. 7.4.2. To review the NCCO functionality, the pseudo-code in Algorithm 7.1 is manifested. The proposed method considers as input vector data the number of solutions N, and the maximum number of iterations. Comparable to some other ECTs, in the first step (line 2), the algorithm is initialized considering random values in the position of the agents. Following the initialization process (line 4), the target position (best individual) is chosen. Then, the whole population is adjusted through the flocking model in Sect. 7.4.2. For the exploration phase (lines 6–7), the separation and alignment procedures are performed. Later, the exploitation stage (lines 9–10), is fulfilled by the cohesion and seek mechanisms.

Algorithm 7.1 Pseudo-code for the proposed NCCO algorithm

1.	**Input:** N, *gen*, *k*=0, *s*=1	
2.	$\mathbf{Z}^k \leftarrow$ **Initialize** (N); // Using Eqs. (7.2)–(7.5).	
3.	**while** *k*<=*gen* **do**	
4.	$\mathbf{B}_G \leftarrow$ **SelectBestPosition** ($\mathbf{Z}^k$);	
5.	**for** (*i*=1; *i*>*N*; *i*++)	
6.	**if** (*s*==1)	Exploration
7.	$\left[\mathbf{a}_i^k, \mathbf{Z}^k\right] \leftarrow$ **Separation** ($\mathbf{a}_i^k, \mathbf{Z}^k$); // Using Eq. (7.10)	
8.	$\mathbf{a}_i^k \leftarrow$ **Alignment** ($\mathbf{a}_i^k, \mathbf{Z}^k$); // Using Eqs. (7.11)–(7.13)	
9.	$\mathbf{a}_i^k \leftarrow$ **Update** ($\mathbf{a}_i^k$); // Using Eq. (7.19)	
10.	**end if**	
11.	**if** (*s*==2)	Exploitation
12.	$\mathbf{a}_i^k \leftarrow$ **Cohesion** ($\mathbf{a}_i^k, \mathbf{Z}^k$); // Using Eqs. (7.14)–(7.16)	
13.	$\mathbf{a}_i^k \leftarrow$ **Seek** ($\mathbf{a}_i^k, \mathbf{B}_G$); // Using Eqs. (7.17)–(7.18)	
14.	$\mathbf{a}_i^k \leftarrow$ **Update** ($\mathbf{a}_i^k$); // Using Eq. (7.19)	
15.	**If Max Distance** ($\mathbf{a}_i^k, \mathbf{Z}^k$) <= *eps*	
16.	$\mathbf{a}_i^k \leftarrow$ **Update** ($\mathbf{a}_i^k$); // Using Eq. (7.20)	
17.	**else**	
18.	$\mathbf{a}_i^k \leftarrow$ **Update** ($\mathbf{a}_i^k$); // Using Eq. (7.21)	
19.	**end if**	
20.	**end for**	
21.	*k*=*k*+1;	
22.	$\mathbf{B}_G \leftarrow$ **Select Best Position** ($\mathbf{a}_i^k$);	
23.	$s \leftarrow$ **Evolution Progress** (*k*,*gen*);	
24.	**end while**	
25.	**Output:** $\mathbf{B}_G$	

7.6 Experimental Study

Evolutionary optimization algorithms are conceived as stochastic search methods where classical mathematical methods are unsuitable when the objective function contains multiple optima. Usually, evolutionary approaches are analyzed evaluating a suite of 23 test functions [45, 46], regarding a pre-defined number of function evaluations (NFE) as a stop mechanism. This section presents the performance results of the comparison among the NCCO versus some state-of-art evolutionary methodologies. The performance study also includes the numerical results from the evaluation of three design problems (Appendix B) commonly used in published works [1, 47], namely: Three bar truss (Table B.7), Tension/compression spring (Table B.8) and

Welded beam (Table B.9) design problems. Tables B.4, B.5 and B.6 in Appendix B present the unimodal, multimodal, and hybrid test functions. In the tables, n represents the dimensionality of the problem, $f(x^*)$ denotes the minimum value of the corresponding evaluated function at location x^*, and $\boldsymbol{S}$ denotes the limits of the search space.

For comparison purposes, 9 evolutionary methods were chosen based on their influence in the current EC investigation field. Particle Swarm Optimization (PSO) [5] consolidates the notion of swarm intelligence supplied by animal organizations to lead the search based on leadership judgements. Differential Evolution (DE) [48] combines the features of direct search approaches with evolutionary schemes to avoid unanticipated convergence. Artificial Bee Colony (ABC) [49] employs the idea of cooperative behavior based on the collaboration between individuals belonging to a population to solve specific tasks. On the other hand, Differential Search (DS) [50] introduces the idea of migration inside a population in union with some features of direct search methods to improve the scalability and versatility of the search process. The Cuckoo Search (CS) [51] method applies the notion of brood parasitism observed in the cuckoo species to manage the intercommunication of two different kinds of populations to complete the exploration and exploitation phases. Additionally, Multi-Verse Optimizer (MVO) [52] consolidates diverse features of cosmology, based on attraction-repulsion elements constrained by physical laws. Also, the Moth-Flame Optimization (MFO) [53] algorithm aims the transverse orientation performance of some natural individuals to navigate through promissory zones in the optimization process. Sine Cosine Algorithm (SCA) [54] exhibits the union of sinusoidal fluctuations expressed by mathematical models to produce evolutionary schemes based on mathematical principles. Ultimately, Covariance Matrix Adaptation Evolutionary Strategy (CMA-ES) [55] is considered as the newest paradigm in EC algorithms, since it connects the principal features of evolutionary strategies with the statistical principles of likelihood.

Section 7.6.1 analyzes the numerical results achieved by NCCO and its opponents, analyzing a set of test functions. Section 7.6.2, describes the computational effort and convergence results for all the algorithms. Lastly, in Sect. 7.6.3, the evaluation of the engineering design problems is shown.

7.6.1 Performance Comparison

In this part, the performance of NCCO is compared to its opponents using a suite composed of 23 test functions. The experimental study is split into three subsections: Sect. 7.6.1.1 reveals the numerical results evaluating unimodal functions, Sect. 7.6.1.2 presents the results over multimodal functions. Lastly, Sect. 7.6.1.3 details the results evaluating hybrid functions. To perform a straight comparison, the stop criterion has been considered as $NFE = 25{,}000$. Such criterion has been chosen to support compatibility with earlier published works [46, 56–59]. The experiments are judged regarding $n = \{10, 30, 100\}$ with the population size of 50 elements.

Each benchmark function is evaluated across 30 times. Additionally, the numerical results are statistically validated by the Wilcoxon rank test [60].

For the study, the configuration of the parameters for each evolutionary scheme has been set according to their references as follows:

1. ABC: The population limit has been to 50 individuals [5].
2. DE: DE/rand/bin has been considered, the $cr = 0.5$ and $dw = 0.2$ [48].
3. PSO: The constants has been set to $c_1 = 2$ and $c_2 = 2$ [5], additionally, the weight factor decreases within the range [0.9–0.2].
4. CS: The $pa = 0.25$ [51].
5. DS: The $sp = 1$ and $cp = 0.5$ [50].
6. MFO: The source code has obtained from [53].
7. MVO: The source code has obtained from [52].
8. SCA: The implementation follows [54].
9. CMA-ES: The source code has been taken from [55].

NCCO: The population size has been configured to 50 individuals.

7.6.1.1 Unimodal Test Functions

In this experiment, the performance of the suggested method is judged regarding functions with just one optimum ($f_1 - f_8$) using $n = \{10, 30, 100\}$ dimensional search spaces. The functions are mathematically described in Table B.4. For this test, the numerical results of the comparison are exposed in Tables 7.1, 7.3 and 7.5. In the tables, the average fitness value is interpreted as $\bar{f}$, the standard deviation as σ_f, the best fitness value as f_{Best}, and the worst fitness value as f_{Worst}. The best entries are highlighted. In Table 7.1, the results using $n = 10$ are manifested. Table 7.1 evidence that NCCO performs better than its opponents for the unimodal functions $f_1 - f_5$ and f_7. This is expected by the hybridization among its double pair of exploration-exploitation components, which include an adequate equilibrium among evolutionary phases, delivering steady solutions. For functions f_6 and f_8, NCCO produces fitness values similar to those achieved by its opponents. These results indicate that NCCO performs competitively with state-of-art methods.

To statistically examine the numerical results of Table 7.1, a non-parametric test is managed to achieve a significance interval among the tested evolutionary techniques. The Wilcoxon rank sum test has been employed at 5% significance value, across the best fitness values for each test function. As every function is judged considering 30 independent executions, it examines the best fitness value of every run to employ the Wilcoxon test. Table 7.2 presents the p-values received by the rank sum for a pair-wise matching between the algorithms. In Table 7.2, the comparison forms 9 groups, namely: NCCO versus ABC, NCCO versus DE, NCCO versus PSO, NCCO versus CS, NCCO versus DS, NCCO versus MFO, NCCO versus MVO, NCCO versus SCA, and NCCO versus CMA-ES. The proposed null hypothesis *H0* symbolizes that there is no significant distinction among a given pair of algorithms. On the other hand, the recommended alternative hypothesis *H1* indicates the presence of a significant

Table 7.1 Results of the minimization of Table B.4 with $n = 10$

		ABC	DE	PSO	CS	DS	MFO	MVO	SCA	CMA-ES	NCCO
f_1	$\bar{f}$	1.43E−16	6.35E−02	1.16E−02	5.04E−01	1.13E−09	8.80E−14	1.49E−02	3.54E−14	8.46E−29	2.42E−167
	σ_f	5.82E−17	3.48E−01	3.38E−02	2.32E−01	2.67E−09	2.05E−13	4.21E−03	9.29E−14	7.82E−29	0.00E+00
	f_{Best}	5.83E−17	7.80E−37	6.86E−06	1.18E−01	5.62E−12	3.16E−16	5.05E−03	5.49E−20	8.69E−30	**1.63E−199**
	f_{Worst}	2.39E−16	1.90E+00	1.51E−01	1.03E+00	1.44E−08	9.77E−13	2.43E−02	4.43E−13	2.63E−28	7.04E−166
f_2	$\bar{f}$	7.78E−12	1.26E−03	9.39E−01	6.07E−01	2.90E−06	3.18E−09	3.75E−02	5.49E−10	1.43E−14	2.87E−103
	σ_f	3.82E−12	6.91E−03	8.12E−01	1.27E−01	4.75E−06	2.98E−09	1.21E−02	1.96E−09	8.15E−15	1.15E−102
	f_{Best}	2.12E−12	5.68E−23	9.01E−02	3.88E−01	4.98E−08	3.67E−10	2.14E−02	9.45E−14	3.84E−15	**1.49E−106**
	f_{Worst}	1.66E−11	3.78E−02	3.94E+00	8.59E−01	2.25E−05	1.56E−08	7.62E−02	1.05E−08	3.49E−14	6.36E−102
f_3	$\bar{f}$	1.74E+02	1.09E+00	8.61E+00	1.85E+01	3.66E+01	1.67E+02	5.67E−02	2.63E−04	3.98E−19	4.77E−03
	σ_f	1.18E+02	2.32E+00	1.42E+01	6.20E+00	6.94E+01	9.13E+02	3.07E−02	1.25E−03	7.85E−19	2.08E−02
	f_{Best}	1.09E+01	7.89E−06	1.33E−03	9.19E+00	5.44E−01	2.42E−05	1.58E−02	6.37E−10	1.39E−21	**4.47E−180**
	f_{Worst}	5.76E+02	1.01E+01	6.76E+01	3.12E+01	3.39E+02	5.00E+03	1.25E−01	6.84E−03	3.62E−18	1.12E−01
f_4	$\bar{f}$	4.98E−01	1.07E+00	2.62E+00	3.05E+00	1.10E−01	5.97E−02	8.04E−02	1.23E−04	2.94E−13	2.66E−02
	σ_f	2.98E−01	1.91E+00	2.33E+00	7.04E−01	1.06E−01	2.06E−01	2.78E−02	2.27E−04	1.66E−13	9.10E−02
	f_{Best}	9.83E−02	5.66E−05	1.74E−01	1.47E+00	1.69E−02	9.17E−05	3.64E−02	1.77E−07	5.70E−14	**3.50E−93**
	f_{Worst}	1.71E+00	8.94E+00	1.11E+01	4.55E+00	4.03E−01	8.20E−01	1.53E−01	1.14E−03	8.12E−13	4.96E−01
f_5	$\bar{f}$	4.75E+00	6.87E+00	9.67E+00	8.71E+00	4.46E+00	5.09E+00	3.95E+00	7.18E+00	6.20E−02	8.11E+00
	σ_f	1.56E+00	8.29E−01	2.48E+00	2.29E+00	1.20E+00	2.34E+00	2.18E+00	3.02E+00	2.38E−02	3.36E−01
	f_{Best}	6.73E−01	5.12E+00	6.78E+00	6.71E+00	7.53E−02	0.00E+00	5.56E−01	1.96E−02	2.06E−02	**7.28E+00**
	f_{Worst}	7.11E+00	9.57E+00	1.86E+01	1.72E+01	5.88E+00	9.60E+00	9.41E+00	1.86E+01	1.09E−01	8.81E+00
f_6	$\bar{f}$	0.00E+00	1.77E+00	5.13E+00	1.53E+00	0.00E+00	0.00E+00	9.00E−01	0.00E+00	0.00E+00	0.00E+00

(continued)

Table 7.1 (continued)

		ABC	DE	PSO	CS	DS	MFO	MVO	SCA	CMA-ES	NCCO
	σ_f	0.00E+00	8.95E+00	5.16E+00	8.19E−01	0.00E+00	0.00E+00	9.95E−01	0.00E+00	0.00E+00	0.00E+00
	f_{Best}	**0.00E+00**	**0.00E+00**	**0.00E+00**	**0.00E+00**	**0.00E+00**	**0.00E+00**	**0.00E+00**	**0.00E+00**	**0.00E+00**	**0.00E+00**
	f_{Worst}	0.00E+00	4.90E+01	2.00E+01	3.00E+00	0.00E+00	0.00E+00	3.00E+00	0.00E+00	0.00E+00	0.00E+00
f_7	$\bar{f}$	2.21E−02	2.48E−03	3.90E−02	2.10E−02	7.07E−03	5.15E−03	2.14E−03	1.52E−03	3.02E−03	2.45E−03
	σ_f	8.70E−03	1.11E−03	4.51E−02	9.28E−03	3.81E−03	2.31E−03	1.56E−03	1.00E−03	1.15E−03	3.27E−03
	f_{Best}	7.83E−03	3.33E−04	4.28E−03	4.12E−03	1.21E−03	1.17E−03	4.23E−04	1.47E−04	8.79E−04	**1.02E−06**
	f_{Worst}	4.10E−02	4.66E−03	2.36E−01	4.30E−02	1.54E−02	9.46E−03	6.44E−03	3.34E−03	5.50E−03	1.37E−02
f_8	$\bar{f}$	2.22E+00	2.22E+00	−3.13E+03	2.22E+00	2.22E+00	2.23E+00	2.22E+00	2.57E+00	2.68E+00	2.22E+00
	σ_f	1.19E−14	6.99E−06	1.17E+04	1.00E−04	2.24E−10	1.05E−02	6.29E−06	1.08E−01	1.00E−01	1.61E−03
	f_{Best}	**2.22E+00**	**2.22E+00**	−4.75E+04	**2.22E+00**	**2.22E+00**	**2.22E+00**	**2.22E+00**	2.34E+00	2.50E+00	**2.22E+00**
	f_{Worst}	2.22E+00	2.22E+00	4.36E+00	2.22E+00	2.22E+00	2.25E+00	2.22E+00	2.67E+00	2.91E+00	2.22E+00

Table 7.2 p-values produced by Wilcoxon rank sum test over the averaged fitness value $\bar{f}$ from Table 7.1

NCCO versus	ABC	DE	PSO	CS	DS	MFO	MVO	SCA	CMA-ES
f_1	6.64E−09▲	6.64E−09▲	6.64E−09▲	6.64E−09▲	6.64E−09▲	6.64E−09▲	6.64E−09▲	6.64E−09▲	6.64E−09▲
f_2	6.64E−09▲	6.64E−09▲	6.64E−09▲	6.64E−09▲	6.64E−09▲	6.64E−09▲	6.64E−09▲	6.64E−09▲	6.64E−09▲
f_3	6.64E−09▲	1.52E−07▲	1.86E−08▲	6.64E−09▲	6.64E−09▲	5.29E−05▲	1.71E−07▲	2.63E−06▲	8.09E−07▲
f_4	7.56E−09▲	2.46E−05▲	6.64E−09▲	6.64E−09▲	5.47E−07▲	4.16E−06▲	6.50E−08▲	4.96E−05▲	2.10E−06▲
f_5	6.64E−09▲	2.74E−07▲	4.82E−05▲	1.19E−07▲	6.64E−09▲	1.19E−07▲	1.19E−07▲	1.50E−06▲	6.64E−09▲
f_6	6.24E−01►	6.78E−01►	3.54E−01►	1.74E−01►	2.57E−01►	2.75E−01►	1.32E−01►	3.41E−01►	7.68E−02►
f_7	9.81E−09▲	9.07E−05▲	2.12E−08▲	3.97E−08▲	1.01E−06▲	9.83E−07▲	2.53E−08▲	7.75E−08▲	3.31E−05▲
f_8	7.31E−01►	9.58E−01►	7.88E−06▲	5.95E−01►	7.95E−01►	2.80E−01►	8.47E−01►	6.64E−09▲	6.64E−09▲
▲	6	6	7	6	6	6	6	6	6
▼	0	0	0	0	0	0	0	0	0
►	2	2	1	2	2	2	2	2	2

difference among a pair of algorithms. To make clear the statistical results of the non-parametric test, Table 7.2 assumes the symbols ▲, ▼, and ►. The ▲ symbol indicates that the proposed method achieves significantly better performanche. While ▼ hints that the proposed methodology performs worse. And ► is utilized while the rank sum test is not capable to discriminate among evolutionary methods. The *p*-values from Table 7.2 manifest that NCCO performs better than its opponents for most of the evaluations. Nevertheless, there is no statistical discrepancy among NCCO and the rest of the techniques for function f_6. Also, for function f_8, the test symbolizes that there is no distinction among NCCO and ABC, DE, CD, DS, MFO and MVO.

The subsequent experiment examines the performance of unimodal functions using a 30-dimensional feasible space. The numerical values, manifested in Table 7.3, show that NCCO provides more reliable and scalable solutions. NCCO outperforms its competitors in functions $f_1 - f_5$ and f_7. Nevertheless, it presents comparable performance in functions f_6 and f_8.

The statistical values of the outcomes from Table 7.3 are exhibited in Table 7.4. Regarding to the p-values from Table 7.4, the proposed NCCO performs better than its competitors. But, in functions f_6 and f_8, the *p*-values indicate that there is no statistical discrepancy among the outcomes achieved by NCCO and most of the remaining algorithms. The last experiment involves the experimental performance considering a 100-dimensional feasible space. The numerical values are manifested in Table 7.5. From the table, it is visible that NCCO beats its competitors by obtaining the best fitness values, with the lowest standard deviations. The exception is in function f_6, where CMA-ES reaches the same fitness value as the proposed method. But, NCCO is capable of producing more steady solutions.

To statistically verify the outcomes of Table 7.5, Table 7.6 displays the results of the rank sum analysis regarding a 100-dimensional search space. Regarding the *p*-values, NCCO presents a dominant performance in most of the functions.

7.6.1.2 Multimodal Test Functions

This section presents the performance results over multimodal scenarios. The presence of multiple optimal values inside a search space usually generates solutions to be caught in local optima, making the optimization process difficult to solve. In this test, the performance of the suggested approach is judged regarding functions with many optima using $n = \{10, 30, 100\}$ dimensional search spaces. These functions are characterized by $f_9 - f_{18}$ in Table B.5. The numerical results of NCCO, and its opponents are sponsored in Tables 7.7, 7.9 and 7.11. In Table 7.7, the results using $n = 10$ dimensions are exhibited. From Table 7.7, NCCO performs considerably similarly to ABC, DE, CS, DS, MFO, MVO and CMA-ES in function f_9. Additionally, for function f_{10}, ABC, DE and SCA obtain identical performance as well as NCCO. Nonetheless, in those situations NCCO exhibits more steady solutions. For functions f_{13} and f_{14} NCCO operates likewise to its opponents. From the table, it is clear that NCCO has some troubles in functions $f_{16} - f_{18}$ since these functions

Table 7.3 Results of the minimization of Table B.4 with $n = 30$

		ABC	DE	PSO	CS	DS	MFO	MVO	SCA	CMA-ES	NCCO
f_1	$\bar{f}$	3.47E−05	3.99E+00	9.16E+02	6.62E+02	1.43E+00	1.20E+03	7.33E−01	4.03E+00	2.80E−07	8.15E−140
	σ_f	6.89E−05	8.94E+00	3.20E+02	1.39E+02	1.37E+00	3.85E+03	1.87E−01	8.11E+00	1.36E−07	4.46E−139
	f_{Best}	2.90E−06	1.37E−06	3.83E+02	3.99E+02	2.15E−01	3.55E−01	4.75E−01	5.18E−03	1.03E−07	**1.70E−184**
	f_{Worst}	3.66E−04	3.34E+01	1.93E+03	9.67E+02	5.35E+00	2.00E+04	1.15E+00	3.89E+01	6.40E−07	2.44E−138
f_2	$\bar{f}$	3.74E−03	2.34E−03	2.21E+01	5.46E+01	2.42E−01	3.15E+01	3.29E+00	1.24E−02	7.22E−04	4.09E−92
	σ_f	1.79E−03	1.20E−02	7.22E+00	1.12E+01	1.10E−01	1.91E+01	1.42E+01	2.07E−02	2.22E−04	1.55E−91
	f_{Best}	6.95E−04	3.92E−10	1.22E+01	3.55E+01	8.04E−02	8.64E−02	3.21E−01	1.05E−04	4.14E−04	**2.34E−97**
	f_{Worst}	9.79E−03	6.57E−02	3.92E+01	8.22E+01	5.73E−01	9.00E+01	7.86E+01	9.31E−02	1.36E−03	8.23E−91
f_3	$\bar{f}$	1.74E+04	9.34E+03	3.14E+03	9.86E+03	1.10E+04	1.78E+04	8.10E+01	6.04E+03	5.21E−02	3.92E−01
	σ_f	3.59E+03	2.63E+03	1.37E+03	1.76E+03	6.82E+03	1.24E+04	3.61E+01	4.30E+03	5.35E−02	1.40E+00
	f_{Best}	1.12E+04	4.24E+03	1.23E+03	6.42E+03	2.99E+03	2.22E+03	3.76E+01	1.53E+03	1.11E−02	**6.70E−53**
	f_{Worst}	2.63E+04	1.37E+04	7.05E+03	1.32E+04	3.47E+04	6.64E+04	1.60E+02	1.67E+04	2.56E−01	7.12E+00
f_4	$\bar{f}$	5.10E+01	1.41E+01	1.69E+01	2.62E+01	1.98E+01	5.92E+01	1.28E+00	2.99E+01	5.32E+01	7.81E−02
	σ_f	5.68E+00	5.74E+00	3.14E+00	3.35E+00	4.80E+00	9.22E+00	4.39E−01	9.83E+00	4.14E+01	9.25E−02
	f_{Best}	3.63E+01	5.51E+00	1.09E+01	1.64E+01	1.02E+01	3.73E+01	6.04E−01	1.30E+01	7.88E−03	**1.16E−22**
	f_{Worst}	5.91E+01	2.62E+01	2.35E+01	3.09E+01	2.80E+01	7.70E+01	2.23E+00	4.65E+01	9.54E+01	3.20E−01
f_5	$\bar{f}$	2.58E+01	2.77E+01	1.20E+02	2.86E+01	3.15E+01	6.40E+01	2.88E+01	2.92E+01	2.17E+01	2.89E+01
	σ_f	1.23E+00	1.26E+00	4.78E+01	2.51E−02	1.80E+01	3.97E+01	9.65E+00	1.87E+00	8.50E−01	7.28E−02
	f_{Best}	2.13E+01	2.51E+01	4.72E+01	2.86E+01	2.24E+01	1.23E+01	1.85E+01	2.77E+01	2.05E+01	**2.87E+01**
	f_{Worst}	2.72E+01	3.26E+01	2.66E+02	2.87E+01	1.10E+02	1.88E+02	7.84E+01	3.75E+01	2.51E+01	2.90E+01
f_6	$\bar{f}$	3.33E−02	1.25E+01	1.35E+03	7.04E+02	1.20E+00	1.21E+03	8.50E+00	6.73E+00	0.00E+00	0.00E+00

(continued)

Table 7.3 (continued)

		ABC	DE	PSO	CS	DS	MFO	MVO	SCA	CMA-ES	NCCO
	σ_f	1.83E−01	2.77E+01	5.16E+02	1.78E+02	1.83E+00	3.85E+03	4.43E+00	8.17E+00	0.00E+00	0.00E+00
	f_{Best}	**0.00E+00**	**0.00E+00**	3.23E+02	3.78E+02	**0.00E+00**	**0.00E+00**	2.00E+00	**0.00E+00**	**0.00E+00**	**0.00E+00**
	f_{Worst}	1.00E+00	1.23E+02	2.58E+03	1.14E+03	9.00E+00	2.00E+04	2.60E+01	3.00E+01	0.00E+00	0.00E+00
f_7	$\bar{f}$	3.04E−01	1.94E−02	4.58E+00	2.88E−01	9.54E−02	2.44E+00	2.56E−02	6.32E−02	1.46E−02	1.63E−02
	σ_f	7.46E−02	8.71E−03	2.85E+00	1.06E−01	3.44E−02	5.36E+00	8.22E−03	4.85E−02	4.68E−03	8.98E−03
	f_{Best}	1.70E−01	5.73E−03	9.74E−01	9.69E−02	4.58E−02	3.63E−02	1.02E−02	7.23E−03	6.30E−03	**3.24E−03**
	f_{Worst}	4.39E−01	4.44E−02	1.34E+01	5.65E−01	1.93E−01	2.69E+01	3.86E−02	1.92E−01	2.31E−02	4.46E−02
f_8	$\bar{f}$	2.07E+00	2.11E+00	8.03E+00	3.01E+00	2.07E+00	2.09E+00	2.07E+00	2.37E+00	2.61E+00	2.08E+00
	σ_f	4.31E−04	7.82E−02	1.76E+00	5.06E−01	4.22E−03	1.23E−02	2.44E−04	7.55E−02	1.64E−01	2.07E−03
	f_{Best}	**2.07E+00**	**2.07E+00**	4.60E+00	2.21E+00	**2.07E+00**	**2.07E+00**	**2.07E+00**	2.26E+00	2.39E+00	**2.07E+00**
	f_{Worst}	2.07E+00	2.38E+00	1.22E+01	4.42E+00	2.09E+00	2.12E+00	2.07E+00	2.67E+00	2.97E+00	2.08E+00

Table 7.4 *p*-values produced by Wilcoxon rank sum test over the averaged fitness value $\bar{f}$ from Table 7.3

NCCO versus	ABC	DE	PSO	CS	DS	MFO	MVO	SCA	CMA-ES
f_1	6.64E−09▲	6.64E−09▲	6.64E−09▲	6.64E−09▲	6.64E−09▲	6.64E−09▲	6.64E−09▲	6.64E−09▲	6.64E−09▲
f_2	6.64E−09▲	6.64E−09▲	6.64E−09▲	6.64E−09▲	6.64E−09▲	6.64E−09▲	6.64E−09▲	6.64E−09▲	6.64E−09▲
f_3	6.64E−09▲	6.64E−09▲	6.64E−09▲	6.64E−09▲	6.64E−09▲	6.64E−09▲	6.64E−09▲	6.64E−09▲	3.02E−03▲
f_4	6.64E−09▲	6.64E−09▲	6.64E−09▲	6.64E−09▲	6.64E−09▲	6.64E−09▲	6.64E−09▲	6.64E−09▲	4.05E−04▲
f_5	6.64E−09▲	1.19E−07▲	6.64E−09▲	6.64E−09▲	1.67E−06▲	4.55E−02▲	1.06E−03▲	2.72E−04▲	6.64E−09▲
f_6	5.98E−01►	3.90E−01►	5.66E−10▲	5.65E−10▲	4.08E−01►	1.19E−01►	5.43E−10▲	2.61E−01►	1.45E−01►
f_7	6.64E−09▲	3.56E−03▲	6.64E−09▲	6.64E−09▲	6.64E−09▲	7.56E−09▲	9.83E−04▲	6.13E−07▲	3.12E−03▲
f_8	7.68E−02►	2.21E−01►	6.64E−09▲	6.64E−09▲	4.94E−01►	2.63E−01►	2.76E−01►	6.64E−09▲	6.64E−09▲
▲	6	6	8	8	6	6	7	7	7
▼	0	0	0	0	0	0	0	0	0
►	2	2	0	0	2	2	1	1	1

Table 7.5 Results of the minimization of Table B.4 with $n = 100$

		ABC	DE	PSO	CS	DS	MFO	MVO	SCA	CMA-ES	NCCO
f_1	$\bar{f}$	5.46E+02	2.58E+02	1.24E+04	1.43E+04	6.00E+03	5.53E+04	9.47E+01	9.60E+03	1.50E+00	1.34E−28
	σ_f	6.67E+02	2.03E+02	2.77E+03	1.56E+03	2.58E+03	1.16E+04	1.03E+01	7.02E+03	2.67E−01	7.33E−28
	f_{Best}	3.16E+01	2.85E+01	7.41E+03	1.17E+04	2.72E+03	3.85E+04	7.39E+01	1.79E+03	1.03E+00	**2.87E−89**
	f_{Worst}	3.13E+03	9.79E+02	2.09E+04	1.88E+04	1.56E+04	8.42E+04	1.08E+02	3.44E+04	2.09E+00	4.02E−27
f_2	$\bar{f}$	2.39E+00	5.94E−01	1.36E+02	1.00E+10	4.47E+01	2.33E+02	1.32E+25	7.08E+00	2.66E+00	8.82E−35
	σ_f	5.62E−01	8.87E−01	4.11E+01	0.00E+00	1.18E+01	3.88E+01	6.25E+25	7.13E+00	3.53E−01	4.83E−34
	f_{Best}	1.21E+00	1.25E−02	8.67E+01	1.00E+10	2.69E+01	1.70E+02	3.08E+02	5.18E−01	2.02E+00	**2.26E−88**
	f_{Worst}	3.50E+00	4.91E+00	2.23E+02	1.00E+10	7.82E+01	3.35E+02	3.40E+26	3.30E+01	3.38E+00	2.65E−33
f_3	$\bar{f}$	2.16E+05	2.60E+05	6.11E+04	1.66E+05	1.50E+05	2.15E+05	5.26E+04	2.27E+05	3.05E+05	2.93E+01
	σ_f	2.33E+04	3.92E+04	1.96E+04	1.90E+04	6.81E+04	4.00E+04	7.84E+03	5.35E+04	4.59E+04	4.66E+01
	f_{Best}	1.61E+05	1.61E+05	3.00E+04	1.37E+05	5.38E+04	1.42E+05	4.19E+04	1.49E+05	1.94E+05	**3.56E−02**
	f_{Worst}	2.47E+05	3.28E+05	1.02E+05	2.07E+05	3.27E+05	3.06E+05	7.22E+04	3.71E+05	3.96E+05	2.10E+02
f_4	$\bar{f}$	9.16E+01	3.68E+01	2.97E+01	4.27E+01	5.78E+01	9.16E+01	5.39E+01	8.86E+01	9.88E+01	3.09E−01
	σ_f	1.47E+00	5.01E+00	3.65E+00	3.29E+00	6.48E+00	2.61E+00	5.01E+00	2.88E+00	1.28E+00	1.95E−01
	f_{Best}	8.86E+01	2.59E+01	2.19E+01	3.85E+01	4.65E+01	8.40E+01	4.59E+01	7.74E+01	9.57E+01	**1.01E−01**
	f_{Worst}	9.46E+01	4.78E+01	3.67E+01	4.96E+01	7.60E+01	9.55E+01	6.65E+01	9.32E+01	1.00E+02	9.45E−01
f_5	$\bar{f}$	2.40E+02	1.09E+02	3.96E+03	9.96E+01	2.60E+02	7.56E+02	1.06E+02	3.60E+02	9.89E+01	1.16E+02
	σ_f	3.73E+01	8.36E+00	8.55E+02	1.87E−01	6.60E+01	1.25E+02	1.89E+01	1.26E+02	4.21E−02	4.34E+01
	f_{Best}	1.69E+02	9.87E+01	2.29E+03	9.93E+01	1.52E+02	5.38E+02	9.52E+01	1.42E+02	9.87E+01	**9.99E+01**
	f_{Worst}	3.14E+02	1.32E+02	6.64E+03	1.00E+02	4.01E+02	9.95E+02	1.54E+02	7.09E+02	9.90E+01	3.08E+02
f_6	$\bar{f}$	6.99E+02	2.90E+02	1.37E+04	1.40E+04	5.26E+03	5.48E+04	2.38E+02	1.14E+04	3.33E−01	1.67E−01

(continued)

Table 7.5 (continued)

		ABC	DE	PSO	CS	DS	MFO	MVO	SCA	CMA-ES	NCCO
	σ_f	6.34E+02	2.39E+02	2.58E+03	2.07E+03	1.47E+03	1.35E+04	6.47E+01	7.54E+03	4.79E−01	3.79E−01
	f_{Best}	3.10E+01	4.10E+01	8.42E+03	1.05E+04	2.80E+03	3.19E+04	1.37E+02	2.11E+03	**0.00E+00**	**0.00E+00**
	f_{Worst}	2.19E+03	1.10E+03	1.88E+04	1.80E+04	8.24E+03	8.19E+04	4.64E+02	3.26E+04	1.00E+00	1.00E+00
f_7	$\bar{f}$	3.93E+00	6.00E−01	4.10E+02	9.92E+00	5.07E+00	1.79E+02	4.13E−01	1.12E+02	8.31E−02	2.28E−02
	σ_f	9.86E−01	4.24E−01	9.05E+01	2.74E+00	2.04E+00	8.75E+01	1.12E−01	6.49E+01	1.14E−02	1.29E−02
	f_{Best}	2.65E+00	2.31E−01	1.88E+02	4.39E+00	2.51E+00	5.70E+01	2.17E−01	2.05E+01	5.77E−02	**2.33E−03**
	f_{Worst}	7.67E+00	2.00E+00	6.12E+02	1.62E+01	1.12E+01	4.35E+02	7.03E−01	2.68E+02	1.01E−01	6.96E−02
f_8	$\bar{f}$	5.27E+00	3.70E+00	3.79E+01	4.01E+01	3.61E+00	3.34E+00	3.00E+00	2.24E+00	1.09E+01	2.03E+00
	σ_f	1.46E+00	8.71E−01	4.09E+00	2.33E+00	7.80E−01	1.33E+00	5.39E−01	5.11E−02	1.40E+00	2.53E−03
	f_{Best}	2.67E+00	2.67E+00	2.69E+01	3.41E+01	2.10E+00	2.04E+00	2.40E+00	2.15E+00	8.29E+00	**2.03E+00**
	f_{Worst}	7.88E+00	5.94E+00	4.50E+01	4.44E+01	4.93E+00	5.00E+00	4.81E+00	2.34E+00	1.34E+01	2.04E+00

Table 7.6 *p*-values produced by Wilcoxon rank sum test over the averaged fitness value $\bar{f}$ from Table 7.5

NCCO versus	ABC	DE	PSO	CS	DS	MFO	MVO	SCA	CMA-ES
f_1	6.64E−09▲	6.64E−09▲	6.64E−09▲	6.64E−09▲	6.64E−09▲	6.64E−09▲	6.64E−09▲	6.64E−09▲	6.64E−09▲
f_2	6.64E−09▲	6.64E−09▲	6.64E−09▲	5.66E−10▲	6.64E−09▲	6.64E−09▲	6.64E−09▲	6.64E−09▲	6.64E−09▲
f_3	6.64E−09▲	6.64E−09▲	6.64E−09▲	6.64E−09▲	6.64E−09▲	6.64E−09▲	6.64E−09▲	6.64E−09▲	6.64E−09▲
f_4	6.64E−09▲	6.64E−09▲	6.64E−09▲	6.64E−09▲	6.64E−09▲	6.64E−09▲	6.64E−09▲	6.64E−09▲	6.64E−09▲
f_5	6.64E−09▲	1.19E−07▲	6.64E−09▲	6.64E−09▲	6.64E−09▲	6.64E−09▲	1.15E−07▲	6.64E−09▲	6.64E−09▲
f_6	1.78E−09▲	1.78E−09▲	1.78E−09▲	1.77E−09▲	1.78E−09▲	1.78E−09▲	1.77E−09▲	1.78E−09▲	1.65E−01►
f_7	6.64E−09▲	6.64E−09▲	6.64E−09▲	6.64E−09▲	6.64E−09▲	6.64E−09▲	6.64E−09▲	6.64E−09▲	9.81E−09▲
f_8	6.64E−09▲	6.64E−09▲	6.64E−09▲	6.64E−09▲	6.64E−09▲	6.64E−09▲	6.64E−09▲	6.64E−09▲	6.64E−09▲
▲	8	8	8	8	8	8	8	8	7
▼	0	0	0	0	0	0	0	0	0
►	0	0	0	0	0	0	0	0	1

Table 7.7 Results of the minimization of Table B.5 with $n = 10$

		ABC	DE	PSO	CS	DS	MFO	MVO	SCA	CMA-ES	NCCO
f_9	$\bar{f}$	2.47E−01	2.49E−01	2.76E+00	2.47E−01	2.47E−01	2.51E−01	2.47E−01	6.87E−01	6.52E−01	2.49E−01
	σ_f	1.07E−16	9.21E−03	5.93E−01	8.08E−05	2.13E−12	1.19E−02	6.83E−06	1.51E−01	2.39E−01	1.34E−03
	f_{Best}	**2.47E−01**	**2.47E−01**	1.41E+00	**2.47E−01**	**2.47E−01**	**2.47E−01**	**2.47E−01**	4.04E−01	**2.47E−01**	**2.47E−01**
	f_{Worst}	2.47E−01	2.83E−01	3.69E+00	2.47E−01	2.47E−01	2.81E−01	2.47E−01	8.99E−01	1.03E+00	2.54E−01
f_{10}	$\bar{f}$	1.21E−13	2.79E−01	1.68E+01	1.78E+01	2.19E−01	1.61E+01	1.63E+01	1.99E−01	1.58E+01	0.00E+00
	σ_f	3.14E−13	6.33E−01	9.87E+00	3.64E+00	5.41E−01	8.68E+00	8.12E+00	1.09E+00	1.10E+01	0.00E+00
	f_{Best}	**0.00E+00**	**0.00E+00**	3.98E+00	1.24E+01	1.72E−08	3.98E+00	4.98E+00	**0.00E+00**	9.95E−01	**0.00E+00**
	f_{Worst}	1.62E−12	2.98E+00	5.47E+01	2.67E+01	1.99E+00	3.58E+01	3.68E+01	5.98E+00	3.24E+01	0.00E+00
f_{11}	$\bar{f}$	7.48E−10	2.32E−05	2.72E+00	5.19E+00	1.01E−05	6.67E−08	2.32E−01	3.27E−05	5.27E+00	8.88E−16
	σ_f	6.25E−10	6.53E−05	1.22E+00	1.12E+00	1.73E−05	5.64E−08	5.27E−01	1.78E−04	8.89E+00	0.00E+00
	f_{Best}	1.20E−10	4.44E−15	4.08E−02	3.61E+00	2.50E−07	7.20E−09	2.14E−02	6.71E−11	4.44E−15	**8.88E−16**
	f_{Worst}	2.37E−09	2.20E−04	5.48E+00	7.93E+00	7.67E−05	2.26E−07	2.02E+00	9.75E−04	2.00E+01	8.88E−16
f_{12}	$\bar{f}$	7.04E−03	3.57E−03	2.72E−01	3.97E−01	3.70E−02	1.58E−01	3.78E−01	4.98E−02	8.22E−04	1.96E−01
	σ_f	6.49E−03	4.59E−03	1.59E−01	6.46E−02	2.77E−02	1.12E−01	1.45E−01	1.15E−01	2.53E−03	2.32E−01
	f_{Best}	7.84E−12	**0.00E+00**	6.82E−02	2.68E−01	1.65E−04	4.43E−02	1.65E−01	1.78E−14	**0.00E+00**	**0.00E+00**
	f_{Worst}	1.87E−02	1.23E−02	6.92E−01	5.03E−01	9.60E−02	6.17E−01	8.23E−01	6.12E−01	9.86E−03	8.20E−01
f_{13}	$\bar{f}$	−1.00E+00	−1.00E+00	−5.80E−01	−9.04E−01	−1.00E+00	−9.85E−01	−9.06E−01	−1.00E+00	−1.00E+00	−1.00E+00
	σ_f	7.98E−17	2.11E−08	1.79E−01	6.39E−02	9.13E−12	5.95E−02	1.06E−01	2.70E−16	0.00E+00	0.00E+00
	f_{Best}	**−1.00E+00**	**−1.00E+00**	**−1.00E+00**	−9.83E−01	**−1.00E+00**	**−1.00E+00**	**−1.00E+00**	**−1.00E+00**	**−1.00E+00**	**−1.00E+00**
	f_{Worst}	−1.00E+00	−1.00E+00	−2.18E−01	−7.45E−01	−1.00E+00	−7.04E−01	−7.04E−01	−1.00E+00	−1.00E+00	−1.00E+00
f_{14}	$\bar{f}$	2.80E+00	2.80E+00	2.80E+00	2.80E+00	2.80E+00	2.80E+00	2.80E+00	2.81E+00	2.80E+00	2.80E+00

(continued)

Table 7.7 (continued)

		ABC	DE	PSO	CS	DS	MFO	MVO	SCA	CMA-ES	NCCO
	σ_f	1.55E−03	7.88E−04	6.90E−05	2.79E−04	4.03E−11	5.02E−16	4.74E−07	1.03E−03	0.00E+00	5.91E−04
	f_{Best}	2.79E+00	**2.80E+00**	**2.80E+00**	**2.80E+00**	**2.80E+00**	**2.80E+00**	**2.80E+00**	**2.80E+00**	**2.80E+00**	**2.80E+00**
	f_{Worst}	2.80E+00	2.80E+00	2.80E+00	2.80E+00	2.80E+00	2.80E+00	2.80E+00	2.81E+00	2.80E+00	2.80E+00
f_{15}	$\bar{f}$	1.23E−16	2.75E−04	1.20E−05	1.78E−04	9.70E−13	1.89E−17	5.49E−06	3.76E−17	2.57E−32	−1.36E−68
	σ_f	5.36E−17	1.51E−03	2.50E−05	7.30E−05	2.51E−12	2.93E−17	2.23E−06	1.10E−16	2.28E−32	0.00E+00
	f_{Best}	6.02E−17	1.10E−40	1.71E−11	7.45E−05	1.00E−17	8.41E−20	2.09E−06	6.44E−23	2.57E−33	**−1.36E−68**
	f_{Worst}	2.44E−16	8.26E−03	9.75E−05	4.25E−04	1.09E−11	1.01E−16	1.01E−05	4.71E−16	9.31E−32	−1.36E−68
f_{16}	$\bar{f}$	−4.19E+03	−4.18E+03	−2.73E+03	−3.31E+03	−4.19E+03	−3.37E+03	−3.03E+03	−2.18E+03	−2.06E+03	−2.60E+03
	σ_f	1.96E+01	3.00E+01	2.62E+02	1.16E+02	1.39E−02	2.81E+02	3.64E+02	1.13E+02	1.08E+02	3.54E+02
	f_{Best}	**−4.19E+03**	**−4.19E+03**	−3.17E+03	−3.58E+03	**−4.19E+03**	−3.83E+03	−3.74E+03	−2.51E+03	−2.33E+03	−3.57E+03
	f_{Worst}	−4.08E+03	−4.07E+03	−1.90E+03	−3.05E+03	−4.19E+03	−2.88E+03	−2.17E+03	−1.97E+03	−1.92E+03	−1.96E+03
f_{17}	$\bar{f}$	1.33E−16	2.75E−02	2.53E+00	8.22E−01	6.49E−11	5.18E−02	1.05E−01	6.04E−02	7.01E−27	6.92E−02
	σ_f	6.34E−17	1.47E−01	2.50E+00	3.31E−01	1.53E−10	1.18E−01	2.52E−01	2.33E−02	7.13E−27	7.45E−02
	f_{Best}	2.43E−17	**4.71E−32**	8.85E−03	2.03E−01	9.23E−14	6.01E−17	1.33E−04	2.89E−02	1.04E−28	1.10E−03
	f_{Worst}	2.69E−16	8.04E−01	1.22E+01	1.58E+00	7.55E−10	3.11E−01	1.26E+00	1.27E−01	2.87E−26	2.60E−01
f_{18}	$\bar{f}$	1.22E−16	8.23E−02	6.82E−01	2.65E−01	2.87E−10	2.20E−03	2.96E−03	2.39E−01	2.12E−26	1.49E−01
	σ_f	5.19E−17	2.44E−01	1.23E+00	1.48E−01	8.97E−10	4.47E−03	3.06E−03	7.13E−02	2.43E−26	1.15E−01
	f_{Best}	6.15E−17	**1.35E−32**	3.00E−03	1.00E−01	4.90E−13	1.32E−17	6.94E−04	1.22E−01	5.63E−28	3.65E−03
	f_{Worst}	2.58E−16	9.03E−01	4.24E+00	9.27E−01	3.80E−09	1.10E−02	1.36E−02	4.08E−01	1.07E−25	4.26E−01

support non-symmetrical shape round the optimal value. For functions f_{11} and f_{15}, NCCO delivers exceptional performance.

Table 7.8 reveals that NCCO has some troubles in functions f_{16}, f_{17}, and f_{18} since this kind of functions present a non-symmetrical shape round the optimal solution. For functions f_9, f_{13} and f_{14}, NCCO works considerably similar to its competitors in terms of robustness.

When a 30-dimensional search space appears to be evaluated, the NCCO approach enhances its performance. For functions $f_9 - f_{12}$ and f_{15}, Table 7.9 demonstrates that NCCO dominates its competitors. For functions f_{13} and f_{14}, the results of the proposed method are considerably similar to those of ABC, DE, DS, MFO, SCA and CMA-ES. On the other hand, DE and CMA-ES knocked the proposed method in functions $f_{16} - f_{18}$. It can be inferred from this analysis that the NCCO works better when it runs within a higher-dimensional space that offers high multimodality because of the accumulative effect. Since multimodal high-dimensional functions lead to rise the distances among agents, the norms of the bias also grow. As a consequence, future movements will lead to be larger (in the exploration phase) or closer (in the exploitation phase).

Regarding the statistical outcomes of Table 7.10, NCCO achieves related results than ABC, DE, DS, MFO and CMAES for functions f_{13} and f_{14}. On the other hand, NCCO achieves high-grade results in functions $f_9 - f_{12}$ and f_{15}. However Nonetheless, it can be observed that in functions $f_{16} - f_{18}$, the recommended NCCO method is not able of delivering better results.

Table 7.11 exhibits the outcomes within a 100-dimensional search space, and proves that NCCO delivers a higher performance index when it works across high-dimensional functions that present high multimodality. Even judging $n = 100$ dimensions, NCCO beats its adversaries in functions $f_9 - f_{12}$ and f_{15}. Nevertheless, the proposed method fails against ABC and CMA-ES in functions $f_{16} - f_{18}$. Based on the content of the table, it can be assumed that the operators of NCCO display a good balance among the exploration and exploitation phases since they raise the population diversity in the stages, even when it faces higher dimensionality.

The p-values summarized in Table 7.12 confirm the scalability of NCCO when judged in high-dimensional search spaces. In functions $f_9 - f_{12}$ and f_{15}, NCCO persists triumphant. Nevertheless, it still manifests problems in non-symmetrical functions.

7.6.1.3 Hybrid Test Functions

In this segment, the ability of NCCO to obtain optimal values in hybrid situations is exhibited. Hybrid functions model most real-life applicability since they consolidate the complexities of both unimodal and multimodal objective functions. In this experiment, the performance of NCCO is judged regarding hybrid functions using $n = \{10, 30, 100\}$ dimensional search spaces. The functions are described by $f_{19} - f_{23}$ in Table B.6. The numerical results of NCCO, and its competitors are detailed in Tables 7.13, 7.15 and 7.17. In Table 7.13, the results using $n = 10$ dimensions are

Table 7.8 p-values produced by Wilcoxon rank sum test over the averaged fitness value $\bar{f}$ from Table 7.7

NCCO versus	ABC	DE	PSO	CS	DS	MFO	MVO	SCA	CMA-ES
f_9	1.61E−01▶	7.37E−01▶	6.64E−09▲	2.37E−01▶	1.54E−01▶	8.37E−01▶	3.14E−01▶	6.64E−09▲	5.01E−01▶
f_{10}	2.85E−01▶	3.41E−01▶	5.66E−10▲	5.66E−10▲	5.66E−10▲	5.63E−10▲	5.66E−10▲	1.43E−01▶	5.62E−10▲
f_{11}	5.66E−10▲	2.52E−10▲	5.66E−10▲	5.66E−10▲	5.66E−10▲	5.66E−10▲	5.66E−10▲	5.66E−10▲	4.42E−10▲
f_{12}	1.35E−07▲	6.43E−02▶	1.38E−06▲	9.91E−05▲	4.68E−01▼	3.79E−06▲	1.54E−08▲	3.67E−09▲	9.73E−02▶
f_{13}	2.78E−01▶	1.32E−01▶	1.65E−01▶	5.66E−10▲	3.13E−01▶	2.45E−01▶	8.12E−02▶	2.87E−01▶	1.34E−01▶
f_{14}	6.64E−09▲	5.42E−01▶	2.39E−01▶	4.96E−01▶	4.70E−01▶	2.34E−01▶	7.24E−02▶	1.21E−01▶	1.65E−01▶
f_{15}	5.66E−10▲	5.66E−10▲	5.66E−10▲	5.66E−10▲	5.66E−10▲	5.66E−10▲	5.66E−10▲	5.66E−10▲	5.66E−10▲
f_{16}	6.64E−09▼	3.58E−09▼	1.19E−08▲	1.52E−07▼	6.64E−09▼	4.83E−07▼	2.89E−04▼	1.51E−05▲	3.45E−07▲
f_{17}	6.64E−09▲	3.75E−07▲	1.19E−07▲	8.62E−09▲	6.64E−09▲	1.58E−04▼	2.36E−05▼	1.77E−07▲	6.64E−09▼
f_{18}	6.64E−09▲	5.14E−05▲	3.02E−06▼	5.29E−05▼	6.64E−09▼	1.27E−08▼	1.12E−08▼	8.41E−05▲	6.64E−09▼
▲	6	4	7	6	4	4	4	7	4
▼	1	1	1	2	3	3	3	0	2
▶	3	5	2	2	3	3	3	3	4

Table 7.9 Results of the minimization of Table B.5 with $n = 30$

		ABC	DE	PSO	CS	DS	MFO	MVO	SCA	CMA-ES	NCCO
f_9	$\bar{f}$	7.14E−02	2.02E−01	1.11E+01	1.46E−01	7.22E−02	9.25E−02	7.21E−02	4.26E−01	5.18E−01	7.88E−02
	σ_f	5.20E−05	1.42E−01	2.22E+00	4.91E−02	1.11E−03	1.01E−02	2.40E−04	1.05E−01	1.05E−01	1.77E−03
	f_{Best}	7.14E−02	8.02E−02	6.37E+00	1.01E−01	7.14E−02	7.43E−02	7.17E−02	2.69E−01	2.96E−01	**7.13E−02**
	f_{Worst}	7.16E−02	8.30E−01	1.53E+01	3.46E−01	7.44E−02	1.25E−01	7.26E−02	6.37E−01	7.45E−01	8.33E−02
f_{10}	$\bar{f}$	4.93E+00	5.28E+01	1.13E+02	1.46E+02	3.25E+01	1.55E+02	1.23E+02	4.01E+01	1.30E+02	3.44E−02
	σ_f	1.76E+00	1.22E+01	2.35E+01	1.14E+01	7.57E+00	3.63E+01	3.28E+01	3.56E+01	6.96E+01	1.88E−01
	f_{Best}	1.26E+00	3.09E+01	6.34E+01	1.24E+02	1.51E+01	9.56E+01	4.42E+01	6.41E−02	1.22E+01	**0.00E+00**
	f_{Worst}	7.40E+00	8.31E+01	1.71E+02	1.80E+02	4.79E+01	2.63E+02	1.74E+02	1.12E+02	1.85E+02	1.03E+00
f_{11}	$\bar{f}$	1.27E−01	5.77E−01	9.61E+00	1.45E+01	1.04E+00	1.24E+01	1.51E+00	1.40E+01	9.31E+00	8.88E−16
	σ_f	1.40E−01	8.20E−01	1.54E+00	1.36E+00	5.44E−01	8.63E+00	5.35E−01	8.74E+00	1.01E+01	0.00E+00
	f_{Best}	1.48E−02	7.15E−06	6.99E+00	1.18E+01	1.83E−01	2.09E−01	3.46E−01	2.51E−02	8.96E−05	**8.88E−16**
	f_{Worst}	5.54E−01	2.97E+00	1.25E+01	1.77E+01	2.18E+00	2.00E+01	2.37E+00	2.03E+01	2.00E+01	8.88E−16
f_{12}	$\bar{f}$	2.49E−02	1.71E−01	9.82E+00	6.63E+00	7.53E−01	2.96E+01	7.44E−01	7.42E−01	2.87E−06	5.19E−01
	σ_f	2.31E−02	2.33E−01	3.67E+00	1.50E+00	2.05E−01	5.87E+01	8.47E−02	3.08E−01	1.27E−06	4.12E−01
	f_{Best}	1.41E−04	1.34E−09	2.98E+00	3.58E+00	2.37E−01	3.97E−01	5.77E−01	9.96E−03	6.73E−07	**0.00E+00**
	f_{Worst}	8.53E−02	9.07E−01	2.21E+01	9.74E+00	1.02E+00	1.81E+02	9.50E−01	1.36E+00	5.72E−06	1.49E+00
f_{13}	$\bar{f}$	−3.00E+00	−3.00E+00	8.74E−01	−1.10E+00	−3.00E+00	−2.56E+00	−1.96E+00	−3.00E+00	−3.00E+00	−3.00E+00
	σ_f	5.60E−06	2.70E−02	9.09E−01	1.96E−01	1.23E−03	3.80E−01	3.72E−01	3.62E−03	4.31E−10	0.00E+00
	f_{Best}	**−3.00E+00**	**−3.00E+00**	−4.97E−01	−1.66E+00	**−3.00E+00**	**−3.00E+00**	−2.70E+00	**−3.00E+00**	**−3.00E+00**	**−3.00E+00**
	f_{Worst}	−3.00E+00	−2.85E+00	3.30E+00	−8.28E−01	−2.99E+00	−1.50E+00	−1.23E+00	−2.99E+00	−3.00E+00	−3.00E+00
f_{14}	$\bar{f}$	2.75E+00	2.77E+00	2.98E+00	3.17E+00	2.75E+00	1.97E+01	2.75E+00	2.78E+00	2.75E+00	2.75E+00

(continued)

Table 7.9 (continued)

		ABC	DE	PSO	CS	DS	MFO	MVO	SCA	CMA-ES	NCCO
	σ_f	5.18E−04	4.63E−02	1.97E−01	2.46E−01	2.75E−04	2.78E+01	6.42E−06	1.01E−01	1.04E−11	2.10E−04
	f_{Best}	**2.75E+00**	**2.75E+00**	2.76E+00	2.86E+00	**2.75E+00**	**2.75E+00**	**2.75E+00**	**2.75E+00**	**2.75E+00**	**2.75E+00**
	f_{Worst}	2.75E+00	2.99E+00	3.61E+00	3.78E+00	2.75E+00	1.53E+02	2.75E+00	3.18E+00	2.75E+00	2.75E+00
f_{15}	$\bar{f}$	4.29E−07	5.16E−03	1.45E+00	3.01E−01	4.67E−04	4.01E−01	3.12E−04	2.53E−03	1.04E−10	1.20E−180
	σ_f	3.25E−07	1.50E−02	6.71E−01	7.14E−02	5.08E−04	1.21E+00	9.22E−05	5.22E−03	4.92E−11	0.00E+00
	f_{Best}	3.35E−08	1.79E−13	4.49E−01	1.84E−01	2.62E−05	9.06E−05	1.90E−04	3.92E−06	3.36E−11	**2.83E−192**
	f_{Worst}	1.32E−06	7.34E−02	3.29E+00	4.88E−01	2.43E−03	4.00E+00	5.50E−04	2.40E−02	2.43E−10	3.49E−179
f_{16}	$\bar{f}$	−1.15E+04	−1.22E+04	−6.52E+03	−7.60E+03	−1.08E+04	−8.72E+03	−7.81E+03	−3.93E+03	−5.44E+03	−5.20E+03
	σ_f	1.89E+02	5.99E+02	6.33E+02	2.08E+02	4.10E+02	9.28E+02	5.85E+02	2.60E+02	5.36E+01	4.64E+02
	f_{Best}	−1.19E+04	**−1.26E+04**	−7.70E+03	−8.01E+03	−1.15E+04	−1.08E+04	−8.74E+03	−4.51E+03	−5.68E+03	−6.64E+03
	f_{Worst}	−1.11E+04	−1.04E+04	−5.24E+03	−7.15E+03	−1.01E+04	−6.65E+03	−6.55E+03	−3.36E+03	−5.42E+03	−4.42E+03
f_{17}	$\bar{f}$	1.74E−05	1.55E+01	1.90E+01	1.83E+01	9.87E−02	4.38E+00	1.94E+00	2.56E+02	1.36E−08	8.65E−01
	σ_f	4.88E−05	8.22E+01	1.30E+01	5.30E+00	1.08E−01	3.28E+00	1.05E+00	1.10E+03	9.40E−09	3.38E−01
	f_{Best}	2.72E−07	1.56E−02	6.43E+00	1.14E+01	2.85E−03	1.40E+00	5.97E−02	6.48E−01	**3.58E−09**	1.79E−01
	f_{Worst}	2.62E−04	4.51E+02	7.74E+01	3.44E+01	4.60E−01	2.03E+01	3.77E+00	5.97E+03	5.58E−08	1.45E+00
f_{18}	$\bar{f}$	1.16E−04	2.22E+01	5.38E+03	7.11E+03	2.18E−01	8.20E+06	1.18E−01	3.05E+04	2.18E−07	2.62E+00
	σ_f	1.93E−04	5.79E+01	1.88E+04	8.46E+03	1.52E−01	5.80E+07	5.32E−02	1.13E+05	8.88E−08	4.88E−01
	f_{Best}	2.27E−06	4.46E−03	3.89E+01	5.75E+01	3.36E−02	5.36E−01	3.93E−02	2.91E+00	**8.76E−08**	1.67E+00
	f_{Worst}	8.20E−04	2.81E+02	1.03E+05	2.93E+04	6.79E−01	4.10E+08	2.65E−01	6.12E+05	3.78E−07	3.73E+00

Table 7.10 *p*-values produced by Wilcoxon rank sum test over the averaged fitness value from Table 7.9

NCCO versus	ABC	DE	PSO	CS	DS	MFO	MVO	SCA	CMA-ES
f_9	6.64E−09▲	9.81E−09▲	6.64E−09▲	6.64E−09▲	6.64E−09▲	9.37E−08▲	6.64E−09▲	6.64E−09▲	6.64E−09▲
f_{10}	7.99E−10▲	7.99E−10▲	7.99E−10▲	7.99E−10▲	7.99E−10▲	7.99E−10▲	7.99E−10▲	9.25E−10▲	7.99E−10▲
f_{11}	5.66E−10▲	5.66E−10▲	5.66E−10▲	5.66E−10▲	5.66E−10▲	5.66E−10▲	5.66E−10▲	5.66E−10▲	5.66E−10▲
f_{12}	1.57E−04▲	3.47E−06▲	6.61E−09▲	6.61E−09▲	3.89E−07▲	1.95E−05▲	1.04E−05▲	4.56E−07▲	1.57E−04▲
f_{13}	1.37E−01►	2.64E−01►	5.66E−10▲	5.66E−10▲	2.89E−01►	4.23E−01►	5.66E−10▲	4.78E−01►	8.37E−01►
f_{14}	1.68E−01►	8.37E−01►	6.64E−09▲	6.64E−09▲	1.51E−01►	5.98E−01►	2.96E−01►	2.11E−01►	4.65E−01►
f_{15}	6.64E−09▲	6.64E−09▲	6.64E−09▲	6.64E−09▲	6.64E−09▲	6.64E−09▲	6.64E−09▲	6.64E−09▲	6.64E−09▲
f_{16}	6.64E−09▼	6.64E−09▼	3.09E−08▼	6.64E−09▼	6.64E−09▼	6.64E−09▼	6.64E−09▼	6.64E−09▲	8.37E−07▲
f_{17}	6.64E−09▼	1.24E−06▲	6.64E−09▲	6.64E−09▲	7.56E−09▼	6.64E−09▲	2.61E−08▼	1.92E−07▼	6.64E−09▼
f_{18}	6.64E−09▼	2.58E−05▼	6.64E−09▲	6.64E−09▲	6.64E−09▼	2.89E−09▼	6.64E−09▼	1.12E−08▲	6.64E−09▼
▲	5	6	9	9	5	6	6	7	6
▼	3	2	1	1	3	2	3	1	2
►	2	2	0	0	2	2	1	2	2

Table 7.11 Results of the minimization of Table B.5 with $n = 100$

		ABC	DE	PSO	CS	DS	MFO	MVO	SCA	CMA-ES	NCCO
f_9	$\bar{f}$	1.35E+00	1.67E+00	5.02E+01	2.50E+01	1.73E+00	2.10E+00	3.34E−02	2.21E−01	3.34E+00	2.65E−02
	σ_f	7.40E−01	4.42E−01	3.97E+00	3.23E+00	5.21E−01	3.63E+00	2.65E−03	5.28E−02	7.20E−01	1.01E−03
	f_{Best}	2.36E−01	9.64E−01	3.99E+01	1.92E+01	7.70E−01	4.29E−02	3.02E−02	1.27E−01	1.91E+00	**2.49E−02**
	f_{Worst}	2.78E+00	2.67E+00	5.71E+01	3.13E+01	2.79E+00	1.60E+01	4.38E−02	3.43E−01	4.79E+00	2.88E−02
f_{10}	$\bar{f}$	2.05E+02	5.70E+02	7.27E+02	7.95E+02	5.02E+02	8.24E+02	6.73E+02	2.60E+02	8.43E+02	1.01E+00
	σ_f	2.38E+01	3.66E+01	5.38E+01	3.31E+01	6.97E+01	6.91E+01	8.03E+01	1.36E+02	2.73E+01	2.89E+00
	f_{Best}	1.58E+02	5.00E+02	6.07E+02	7.14E+02	3.85E+02	7.10E+02	5.42E+02	3.72E+01	7.76E+02	**0.00E+00**
	f_{Worst}	2.51E+02	6.46E+02	8.39E+02	8.41E+02	6.32E+02	1.07E+03	8.11E+02	6.12E+02	8.89E+02	1.12E+01
f_{11}	$\bar{f}$	1.11E+01	2.18E+00	1.33E+01	1.72E+01	1.23E+01	1.99E+01	7.07E+00	1.93E+01	1.42E+01	1.54E−06
	σ_f	1.04E+00	7.26E−01	7.78E−01	9.76E−01	3.75E+00	1.66E−01	5.93E+00	3.60E+00	8.93E+00	8.43E−06
	f_{Best}	9.20E+00	9.64E−01	1.20E+01	1.48E+01	7.96E+00	1.93E+01	3.54E+00	5.96E+00	3.46E−01	**8.88E−16**
	f_{Worst}	1.29E+01	3.63E+00	1.48E+01	1.89E+01	2.00E+01	2.00E+01	2.02E+01	2.06E+01	2.00E+01	4.62E−05
f_{12}	$\bar{f}$	1.10E+01	4.11E+00	1.12E+02	1.25E+02	4.60E+01	4.35E+02	1.86E+00	8.08E+01	6.91E−01	1.74E+00
	σ_f	1.19E+01	2.55E+00	2.07E+01	2.01E+01	1.46E+01	1.16E+02	1.22E−01	7.01E+01	1.02E−01	8.99E−01
	f_{Best}	1.47E+00	1.12E+00	6.70E+01	8.57E+01	2.28E+01	1.95E+02	1.66E+00	1.53E+00	4.73E−01	**4.36E−01**
	f_{Worst}	4.19E+01	1.13E+01	1.46E+02	1.70E+02	7.70E+01	7.26E+02	2.06E+00	3.29E+02	8.77E−01	3.94E+00
f_{13}	$\bar{f}$	−9.00E+00	−9.77E+00	1.99E+01	1.78E−01	−4.94E+00	1.74E+00	−4.04E+00	−7.79E+00	−9.85E+00	−1.00E+01
	σ_f	3.90E−01	1.82E−01	2.53E+00	5.01E−01	8.71E−01	1.55E+00	8.85E−01	1.31E+00	1.26E−01	0.00E+00
	f_{Best}	−9.60E+00	**−1.00E+01**	1.54E+01	−7.72E−01	−6.32E+00	−2.31E+00	−5.87E+00	−9.59E+00	**−1.00E+01**	**−1.00E+01**
	f_{Worst}	−8.25E+00	−9.39E+00	2.56E+01	1.06E+00	−2.16E+00	5.46E+00	−1.79E+00	−4.99E+00	−9.55E+00	−1.00E+01
f_{14}	$\bar{f}$	8.15E+00	3.33E+00	1.54E+01	2.87E+01	8.09E+00	3.66E+02	2.73E+00	1.74E+02	2.73E+00	2.73E+00

(continued)

Table 7.11 (continued)

		ABC	DE	PSO	CS	DS	MFO	MVO	SCA	CMA-ES	NCCO
	σ_f	4.97E+00	2.53E−01	5.86E+00	5.52E+00	3.26E+00	1.76E+02	1.11E−03	2.74E+02	1.45E−05	9.60E−06
	f_{Best}	2.90E+00	2.93E+00	6.69E+00	2.02E+01	3.64E+00	1.38E+02	**2.73E+00**	2.17E+01	**2.73E+00**	**2.73E+00**
	f_{Worst}	2.18E+01	3.91E+00	3.33E+01	3.77E+01	1.83E+01	8.39E+02	2.73E+00	1.30E+03	2.73E+00	2.73E+00
f_{15}	$\bar{f}$	5.38E−01	1.22E−01	2.42E+01	5.55E+00	2.47E+00	2.10E+01	4.01E−02	4.13E+00	6.36E−04	4.28E−82
	σ_f	4.78E−01	8.86E−02	3.64E+00	9.98E−01	9.54E−01	5.07E+00	4.85E−03	2.48E+00	1.04E−04	2.35E−81
	f_{Best}	1.28E−02	1.33E−02	1.92E+01	3.92E+00	1.02E+00	1.17E+01	3.09E−02	7.38E−01	4.80E−04	**6.40E−173**
	f_{Worst}	1.71E+00	3.89E−01	3.35E+01	7.31E+00	5.24E+00	3.01E+01	4.83E−02	1.00E+01	8.98E−04	1.29E−80
f_{16}	$\bar{f}$	−3.03E+04	−1.53E+04	−1.45E+04	−1.70E+04	−2.35E+04	−2.33E+04	−2.38E+04	−7.18E+03	−1.49E+04	−1.01E+04
	σ_f	6.92E+02	8.24E+02	1.43E+03	3.88E+02	1.14E+03	2.57E+03	1.45E+03	4.83E+02	6.67E+02	1.36E+03
	f_{Best}	**−3.13E+04**	−1.74E+04	−1.72E+04	−1.79E+04	−2.55E+04	−2.80E+04	−2.69E+04	−8.18E+03	−1.66E+04	−1.24E+04
	f_{Worst}	−2.87E+04	−1.37E+04	−1.14E+04	−1.62E+04	−2.14E+04	−1.88E+04	−2.03E+04	−6.40E+03	−1.37E+04	−7.59E+03
f_{17}	$\bar{f}$	1.76E+02	5.61E+04	9.56E+03	3.38E+05	1.44E+05	1.78E+08	1.61E+01	2.30E+08	1.56E−02	1.15E+00
	σ_f	8.31E+02	1.27E+05	2.00E+04	2.96E+05	3.87E+05	1.32E+08	5.29E+00	1.23E+08	3.25E−03	9.00E−02
	f_{Best}	2.76E−02	2.04E+00	3.41E+01	8.95E+02	6.65E+01	3.39E+07	1.06E+01	4.84E+07	**1.11E−02**	9.83E−01
	f_{Worst}	4.52E+03	6.35E+05	8.44E+04	1.24E+06	2.02E+06	4.73E+08	2.84E+01	5.17E+08	2.70E−02	1.29E+00
f_{18}	$\bar{f}$	7.20E+03	6.79E+05	1.78E+06	5.36E+06	2.42E+06	4.68E+08	1.39E+02	4.85E+08	3.28E−01	1.00E+01
	σ_f	2.63E+04	8.26E+05	1.26E+06	2.56E+06	2.42E+06	3.11E+08	2.44E+01	2.55E+08	9.77E−02	3.30E−01
	f_{Best}	9.00E−01	6.36E+03	3.59E+05	1.23E+06	8.24E+04	6.65E+07	9.78E+01	5.68E+07	**1.94E−01**	9.69E+00
	f_{Worst}	1.29E+05	2.82E+06	5.08E+06	1.14E+07	1.01E+07	1.34E+09	1.99E+02	9.88E+08	5.81E−01	1.09E+01

Table 7.12 p-values produced by Wilcoxon rank sum test over the averaged fitness value from Table 7.11

NCCO versus	ABC	DE	PSO	CS	DS	MFO	MVO	SCA	CMA-ES
f_9	6.64E−09▲	6.64E−09▲	6.64E−09▲	6.64E−09▲	6.64E−09▲	6.64E−09▲	6.64E−09▲	6.64E−09▲	6.64E−09▲
f_{10}	1.80E−09▲	1.80E−09▲	1.80E−09▲	1.80E−09▲	1.80E−09▲	1.80E−09▲	1.80E−09▲	1.80E−09▲	1.80E−09▲
f_{11}	1.08E−09▲	1.08E−09▲	1.08E−09▲	1.08E−09▲	1.08E−09▲	1.08E−09▲	1.08E−09▲	1.08E−09▲	1.08E−09▲
f_{12}	6.68E−06▲	7.00E−05▲	6.64E−09▲	6.64E−09▲	6.64E−09▲	6.64E−09▲	3.31E−02▲	3.97E−08▲	1.35E−07▲
f_{13}	5.66E−10▲	8.37E−02►	5.66E−10▲	5.66E−10▲	5.66E−10▲	5.66E−10▲	5.66E−10▲	5.66E−10▲	6.14E−02►
f_{14}	6.64E−09▲	6.64E−09▲	6.64E−09▲	6.64E−09▲	6.64E−09▲	6.64E−09▲	1.94E−01►	6.64E−09▲	1.15E−01►
f_{15}	6.64E−09▲	6.64E−09▲	6.64E−09▲	6.64E−09▲	6.64E−09▲	6.64E−09▲	6.64E−09▲	6.64E−09▲	6.64E−09▲
f_{16}	6.64E−09▼	6.64E−09▼	1.27E−08▼	6.64E−09▼	6.64E−09▼	6.64E−09▼	6.64E−09▼	1.12E−08▼	6.08E−09▼
f_{17}	2.16E−07▼	6.64E−09▲	6.64E−09▲	6.64E−09▲	6.64E−09▲	6.64E−09▲	6.64E−09▲	6.64E−09▲	6.64E−09▼
f_{18}	1.58E−04▼	6.64E−09▲	6.64E−09▲	6.64E−09▲	6.64E−09▲	6.64E−09▲	6.64E−09▲	6.64E−09▲	6.64E−09▼
▲	7	8	9	9	9	9	8	9	5
▼	3	1	1	1	1	1	1	1	3
►	0	1	0	0	0	0	1	0	2

Table 7.13 Results of the minimization of Table B.6 with $n = 10$

		ABC	DE	PSO	CS	DS	MFO	MVO	SCA	CMA-ES	NCCO
f_{19}	$\bar{f}$	1.03E−11	3.82E−03	4.52E−02	1.45E+00	2.96E−06	9.18E−09	5.07E−02	1.14E−09	7.34E−15	8.32E−83
	σ_f	5.91E−12	1.98E−02	6.25E−02	6.06E−01	3.35E−06	1.14E−08	1.40E−02	2.79E−09	3.54E−15	4.56E−82
	f_{Best}	2.78E−12	1.18E−22	3.01E−03	4.19E−01	1.90E−07	5.39E−10	3.06E−02	2.18E−12	1.68E−15	**8.82E−101**
	f_{Worst}	2.86E−11	1.08E−01	2.82E−01	3.18E+00	1.09E−05	4.53E−08	8.12E−02	1.53E−08	1.48E−14	2.50E−81
f_{20}	$\bar{f}$	9.00E+00	9.54E+00	5.06E+01	3.37E+01	9.00E+00	2.05E+01	2.58E+01	9.00E+00	9.16E+00	9.00E+00
	σ_f	2.57E−10	2.04E+00	2.10E+01	4.76E+00	3.66E−07	8.89E+00	9.64E+00	1.21E−05	8.96E−01	1.08E−02
	f_{Best}	**9.00E+00**	**9.00E+00**	1.75E+01	2.42E+01	**9.00E+00**	**9.00E+00**	9.07E+00	**9.00E+00**	**9.00E+00**	**9.00E+00**
	f_{Worst}	9.00E+00	1.72E+01	1.31E+02	4.28E+01	9.00E+00	3.88E+01	4.05E+01	9.00E+00	1.39E+01	9.06E+00
f_{21}	$\bar{f}$	1.49E+01	1.48E+01	2.30E+01	2.04E+01	1.08E+01	1.00E+01	1.02E+01	1.00E+01	1.00E+01	1.00E+01
	σ_f	2.29E+00	9.84E+00	1.21E+01	2.61E+00	9.49E−01	1.22E−05	1.34E−01	1.75E−05	2.34E−12	7.76E−06
	f_{Best}	1.14E+01	**1.00E+01**	1.04E+01	1.62E+01	**1.00E+01**	**1.00E+01**	1.01E+01	**1.00E+01**	**1.00E+01**	**1.00E+01**
	f_{Worst}	2.22E+01	5.50E+01	6.68E+01	2.65E+01	1.32E+01	1.00E+01	1.07E+01	1.00E+01	1.00E+01	1.00E+01
f_{22}	$\bar{f}$	9.00E+00	9.00E+00	6.06E+01	4.60E+01	9.00E+00	1.45E+01	3.39E+01	9.00E+00	9.00E+00	9.00E+00
	σ_f	7.36E−09	4.79E−03	1.79E+01	5.56E+00	1.63E−05	8.21E+00	1.46E+01	4.26E−07	1.54E−14	0.00E+00
	f_{Best}	**9.00E+00**	**9.00E+00**	2.56E+01	3.58E+01	**9.00E+00**	**9.00E+00**	9.15E+00	**9.00E+00**	**9.00E+00**	**9.00E+00**
	f_{Worst}	9.00E+00	9.03E+00	1.10E+02	5.66E+01	9.00E+00	3.54E+01	5.64E+01	9.00E+00	9.00E+00	9.00E+00
f_{23}	$\bar{f}$	−2.23E+01	−2.13E+01	−1.14E+01	−1.59E+01	−2.21E+01	−2.15E+01	−2.13E+01	−1.48E+01	−2.19E+01	−2.03E+01
	σ_f	2.12E−02	4.53E+00	2.12E+01	1.95E+00	1.33E−01	5.93E−01	6.78E−01	1.95E+00	3.00E−01	9.33E−01
	f_{Best}	**−2.23E+01**	**−2.23E+01**	−2.18E+01	−1.95E+01	**−2.23E+01**	−2.22E+01	−2.21E+01	−1.77E+01	−2.22E+01	−2.17E+01
	f_{worst}	−2.22E+01	2.70E+00	5.16E+01	−8.38E+00	−2.17E+01	−2.04E+01	−2.03E+01	−1.09E+01	−2.10E+01	−1.69E+01

shown. According to the table, NCCO obtains dominant performance only in function f_{19}. NCCO only declines to ABC, DE, and DS in function f_{23}. For functions $f_{20} - f_{22}$, NCCO works similar to its opponents.

Regarding to Table 7.14, NCCO provides more reliable statistical outcomes than its competitors in function f_{19}. Nevertheless, in functions $f_{20} - f_{22}$, the performance of all the tested algorithms is considerably similar. In function f_{23}, almost all the competitors beat NCCO since the optimal value for this kind of functions is located near the limits of the search space.

The following analysis includes the evaluation of hybrid function in $n = 30$ dimensions. In this instance, Table 7.15 shows that the performance of the tested methods are quite similar as those contained in Table 7.13. Consequently, in an attempt to give a clear conclusion of hybrid scenarios, the subsequent analysis is carried using n = 100 dimensions.

The statistical outcomes summarized in Table 7.16 hint that ABC, DE, SCA and CMA-ES, as well as NCCO, are competent since they obtain the optimal value in functions $f_{20} - f_{22}$. Nevertheless, CMA-ES and DE lead the measurement results in functions f_{21} and f_{23}.

Table 7.17 summarizes the results achieved regarding $n = 100$ dimensions over hybrid functions. The experimental results confirm that the NCCO algorithm is competent in beating its opponents considering functions presenting high multimodality. NCCO achieves exceptional outcomes for functions $f_{19} - f_{22}$. Nevertheless, in function f_{23}, ABC strikes the proposed method. It can be inferred that NCCO achieves better results when analyzing a high-dimensional search space, although it presents some disadvantages in functions where the optimum is located in the boundaries of the search space.

Regarding Table 7.18, ABC gets the best results for function f_{23}, while NCCO defeats its rivals in functions $f_{19} - f_{22}$. Although some tested algorithms strike the NCCO method in some functions, NCCO preserves scalability in low and high dimensions for the majority of the test functions because of the versatility of its operators, which increases population diversity in the evolutionary phases.

When examining all of the experimental tests, it is visible that NCCO carries the advantage of local consensus agreements to deliver superior performance than the rest of the algorithms. Since most of the test functions include high dimensionality and high multimodality, test results hint that the hybridization among the double pair of operators encourages a balance among evolutionary phases. With such features, many problems generally found by classical EC algorithms are raised. Nevertheless, NCCO exhibits some deficiencies when judging low-dimensional multimodal or hybrid surfaces with high multi-modality. Such objective functions lead to creating zero movement or small perturbations in the location of the individuals, because of the mathematical definition of the bias vectors. Under these conditions, the search procedure will converge into a non-optimal value. Additionally, NCCO exhibits some problems in functions where the global optimum is placed at the borders of the search space, or where a non-symmetrical shape surrounds the optimal solution. These weaknesses are affected by the characterization of the consensus formulations, since the consensus agreements may lead agents to be positioned outside the search

Table 7.14 p-values produced by Wilcoxon rank sum test over the averaged fitness value from Table 7.13

NCCO versus	ABC	DE	PSO	CS	DS	MFO	MVO	SCA	CMA-ES
f_{19}	6.64E−09▲	6.64E−09▲	6.64E−09▲	6.64E−09▲	6.64E−09▲	6.64E−09▲	6.64E−09▲	6.64E−09▲	6.64E−09▲
f_{20}	1.36E−01►	7.36E−02►	6.64E−09▲	6.64E−09▲	1.37E−01►	7.02E−02►	6.64E−09▲	3.62E−01►	9.22E−02►
f_{21}	2.99E−09▲	2.25E−01►	2.99E−09▲	2.99E−09▲	4.62E−01►	6.10E−02►	2.99E−09▲	7.71E−02►	1.59E−01►
f_{22}	8.37E−02►	1.73E−01►	5.66E−10▲	5.66E−10▲	1.63E−01►	1.62E−01►	5.66E−10▲	4.33E−01►	8.69E−02►
f_{23}	6.64E−09▼	1.18E−07▼	8.26E−07▼	2.12E−08▲	6.64E−09▼	1.87E−06▼	4.77E−04▼	9.81E−09▲	1.27E−08▼
▲	2	1	4	5	1	1	4	2	1
▼	1	1	1	0	1	1	1	0	1
►	2	3	0	0	3	3	0	3	3

Table 7.15 Results of the minimization of Table B.6 with $n = 30$

		ABC	DE	PSO	CS	DS	MFO	MVO	SCA	CMA-ES	NCCO
f_{19}	$\bar{f}$	1.27E−03	3.42E+00	2.14E+03	1.27E+04	1.67E+00	2.55E+04	1.73E+00	5.61E+00	4.31E−04	1.02E−30
	σ_f	6.32E−04	9.43E+00	1.78E+03	4.74E+03	1.06E+00	2.00E+04	3.23E−01	8.89E+00	1.24E−04	5.59E−30
	f_{Best}	5.54E−04	1.11E−09	8.63E+02	5.19E+03	3.72E−01	7.85E−01	1.09E+00	5.26E−03	2.09E−04	**1.15E−93**
	f_{Worst}	3.22E−03	4.64E+01	1.03E+04	2.52E+04	4.34E+00	9.04E+04	2.32E+00	4.32E+01	7.26E−04	3.06E−29
f_{20}	$\bar{f}$	2.94E+01	3.65E+01	2.60E+02	2.77E+02	4.57E+01	1.91E+02	1.53E+02	3.49E+01	3.13E+01	2.90E+01
	σ_f	4.57E−01	7.58E+00	5.00E+01	2.26E+01	1.09E+01	1.22E+02	4.02E+01	2.73E+01	4.42E+00	1.40E−01
	f_{Best}	**2.90E+01**	**2.90E+01**	1.50E+02	2.09E+02	3.09E+01	6.90E+01	1.07E+02	**2.90E+01**	**2.90E+01**	**2.90E+01**
	f_{Worst}	3.05E+01	5.28E+01	3.52E+02	3.15E+02	6.64E+01	5.13E+02	2.90E+02	1.79E+02	4.49E+01	2.98E+01
f_{21}	$\bar{f}$	3.87E+02	1.99E+02	1.22E+04	1.82E+04	4.22E+02	4.48E+02	8.43E+01	6.14E+04	3.20E+01	3.21E+01
	σ_f	5.20E+01	8.84E+01	1.41E+04	1.10E+04	1.17E+02	1.57E+02	2.42E+01	1.58E+05	3.02E−03	2.38E−01
	f_{Best}	2.89E+02	8.63E+01	7.90E+02	3.54E+03	2.05E+02	1.96E+02	5.07E+01	1.04E+02	**3.20E+01**	**3.20E+01**
	f_{Worst}	4.71E+02	5.18E+02	5.71E+04	5.00E+04	7.70E+02	1.11E+03	1.64E+02	7.35E+05	3.20E+01	3.33E+01
f_{22}	$\bar{f}$	2.90E+01	3.08E+01	3.79E+02	7.62E+02	3.66E+01	9.19E+02	2.01E+02	3.06E+01	3.22E+01	2.90E+01
	σ_f	4.04E−02	5.32E+00	1.03E+02	1.65E+02	7.30E+00	5.67E+02	1.10E+02	3.74E+00	5.53E+00	0.00E+00
	f_{Best}	**2.90E+01**	**2.90E+01**	2.02E+02	5.43E+02	2.99E+01	4.59E+01	9.26E+01	**2.90E+01**	**2.90E+01**	**2.90E+01**
	f_{Worst}	2.92E+01	5.15E+01	7.22E+02	1.16E+03	5.81E+01	2.25E+03	7.34E+02	4.96E+01	4.35E+01	2.90E+01
f_{23}	$\bar{f}$	−8.31E+01	−5.80E+01	2.19E+03	8.66E+05	−7.57E+01	2.36E+08	−6.88E+01	1.31E+03	−8.28E+01	−3.86E+01
	σ_f	2.13E−01	4.37E+01	8.84E+02	1.53E+06	5.15E+00	3.54E+08	2.26E+01	6.56E+03	3.11E−01	7.68E+00
	f_{Best}	−8.35E+01	**−8.38E+01**	6.71E+02	5.48E+03	−8.12E+01	−7.96E+01	−8.05E+01	−2.33E+01	−8.34E+01	−5.93E+01
	f_{Worst}	−8.25E+01	1.16E+02	3.73E+03	6.28E+06	−6.58E+01	1.28E+09	−1.14E+01	3.60E+04	−8.22E+01	−2.25E+01

Table 7.16 *p*-values produced by Wilcoxon rank sum test over the averaged fitness value from Table 7.15

NCCO versus	ABC	DE	PSO	CS	DS	MFO	MVO	SCA	CMA-ES
f_{19}	6.64E−09▲	6.64E−09▲	6.64E−09▲	6.64E−09▲	6.64E−09▲	6.64E−09▲	6.64E−09▲	6.64E−09▲	6.64E−09▲
f_{20}	1.36E−01►	1.06E−01►	6.64E−09▲	6.64E−09▲	6.64E−09▲	6.64E−09▲	6.64E−09▲	3.87E−01►	2.36E−01►
f_{21}	6.64E−09▲	6.64E−09▲	6.64E−09▲	6.64E−09▲	6.64E−09▲	6.64E−09▲	6.64E−09▲	6.64E−09▲	4.92E−01►
f_{22}	4.57E−01►	6.75E−01►	5.66E−10▲	5.66E−10▲	5.66E−10▲	5.66E−10▲	5.66E−10▲	2.54E−01►	1.33E−01►
f_{23}	6.64E−09▼	5.27E−03▼	6.64E−09▲	6.64E−09▲	6.64E−09▼	1.81E−03▼	1.84E−05▼	8.62E−09▲	6.64E−09▼
▲	2	2	5	5	4	4	4	3	1
▼	1	1	0	0	1	1	1	0	1
►	2	2	0	0	0	0	0	2	3

Table 7.17 Results of the minimization of Table B.6 with $n = 100$

		ABC	DE	PSO	CS	DS	MFO	MVO	SCA	CMA-ES	NCCO
f_{19}	$\bar{f}$	3.91E+01	1.54E+02	6.24E+04	1.00E+10	5.45E+03	1.56E+05	4.25E+24	8.82E+03	1.23E+01	7.03E−07
	σ_f	1.19E+02	1.74E+02	4.48E+04	0.00E+00	1.93E+03	4.22E+04	2.31E+25	7.03E+03	3.59E+00	2.87E−06
	f_{Best}	6.73E+00	7.33E+00	1.33E+04	1.00E+10	2.79E+03	8.28E+04	2.17E+05	8.37E+02	8.12E+00	**1.37E−41**
	f_{Worst}	6.65E+02	8.43E+02	1.76E+05	1.00E+10	1.05E+04	2.63E+05	1.26E+26	3.01E+04	2.42E+01	1.45E−05
f_{20}	$\bar{f}$	5.36E+02	1.86E+02	1.28E+03	1.41E+03	8.89E+02	2.54E+03	8.52E+02	7.71E+02	1.67E+02	9.90E+01
	σ_f	6.69E+01	2.69E+01	1.22E+02	7.47E+01	1.26E+02	3.95E+02	1.14E+02	3.37E+02	1.33E+01	9.91E−05
	f_{Best}	3.64E+02	1.40E+02	1.04E+03	1.27E+03	6.60E+02	1.80E+03	6.00E+02	1.38E+02	1.47E+02	**9.90E+01**
	f_{Worst}	7.06E+02	2.42E+02	1.55E+03	1.56E+03	1.09E+03	3.40E+03	1.09E+03	1.71E+03	2.01E+02	9.90E+01
f_{21}	$\bar{f}$	1.22E+05	4.10E+05	2.36E+06	6.99E+06	2.56E+06	5.10E+08	3.74E+03	4.62E+08	1.92E+03	1.18E+02
	σ_f	3.61E+05	9.17E+05	1.55E+06	3.11E+06	1.83E+06	3.16E+08	7.54E+02	2.22E+08	1.46E+02	2.19E+01
	f_{Best}	5.83E+03	9.59E+03	2.28E+05	1.21E+06	5.07E+05	1.39E+08	2.42E+03	7.59E+07	1.41E+03	**1.09E+02**
	f_{Worst}	1.84E+06	5.18E+06	6.85E+06	1.76E+07	5.98E+06	1.33E+09	5.06E+03	8.80E+08	2.17E+03	2.29E+02
f_{22}	$\bar{f}$	1.86E+02	1.34E+02	3.11E+03	1.00E+10	8.38E+02	5.49E+03	6.28E+22	7.39E+02	1.81E+02	9.90E+01
	σ_f	3.08E+01	1.71E+01	1.36E+03	0.00E+00	1.38E+02	1.39E+03	3.04E+23	3.44E+02	2.63E+01	0.00E+00
	f_{Best}	1.17E+02	9.96E+01	1.54E+03	1.00E+10	5.68E+02	3.19E+03	6.39E+03	2.07E+02	1.35E+02	**9.90E+01**
	f_{Worst}	2.43E+02	1.76E+02	5.98E+03	1.00E+10	1.11E+03	8.72E+03	1.66E+24	1.59E+03	2.32E+02	9.90E+01
f_{23}	$\bar{f}$	−2.70E+02	3.43E+03	1.93E+08	1.00E+10	5.50E+04	3.37E+09	1.89E+24	5.61E+07	−6.13E−02	−1.33E+01
	σ_f	4.70E+01	1.18E+04	2.75E+08	0.00E+00	1.28E+05	1.52E+09	1.03E+25	5.05E+07	4.71E+01	1.30E+01
	f_{Best}	**−2.87E+02**	−4.89E+01	3.12E+04	1.00E+10	4.98E+03	6.36E+08	1.85E+09	3.91E+06	−1.07E+02	−4.49E+01
	f_{Worst}	−2.42E+01	5.70E+04	8.79E+08	1.00E+10	6.03E+05	6.61E+09	5.65E+25	2.58E+08	7.70E+01	2.97E+01

Table 7.18 p-values produced by Wilcoxon rank sum test over the averaged fitness value from Table 7.17

NCCO versus	ABC	DE	PSO	CS	DS	MFO	MVO	SCA	CMA-ES
f_{19}	6.64E−09▲	6.64E−09▲	6.64E−09▲	5.66E−10▲	6.64E−09▲	6.64E−09▲	6.64E−09▲	6.64E−09▲	6.64E−09▲
f_{20}	6.64E−09▲	6.64E−09▲	6.64E−09▲	6.64E−09▲	6.64E−09▲	6.64E−09▲	6.64E−09▲	6.64E−09▲	6.64E−09▲
f_{21}	6.64E−09▲	6.64E−09▲	6.64E−09▲	6.64E−09▲	6.64E−09▲	6.64E−09▲	6.64E−09▲	6.64E−09▲	6.64E−09▲
f_{22}	5.66E−10▲	5.66E−10▲	5.66E−10▲	2.15E−11▲	5.66E−10▲	5.66E−10▲	5.66E−10▲	5.66E−10▲	5.66E−10▲
f_{23}	6.64E−09▼	1.67E−06▼	6.64E−09▲	5.66E−10▲	6.64E−09▲	6.64E−09▲	6.64E−09▲	6.64E−09▲	7.88E−07▼
▲	4	4	5	5	5	5	5	5	4
▼	1	1	0	0	0	0	0	0	1
►	0	0	0	0	0	0	0	0	0

space, twisting the sensed knowledge of the consensus arrangement, enabling future iterations to perform badly.

7.6.2 Computational Time and Convergence Analysis

The performance results of the comparison study in Sect. 7.6.1 are based on the robustness obtained by the evaluation of the test functions for each evolutionary method. These results cannot fully characterize the performance of evolutionary techniques. Consequently, computational effort and convergence tests must be incorporated. In this segment, the computational time and convergence analysis of the evolutionary approaches is detailed. The principal aim of these experiments is to judge two different criteria: the average computational effort needed for the execution of the benchmark set, and how fast each algorithm leads the optimal solution. Both criteria are crucial performance indexes because a fast convergence does not ever imply acceptable performance.

In this test, the performance of each EC method is judged versus all the test functions from Tables B.4, B.5 and B.6, using 30 dimensions and the stop criterion definition as $NFE = 25{,}000$. To build the computational time table and converge graphs, each test function is evaluated on 30 independent executions. Table 7.19 summarizes the average execution time for each EC technique.

Regarding the table, the proposed NCCO needs around the same quantity of time to perform each function as the competitors. Nonetheless, for functions f_3, f_4, f_5 and $f_{20} - f_{23}$, NCCO surrenders the best execution values. For the rest of the test functions, the DS scheme beats the rest of EC methods. On the other hand, the convergence information was chosen regarding the average fitness value from Tables 7.3, 7.9 and 7.15.

Figures 7.5, 7.6 and 7.7, display the converge plots for unimodal, multimodal and hybrid test functions. As presented in Fig. 7.5, NCCO converges to the optimum solution faster than its competitors in functions f_1, f_2, $f_4 - f_8$. The exception to this rule is CMA-ES, which defeats the proposed method in function f_3. These plots prove the evolutionary operators of NCCO. Since NCCO includes two pairs of evolutionary operators, it is capable of delivering the optimum value quicker than its adversaries. Additionally, Fig. 7.6 records that NCCO gathers faster toward the optimal value in functions $f_9 - f_{11}$ and $f_{13} - f_{15}$, whereas CMA-ES strikes the proposed NCCO method in functions f_{12}, $f_{16} - f_{18}$. Ultimately, Fig. 7.7 confirms that for functions $f_{19} - f_{22}$ NCCO gives the optimum value employing a few number function evaluations.

The data demonstrates that the converging velocity of NCCO when judging the set of benchmark functions is quicker than the converging velocity of the competitors. The outstanding performance of NCCO is produced by its well-adjusted operators since they promote scalability, population diversity, and robustness.

Table 7.19 Computational time used in the experimental study

	ABC	DE	PSO	CS	DS	MFO	MVO	SCA	CMA-ES	NCCO
f_1	9.3650 (0.0099)	9.6971 (0.0160)	9.6575 (0.0534)	9.2804 (0.0054)	**9.1033** (0.0100)	9.2621 (0.0130)	9.2341 (0.0165)	9.1525 (0.0107)	10.4917 (0.0340)	9.2484 (0.2659)
f_2	9.3939 (0.0049)	9.7346 (0.0093)	9.6419 (0.0189)	9.3107 (0.0045)	**9.1265** (0.0020)	9.2876 (0.0033)	9.2506 (0.0042)	9.1815 (0.0047)	10.5242 (0.0138)	10.3050 (0.0977)
f_3	9.5704 (0.0077)	9.9151 (0.0112)	9.8422 (0.0312)	9.4840 (0.0066)	9.3031 (0.0048)	9.4758 (0.0042)	9.4240 (0.0066)	9.3964 (0.0250)	10.7225 (0.0129)	**9.1929** (0.0466)
f_4	9.3570 (0.0077)	9.6935 (0.0086)	9.5878 (0.0121)	9.2819 (0.0029)	9.1033 (0.0010)	9.2609 (0.0031)	9.2313 (0.0060)	9.1593 (0.0136)	10.4828 (0.0120)	**8.9838** (0.0524)
f_5	9.4108 (0.0044)	9.7542 (0.0108)	9.6648 (0.0327)	9.3371 (0.0048)	**9.1403** (0.0043)	9.3094 (0.0047)	9.2839 (0.0128)	9.1961 (0.0136)	10.5457 (0.0157)	10.9823 (0.3576)
f_6	9.3631 (0.0061)	9.6925 (0.0089)	9.6115 (0.0288)	9.2859 (0.0041)	9.1048 (0.0011)	9.2690 (0.0030)	9.2403 (0.0155)	9.1521 (0.0052)	10.4879 (0.0147)	**8.9597** (0.0461)
f_7	9.4124 (0.0067)	9.7510 (0.0097)	9.6656 (0.0343)	9.3394 (0.0052)	**9.1497** (0.0031)	9.3188 (0.0036)	9.2864 (0.0145)	9.2050 (0.0063)	10.5592 (0.0136)	10.8941 (0.0542)
f_8	9.3332 (0.0069)	9.6636 (0.0113)	9.6470 (0.0488)	9.2658 (0.0039)	**9.0802** (0.0022)	9.2396 (0.0032)	9.2176 (0.0066)	9.1314 (0.0081)	10.4983 (0.0104)	10.3364 (0.0748)
f_9	9.3705 (0.0052)	9.7027 (0.0096)	9.6529 (0.0510)	9.3061 (0.0036)	**9.1132** (0.0025)	9.2770 (0.0046)	9.2539 (0.0221)	9.1700 (0.0093)	10.5366 (0.0291)	10.3728 (0.0599)
f_{10}	9.3748 (0.0043)	9.7078 (0.0079)	9.6319 (0.0439)	9.2948 (0.0047)	**9.1215** (0.0039)	9.2710 (0.0034)	9.2749 (0.0256)	9.1626 (0.0128)	10.5072 (0.0139)	10.8546 (0.0509)
f_{11}	9.4055 (0.0052)	9.7464 (0.0093)	9.6869 (0.0387)	9.3310 (0.0048)	**9.1479** (0.0065)	9.2983 (0.0028)	9.3019 (0.0079)	9.1902 (0.0165)	10.5518 (0.0259)	10.9628 (0.6366)
f_{12}	9.4666 (0.0081)	9.7998 (0.0088)	9.7468 (0.0387)	9.3592 (0.0060)	**9.1882** (0.0049)	9.3441 (0.0047)	9.3229 (0.0092)	9.2260 (0.0113)	10.5905 (0.0125)	10.8305 (0.0623)

(continued)

Table 7.19 (continued)

	ABC	DE	PSO	CS	DS	MFO	MVO	SCA	CMA-ES	NCCO
f_{13}	9.3444 (0.0047)	9.6605 (0.0075)	9.6014 (0.0455)	9.2692 (0.0048)	**9.0867** (0.0043)	9.2439 (0.0028)	9.1844 (0.0055)	9.1332 (0.0070)	10.4663 (0.0109)	10.8050 (0.0839)
f_{14}	9.3457 (0.0064)	9.6809 (0.0082)	9.6289 (0.0567)	9.2757 (0.0085)	**9.0933** (0.0022)	9.2529 (0.0037)	9.2269 (0.0051)	9.1458 (0.0044)	10.4704 (0.0188)	10.4231 (0.1723)
f_{15}	9.3408 (0.0055)	9.6754 (0.0087)	9.5769 (0.0266)	9.2659 (0.0129)	**9.0836** (0.0045)	9.2439 (0.0041)	9.2195 (0.0059)	9.1420 (0.0042)	10.4611 (0.0131)	10.2987 (0.4409)
f_{16}	9.4310 (0.0066)	9.7587 (0.0129)	9.6612 (0.0231)	9.3396 (0.0091)	**9.1586** (0.0012)	9.3100 (0.0035)	9.2317 (0.0052)	9.2020 (0.0079)	10.5873 (0.0141)	9.2305 (0.2192)
f_{17}	9.7091 (0.0100)	10.1207 (0.0922)	10.0217 (0.0482)	9.5620 (0.0082)	**9.4034** (0.0179)	9.5799 (0.0101)	9.5535 (0.0324)	9.4646 (0.0154)	10.8527 (0.0219)	10.5970 (0.1127)
f_{18}	9.7095 (0.0085)	10.0631 (0.0119)	10.0113 (0.0519)	9.5575 (0.0057)	**9.4079** (0.0147)	9.5831 (0.0062)	9.5562 (0.0390)	9.4644 (0.0154)	10.8517 (0.0151)	10.6636 (0.3350)
f_{19}	9.5056 (0.0065)	9.8290 (0.0293)	9.7279 (0.0245)	9.3793 (0.0055)	**9.2010** (0.0033)	9.3718 (0.0053)	9.3457 (0.0212)	9.2556 (0.0157)	10.6226 (0.0172)	9.3509 (0.2241)
f_{20}	9.6069 (0.0081)	9.9473 (0.0110)	9.8648 (0.0322)	9.4914 (0.0069)	9.2932 (0.0037)	9.4800 (0.0063)	9.4706 (0.0200)	9.3551 (0.0106)	10.7558 (0.0315)	**9.2372** (0.0690)
f_{21}	10.1177 (0.0123)	10.4880 (0.0144)	10.5070 (0.0685)	9.9497 (0.0116)	9.7622 (0.0090)	10.0082 (0.0152)	9.9693 (0.0343)	9.8645 (0.0251)	11.2874 (0.0179)	**9.7170** (0.1030)
f_{22}	9.7179 (0.0093)	10.0924 (0.0098)	10.0074 (0.0296)	9.6111 (0.0078)	9.4541 (0.0151)	9.6083 (0.0100)	9.6196 (0.0475)	9.4950 (0.0231)	10.9139 (0.0141)	**9.3578** (0.0387)
f_{23}	9.9643 (0.0089)	10.3992 (0.0546)	10.3061 (0.0375)	9.8117 (0.0086)	9.6201 (0.0077)	9.8569 (0.0088)	9.8618 (0.0718)	9.7532 (0.0430)	11.1580 (0.0136)	**9.5389** (0.0616)

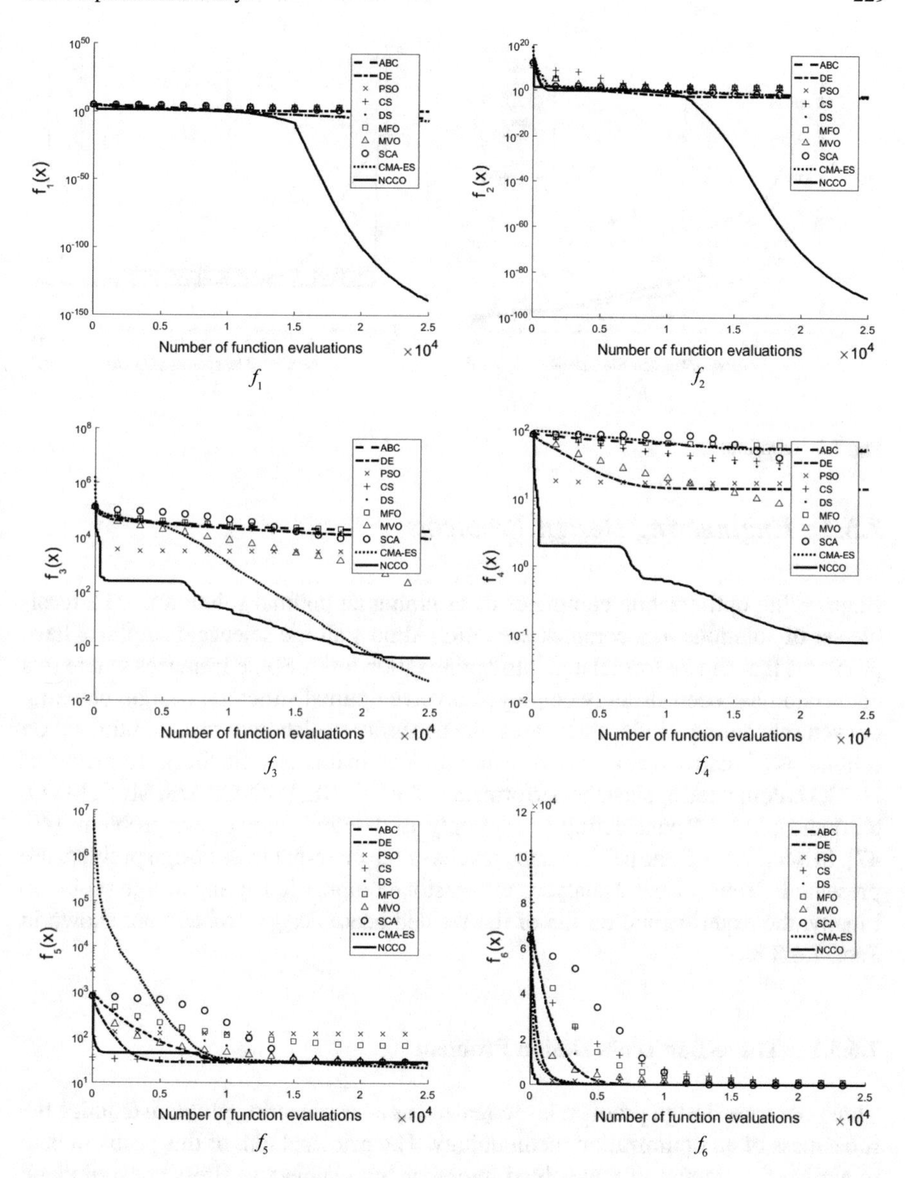

Fig. 7.5 Convergence test graph for functions of Table B.4

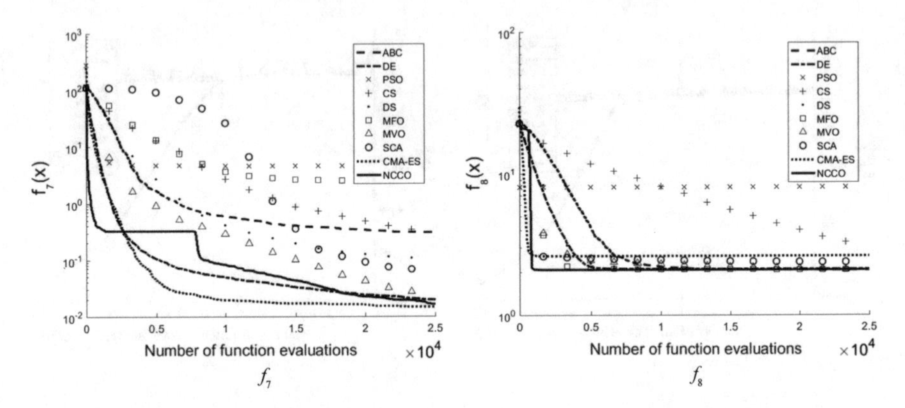

Fig. 7.5 (continued)

7.6.3 Engineering Design Problems

Engineering optimization comprises determining an optimal solution from a feasible set of solutions in a complex scenario. Almost all the science disciplines have problems that can be formulated into optimization tasks. Some instances where this conversion has been strongly employed involve optimal structural design, planning or even scheduling. Under such cases, the necessity to elaborate strong optimization routines is a broad and attractive research subject. In this section, the performance of NCCO is compared against the performance of ABC, DE, PSO, CS, DS, MFO, MVO, SCA and CMA-ES considering 3 commonly used optimization design problems [46, 47]. In Sect. 7.6.3.1, the performance results of the three-bar truss design problem are presented. Section 7.6.3.2 analyzes the tension/compression spring design problem. Finally, the experimental results of the welded beam design problem are shown in Sect. 7.6.3.3.

7.6.3.1 Three-Bar Truss Design Problem

Three-bar truss design problem is recognized as an engineering problem to judge the robustness of an optimization methodology. The principal aim of this problem is to minimize the amount of a weighted three-bar truss subject to stress constraints on each of the truss segments by modifying cross-sectional areas x_1, x_2. The problem examines an objective function with three inequalities and two decision variables as specified in Table B.7. The numeric simulation of the evaluation for the constrained objective function is accomplished regarding the parameter configuration detailed in Sect. 7.6.1 using 1000 iterations as stop criterion. The numerical results are displayed in Table 7.20 and Fig. 7.8 displays the three-bar truss.

Table 7.20, designates that the proposed method achieves outcomes considerably similar to the results of CS, DS, and MVO for the three-bar truss problem, evaluated

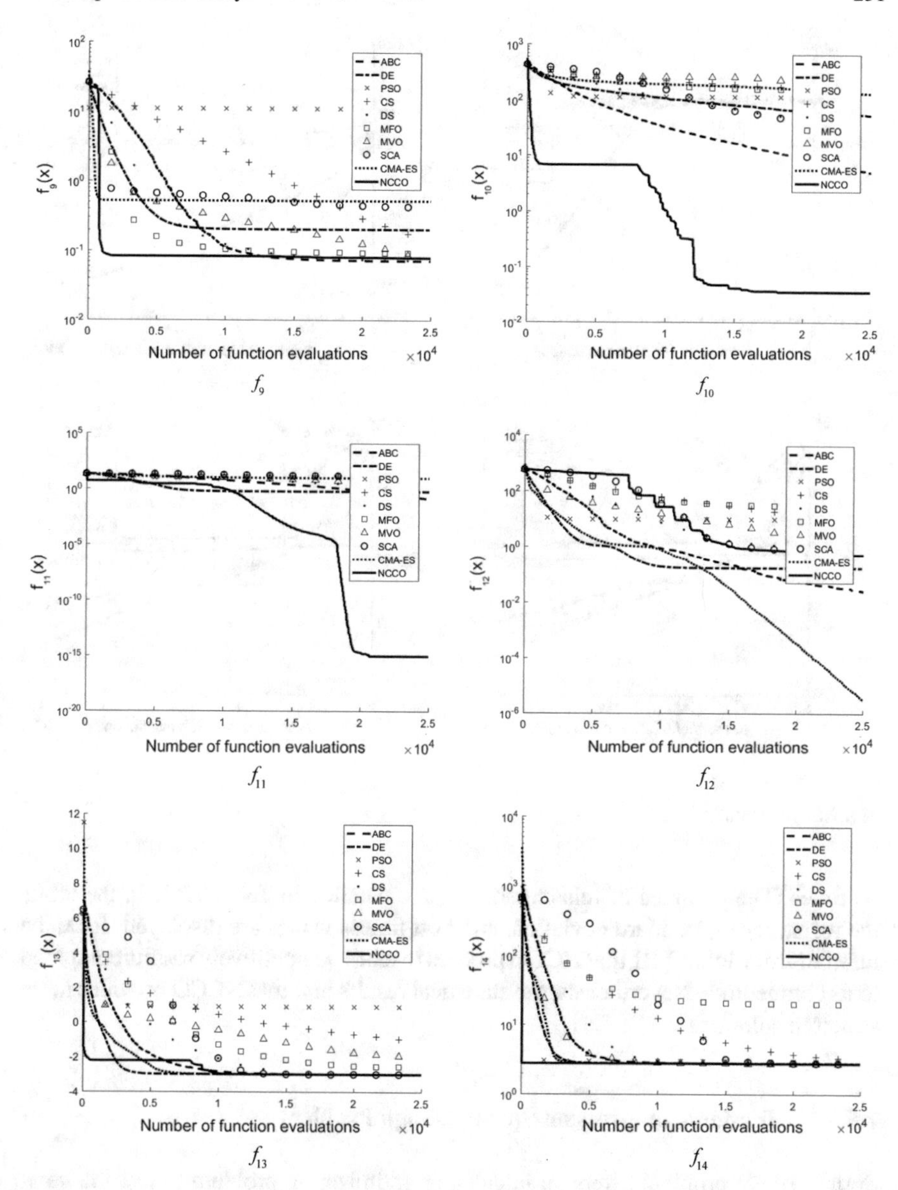

Fig. 7.6 Convergence test graph for functions of Table B.5

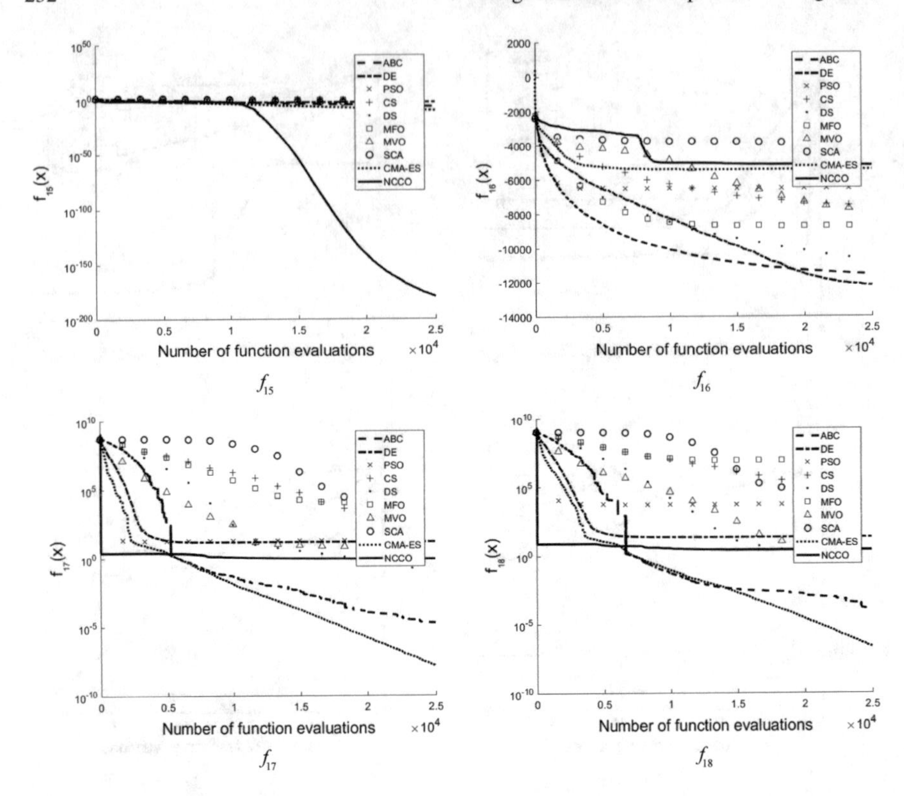

Fig. 7.6 (continued)

30 times. The *p*-values of this experiment are detailed in Table 7.21. In the table, the worst, mean, standard deviation, and best fitness values are displayed. It can be inferred from Table 7.21 that NCCO process reaches competitively results compared to its competitors. Nevertheless, the statistical results hint that NCCO provides more scattered solutions.

7.6.3.2 Tension/Compression Spring Design Problem

Spring design problem offers an interesting optimization problem to test the flexibility of an optimization technique. The main objective is to minimize the weight tension/compression spring. The problem is described in Table B.8. In this optimization problem, three decision variables: wire diameter $w(x_1)$, mean coil diameter $d(x_2)$ and the number of active coils $L(x_3)$ are involved in the process while it contains one linear constraint and three non-linear inequality constraints. The experimental outcomes are exhibited in Table 7.22 and Fig. 7.9 represents a schematic of the problem.

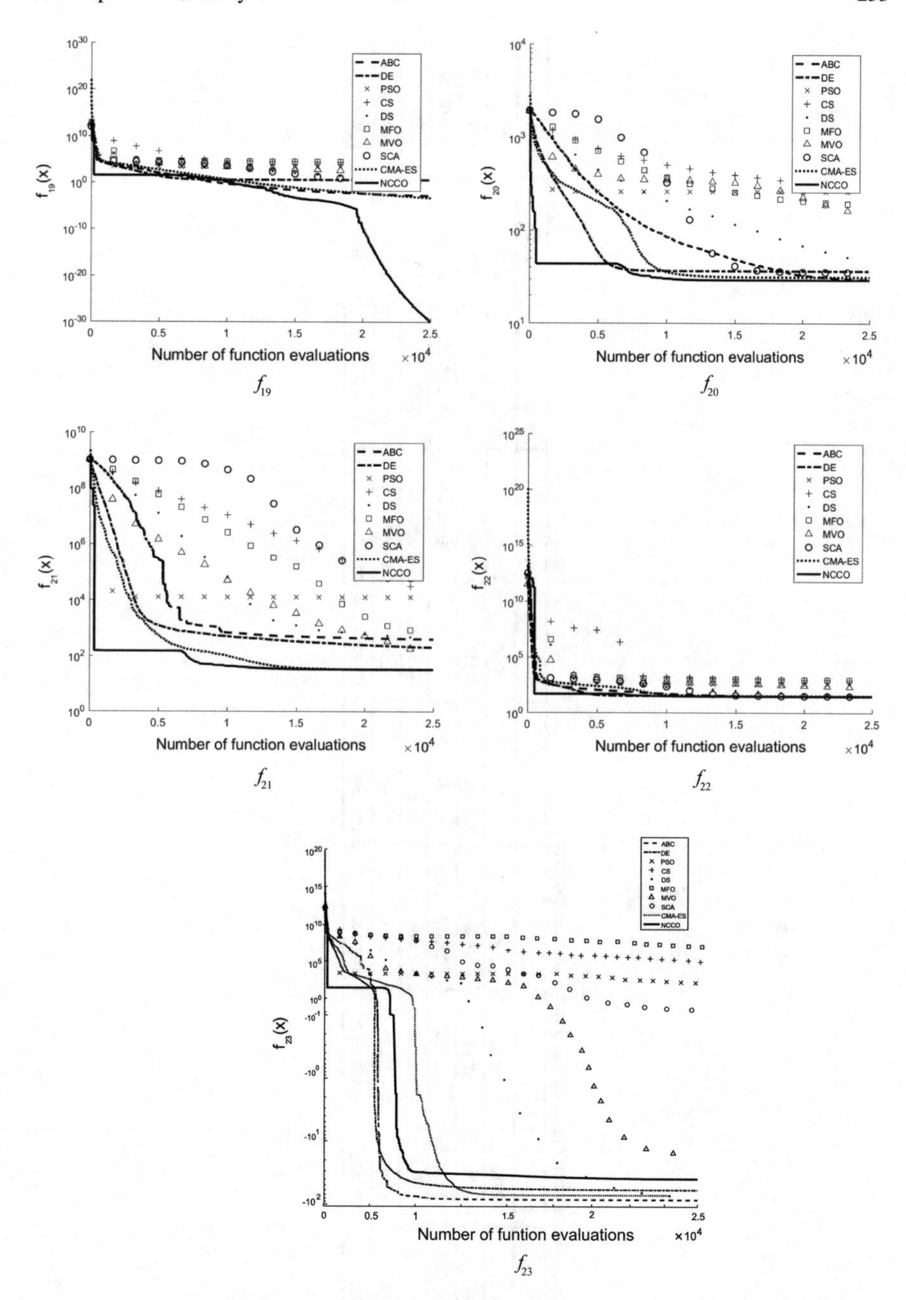

Fig. 7.7 Convergence test graph for functions of Table B.6

Table 7.20 Results of the minimization of the three-bar truss problem

Parameter	ABC	DE	PSO	CS	DS	MFO	MVO	SCA	CMA-ES	NCCO
x_1	0.8643	0.8708	0.8573	0.8698	0.8700	0.8702	0.8697	0.8730	0.8721	0.8699
x_2	0.2278	0.2149	0.2710	0.2166	0.2162	0.2157	0.2167	0.2101	0.2122	0.2164
$g(x_1)$	−1.49E−04	−3.60E−04	−0.0275	−2.2204E−16	−6.2774E−10	0	−2.5611E−07	−5.7400E−05	−2.43E−04	−5.8363E−08
$g(x_2)$	−1.6858	−1.7029	−1.6396	−1.7005	−1.7010	−1.7018	−1.7005	−1.7091	−1.7064	−1.7008
$g(x_3)$	−0.3143	−0.2975	−0.3878	−0.2995	−0.2990	−0.2982	−0.2995	−0.2909	−0.2938	−0.2992
$f(x)$	279.7452	279.7472	280.5426	**279.7245**	**279.7245**	279.7246	**279.7245**	279.7312	279.7411	**279.7245**

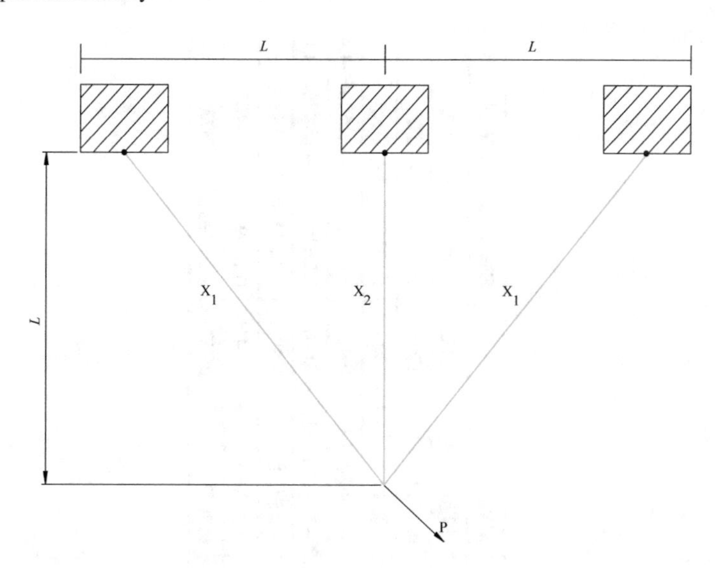

Fig. 7.8 Three-bar truss problem

Table 7.21 Statistical results for the three-bar truss problem

Algorithm	Worst	Mean	Std	Best
ABC	279.7849	279.7372	0.0125	279.7452
DE	280.1867	279.8096	0.0902	279.7472
PSO	282.8226	93.2079	29.9639	280.5426
CS	279.7245	279.7245	1.3843E−13	**279.7245**
DS	280.0971	279.7484	0.0677	**279.7245**
MFO	280.1945	279.7763	0.1028	279.7246
MVO	279.7270	279.7251	5.7728E−04	**279.7245**
SCA	282.8427	281.4059	1.5624	279.7312
CMA-ES	279.7884	191.7808	88.7932	279.7411
NCCO	279.7585	279.7341	0.0106	**279.7245**

Table 7.22 exhibits that the NCCO method operates similarly to CS, DS, MFO, MVO, and SCA, over 30 independent executions for the spring problem. Table 7.23 shows the p-values. From Table 7.23, it can be recognized that NCCO provides more steady results than its opponents.

7.6.3.3 Welded Beam Design Problem

Welded beam problem describes a complex optimization task whose aim is to obtain the minimal cost of a welded beam. The graphical description of the welded beam is

Table 7.22 Results of the minimization of the tension/compression spring problem

Parameter	ABC	DE	PSO	CS	DS	MFO	MVO	SCA	CMA-ES	NCCO
w	0.0500	0.0500	0.0516	0.0500	0.0500	0.0500	0.0500	0.0500	1.6858	0.0500
d	0.4532	0.4278	0.3560	0.4800	0.4800	0.4800	0.4800	0.4796	0.9031	0.4800
L	4.8539	5.8778	11.3295	4.0563	4.0577	4.0563	4.0580	4.0806	6.4315	4.0580
$g(x_1)$	−0.0070	−0.0257	−0.000006	−1.5477E−13	−4.8364E−06	−0.0025	−6.6620E−05	−0.0035	1.0000	−4.1608E−04
$g(x_2)$	−0.1001	−0.1896	−0.000013	−2.0761E−14	−2.1545E−04	−1.6653E−15	−2.1897E−04	−0.0015	2.2629	−5.9926E−06
$g(x_3)$	−6.0440	−5.5282	−4.0523	−6.5135	−6.5126	−6.4950	−6.5121	−6.4810	−44.1382	−6.5103
$g(x_4)$	−0.6645	−0.6815	−0.7282	−0.6467	−0.6467	−0.6467	−0.6467	−0.6469	0.7259	−0.6467
$f(x)$	0.0078	0.0084	0.0127	**0.0073**	**0.0073**	**0.0073**	**0.0073**	**0.0073**	1.00E+06	**0.0073**

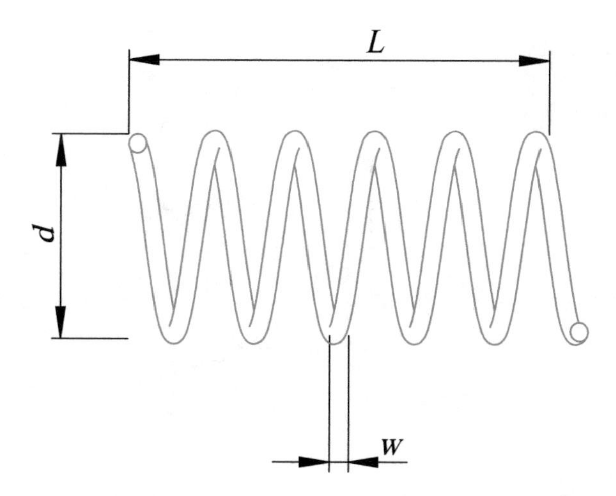

Fig. 7.9 Tension/compression spring problem

Table 7.23 Statistical results for the tension/compression spring problem

Algorithm	Worst	Mean	Std	Best
ABC	0.0084	0.0076	0.0002	0.0078
DE	1.00E+06	2.00E+05	4.07E+05	0.0084
PSO	1.00E+06	1.67E+05	3.79E+05	0.0127
CS	0.0073	0.0073	2.3681E−12	**0.0073**
DS	0.0081	0.0074	2.2511E−04	**0.0073**
MFO	0.0073	0.0073	1.5149E−07	**0.0073**
MVO	1.00E+06	4.0000E+05	4.9827E+05	**0.0073**
SCA	0.0077	0.0074	9.4466E−05	**0.0073**
CMA-ES	1.00E+06	6.00E+05	4.98E+05	1.00E+06
NCCO	0.0073	0.0073	6.4271E−13	**0.0073**

presented in Fig. 7.10 and its mathematical definition is exposed in Table B.9. The design method requires four decision variables: the width $h(x_1)$, the length $l(x_2)$ of the welded area, the depth $t(x_3)$ and the thickness $b(x_4)$. The experimental results are manifested in Table 7.24.

Table 7.24 confirms that the NCCO method provides similar results to CS and MFO, over 30 independent executions for this design problem. Table 7.25 manifests the p-values results achieved by the rest of tested evolutionary methods. The results reveal that NCCO delivers more steady results than its competitors.

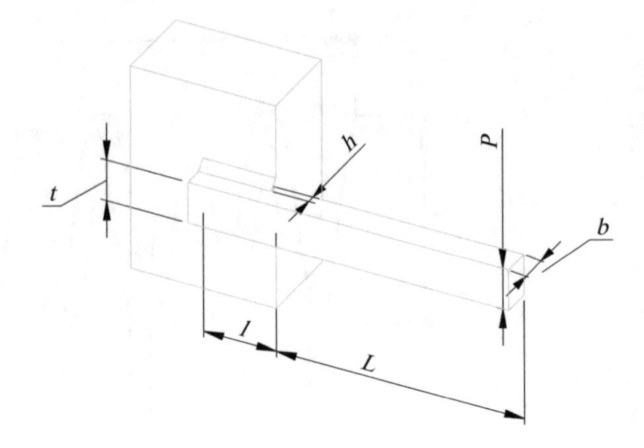

Fig. 7.10 Welded beam problem

7.7 Conclusions

The combination of simple local rules adopted from a swarm of coordinated agents within a multi-agent system model complex intercommunications where the collaboration and interaction among agents defeat the constraints of each agent to solve certain tasks. In this chapter, a reactive model based on a swarm strategy for solving continuous optimization problems is presented. The procedure, called NCCO, combines the simplicity of local consensus agreements with neighborhood decision-based movement to lead the search strategy.

In the optimization method, NCCO identifies each agent of the population to be a self-sufficient agent that works based on neighborhood agreements, to lead the search during the entire optimization process. NCCO employs a double pair of evolutionary operations to adjust the position, velocity, and acceleration vectors of each agent. The first pair of operators; separation and alignment, explore the entire search space. These operations enable the search procedure to be carried into wider search zones until each agent has positioned far apart the rest of agents, while they adjust their heading vector towards promissory search areas. The second pair of operators; cohesion and seek, locally explores and exploits promissory zones, collecting all agents into small groups. In the starting point, this procedure considers each agent as a singleton cluster; later, based on the sensed neighborhood, each agent executes a movement decided to organize with the most related agents in its area. This process proceeds until all agents relate to a cluster. When all the clusters are formed, the final process of this operation intends to converge the clusters into the best position being so far.

The NCCO algorithm has been formulated as an alternative optimization technique since it enhances synergy between operators for its process, as well as leaderless movement judgments. NCCO defines two reactive models for the exploration phase and two reactive models for the exploitation phase. These double pairs help to maintain a complex search strategy, based on neighborhood agreements. NCCO does not contemplate an individual leader to lead the search. This mechanism permits each

Table 7.24 Results of the minimization of the welded beam problem

Parameter	ABC	DE	PSO	CS	DS	MFO	MVO	SCA	CMA-ES	NCCO
h	0.1993	0.1626	0.8120	0.2057	0.2057	0.2057	0.2045	0.2004	0.2031	0.2057
l	3.8115	6.8239	6.4319	3.4705	3.4706	3.4705	3.5040	3.7893	3.5430	3.4705
t	8.6442	8.7650	9.2837	9.0366	9.0385	9.0366	9.0375	9.0748	9.0335	9.0366
b	0.2259	0.2346	0.7008	0.2057	0.2057	0.2057	0.2058	0.2062	0.2062	0.2057
$g(x_1)$	−81.6604	−2977.1639	−1.16E+04	−0.0062	−0.0322	−0.0062	−23.2984	−616.4397	−44.5785	−0.0062
$g(x_2)$	−135.7300	−2035.9101	−2.17E+04	−0.1334	−14.1393	−0.1334	−9.1686	−315.5598	−44.6635	−0.1188
$g(x_3)$	−0.0265	−0.0720	0.1112	−1.2836E−06	−5.0225E−05	−1.2836E−06	−0.0012	−0.0058	−0.0030	−1.1836E−06
$g(x_4)$	−3.3229	−2.9372	1.4640	−3.4330	−3.4326	−3.4406	−3.4297	−3.3945	−3.4237	−3.4330
$g(x_5)$	−0.0743	−0.0376	−0.6870	−0.0807	−0.0807	−0.0807	−0.0795	−0.0754	−0.0781	−0.0807
$g(x_6)$	−0.2350	−0.2361	−0.2461	−0.2355	−0.2356	−0.2355	−0.2355	−0.2358	−0.2355	−0.2355
$g(x_7)$	−1706.7356	−2718.3738	−2.35E+05	−0.0157	−1.9242	−0.0157	−2.5351	−55.4759	−38.0282	−0.0069
$f(x)$	1.8403	2.2593	2.0118	**1.7249**	1.7252	**1.7249**	1.7278	1.7694	1.7334	**1.7249**

Table 7.25 Statistical results of the welded beam problem

Algorithm	Worst	Mean	Std	Best
ABC	2.0848	1.9269	0.0849	1.8403
DE	3.6161	2.6481	0.3445	2.2593
PSO	3.3931	2.1544	0.3594	2.0118
CS	1.7250	1.7249	2.2364E−05	**1.7249**
DS	2.2552	1.8165	0.1205	1.7252
MFO	2.3171	1.7854	0.1282	**1.7249**
MVO	1.8169	1.7412	0.0182	1.7278
SCA	1.9711	1.8519	0.0493	1.7694
CMA-ES	1.11E+172	3.69E+170	NA	1.7334
NCCO	1.7249	1.7249	5.6400E−06	**1.7249**

NA means not available

individual to explore broadly and exploit narrower promissory search zones without following into false-positive positions.

The performance of the proposal has been analyzed to the performance of 9 state-of-art evolutionary methodologies: Particle Swarm Optimization (PSO), Differential Search (DS), Artificial Bee Colony (ABC), Cuckoo Search (CS), Differential Evolution (DE), Moth-Flame Optimization (MFO), Sine Cosine Algorithm (SCA), Multi-Verse Optimizer (MVO), and the Covariance Matrix Adaptation Evolutionary Strategy (CMA-ES). In the experimental comparison, a collection of 23 test functions as well as three real-world optimization problems have been solved and statistically analyzed.

The performance of the algorithms considering the evaluation of the test functions have demonstrated that NCCO achieves the best results in terms of efficiency, consistency, and robustness. This study was statistically verified by the Wilcoxon rank sum test. In the case of the real-world optimization problems, NCCO achieves considerably similar results than its competitors.

For future improvement of the proposed method, NCCO can be modified to solve large scale optimization problems. In addition, NCCO can be extended to include several niching capabilities to construct clusters between agents for classification tasks. On the other hand, NCCO can be reimplemented within a parallel scenario to decrease the computational time needed for real-time optimization tasks. Furthermore, NCCO can be employed to solve energy overconsumption issues commonly encountered in electrical networks. Moreover, NCCO can be helpful in medical image segmentation, where images can be used to detect some kind of diseases.

References

1. X.-S. Yang, *Engineering Optimization: An Introduction with Metaheuristic Application* (Wiley, Hoboken, USA, 2010)
2. P.M. Pardalos, H.E. Romeijn, H. Tuy, Recent developments and trends in global optimization. J. Comput. Appl. Math. **124**, 209–228 (2000)
3. E. Cuevas, J. Gálvez, S. Hinojosa, O. Avalos, D. Zaldívar, M. Pérez-Cisneros, A comparison of evolutionary computation techniques for IIR model identification. J. Appl. Math. **2014** (2014)
4. Y. Ji, K.-C. Zhang, S.-J. Qu, A deterministic global optimization algorithm. Appl. Math. Comput. **185**, 382–387 (2007)
5. J. Kennedy, R.C. Eberhart, Particle swarm optimization. Proc. IEEE Int. Conf. Neural Netw. **4**, 1942–1948 (1995)
6. J.H. Holland, *Adaptation in Natural and Artificial Systems* (University of Michigan Press, 1975)
7. D.E. Goldberg, *Genetic Algorithms in Search, Optimization and Machine Learning* (Addison-Wesley, 1989)
8. E. Rashedi, H. Nezamabadi-Pour, S. Saryazdi, GSA: a gravitational search algorithm. Inf. Sci. (NY) **179**(13), 2232–2248 (2009)
9. Ş.I. Birbil, S.-C. Fang, An electromagnetism-like mechanism for global optimization. J. Glob. Optim. **25**(3), 263–282 (2003)
10. D.H. Wolpert, W.G. Macready, No free lunch theorems for optimization. IEEE Trans. Evol. Comput. **1**(1), 67–82 (1997)
11. K.C. Tan, S.C. Chiam, A.A. Mamun, C.K. Goh, Balancing exploration and exploitation with adaptive variation for evolutionary multi-objective optimization. Eur. J. Oper. Res. **197**(2), 701–713 (2009)
12. E. Alba, B. Dorronsoro, The exploration/exploitation tradeoff in dynamic cellular genetic algorithms. IEEE Trans. Evol. Comput. **9**(2), 126–142 (2005)
13. I. Paenke, Y. Jin, J. Branke, Balancing population- and individual-level adaptation in changing environments. Adapt. Behav. **17**(2), 153–174 (2009)
14. E. Cuevas, A. Echavarría, M.A. Ramírez-Ortegón, An optimization algorithm inspired by the States of Matter that improves the balance between exploration and exploitation. Appl. Intell. **40**(2), 256–272 (2014)
15. M. Brambilla, E. Ferrante, M. Birattari, M. Dorigo, Swarm robotics: a review from the swarm engineering perspective. Swarm Intell. **7**(1), 1–41 (2013)
16. E. Şahin, *Swarm Robotics: From Sources of Inspiration to Domains of Application* (Springer, Berlin, Heidelberg, 2005), pp. 10–20
17. M. Duarte, J. Gomes, V. Costa, T. Rodrigues, F. Silva, V. Lobo, M.M. Marques, S.M. Oliveira, A.L. Christensen, Application of swarm robotics systems to marine environmental monitoring, in *OCEANS 2016—Shanghai* (2016), pp. 1–8
18. Y. Tan, Definition of swarm robotics characteristics of swarm robotics. J. Comput. Sci. Syst. Biol. **6**(6) (2013)
19. F. Mondada, D. Floreano, A. Guignard, J.-L. Deneubourg, L. Gambardella, S. Nolfi, M. Dorigo, *Search for Rescue: An Application for the SWARM-BOT Self-assembling Robot Concept* (2002)
20. S. Camazine, *Self-organization in Biological Systems* (Princeton University Press, 2003)
21. C.W. Reynolds, *Flocks, Herds, and Schools: A Distributed Behavioral Model*, vol. 21, no. 4 (1987)
22. P. De Meo, D. Rosaci, G.M. Sarnè, D. Ursino, G. Terracina, EC-XAMAS: supporting e-commerce activities by an XML-based adaptive multi-agent system. Appl. Artif. Intell. **21**(6), 529–562 (2007)
23. L. Ardissono, A. Goy, G. Petrone, M. Segnan, L. Console, L. Lesmo, C. Simone, P. Torasso, *Agent Technologies for the Development of Adaptive Web Stores* (Springer, Berlin, Heidelberg, 2001), pp. 194–213
24. D. Ursino, D. Rosaci, G.M.L. Sarnè, G. Terracina, An agent-based approach for managing e-commerce activities. Int. J. Intell. Syst. **19**(5), 385–416 (2004)

25. S.M. Aynur, A.A. Dayanik, H. Hirsh, *Information Valets for Intelligent Information Access* (2000)
26. S. Howell, Y. Rezgui, J.-L. Hippolyte, B. Jayan, H. Li, Towards the next generation of smart grids: semantic and holonic multi-agent management of distributed energy resources. Renew. Sustain. Energy Rev. **77**, 193–214 (2017)
27. V.N. Coelho, M. Weiss Cohen, I.M. Coelho, N. Liu, F.G. Guimarães, Multi-agent systems applied for energy systems integration: state-of-the-art applications and trends in microgrids. Appl. Energy **187**, 820–832 (2017)
28. H.S.V.S.K. Nunna, A.M. Saklani, A. Sesetti, S. Battula, S. Doolla, D. Srinivasan, Multi-agent based demand response management system for combined operation of smart microgrids. Sustain. Energy Grids Netw. **6**, 25–34 (2016)
29. A. Anvari-Moghaddam, A. Rahimi-Kian, M.S. Mirian, J.M. Guerrero, A multi-agent based energy management solution for integrated buildings and microgrid system. Appl. Energy **203**, 41–56 (2017)
30. V. Loia, S. Tomasiello, A. Vaccaro, Using fuzzy transform in multi-agent based monitoring of smart grids. Inf. Sci. (NY) **388–389**, 209–224 (2017)
31. X. Zhang, L. Liu, G. Feng, Leader–follower consensus of time-varying nonlinear multi-agent systems. Automatica **52**, 8–14 (2015)
32. X. Zhang, Q. Liu, L. Baron, E.-K. Boukas, Feedback stabilization for high order feedforward nonlinear time-delay systems. Automatica **47**(5), 962–967 (2011)
33. X. Zhang, L. Baron, Q. Liu, E.-K. Boukas, Design of stabilizing controllers with a dynamic gain for feedforward nonlinear time-delay systems. IEEE Trans. Automat. Contr. **56**(3), 692–697 (2011)
34. X. Zhang, G. Feng, Y. Sun, Finite-time stabilization by state feedback control for a class of time-varying nonlinear systems. Automatica **48**(3), 499–504 (2012)
35. J. Alonso-Mora, T. Naegeli, R. Siegwart, P. Beardsley, Collision avoidance for aerial vehicles in multi-agent scenarios. Auton. Robots **39**(1), 101–121 (2015)
36. W. Hönig, T.K.S. Kumar, S. Koenig, L. Cohen, H. Ma, H. Xu, N. Ayanian, S. Koenig, Multi-agent path finding with kinematic constraints, in *Proceedings of the 26th International Conference on Automated Planning and Scheduling* (2016), p. 9
37. S. Shalev-Shwartz, S. Shammah, A. Shashua, Safe, multi-agent, reinforcement learning for autonomous driving. arXiv Prepr. (2016)
38. L. Zhao, Y. Jia, Neural network-based adaptive consensus tracking control for multi-agent systems under actuator faults. Int. J. Syst. Sci. **47**(8), 1931–1942 (2016)
39. A. Nikou, J. Tumova, D.V. Dimarogonas, Cooperative task planning of multi-agent systems under timed temporal specifications, in *2016 American Control Conference (ACC)* (2016), pp. 7104–7109
40. Z. Yang, Large scale evolutionary optimization using cooperative coevolution. Inf. Sci. (NY) **178**(15), 2985–2999 (2008)
41. N. Mladenović, P. Hansen, Variable neighborhood search. Comput. Oper. Res. **24**(11), 1097–1100 (1997)
42. M.A. Potter, K.A. De Jong, Cooperative coevolution: an architecture for evolving coadapted subcomponents. Evol. Comput. **8**(1), 1–29 (2000)
43. F. Glover, M. Laguna, Tabu search, in *Handbook of Combinatorial Optimization* (Springer US, Boston, MA, 1998), pp. 2093–2229
44. P. Hansen, N. Mladenović, J. Brimberg, J.A.M. Pérez, *Variable Neighborhood Search* (Springer, Boston, MA, 2010), pp. 61–86
45. G. Anescu, Further scalable test functions for multidimensional continuous optimization (2017)
46. M.D. Li, H. Zhao, X.W. Weng, T. Han, A novel nature-inspired algorithm for optimization: virus colony search. Adv. Eng. Softw. **92**, 65–88 (2016)
47. A. Askarzadeh, A novel metaheuristic method for solving constrained engineering optimization problems: crow search algorithm. Comput. Struct. **169**, 1–12 (2016)
48. R. Storn, K. Price, Differential evolution—a simple and efficient heuristic for global optimization over continuous spaces. J. Glob. Optim. (1997)

49. D. Karaboga, An idea based on honey bee swarm for numerical optimization. Comput. Eng. Dep. Eng. Fac. Erciyes Univ. (2005)
50. P. Civicioglu, Transforming geocentric cartesian coordinates to geodetic coordinates by using differential search algorithm. Comput. Geosci. **46**, 229–247 (2012)
51. X.-S. Yang, S. Deb, Cuckoo search via Lévy flights, in *Proceedings of World Congress on Nature and Biologically Inspired Computing (NABIC'09)* (2009), pp. 210–214
52. S. Mirjalili, S.M. Mirjalili, A. Hatamlou, Multi-verse optimizer: a nature-inspired algorithm for global optimization. Neural Comput. Appl. **27**(2), 495–513 (2016)
53. S. Mirjalili, Moth-flame optimization algorithm: a novel nature-inspired heuristic paradigm. Knowl.-Based Syst. **89**, 228–249 (2015)
54. S. Mirjalili, SCA: a sine cosine algorithm for solving optimization problems. Knowl.-Based Syst. **96**, 120–133 (2016)
55. N. Hansen, S. Kern, Evaluating the CMA evolution strategy on multimodal test functions, in *Proceedings of the 8th International Conference on Parallel Problem Solving from Nature—PPSN VIII*, vol. 3242/2004 (2004), pp. 282–291
56. J.J.Q. Yu, V.O.K. Li, A social spider algorithm for global optimization. Appl. Soft Comput. J. **30**, 614–627 (2015)
57. M. Han, C. Liu, J. Xing, An evolutionary membrane algorithm for global numerical optimization problems. Inf. Sci. (NY) **276**, 219–241 (2014)
58. Z. Meng, J.S. Pan, Monkey king evolution: a new memetic evolutionary algorithm and its application in vehicle fuel consumption optimization. Knowl.-Based Syst. **97**, 144–157 (2015)
59. X.-S. Yang, M. Karamanoglu, X. He, Flower pollination algorithm: a novel approach for multiobjective optimization. Eng. Optim. **46**(9), 1222–1237 (2014)
60. F. Wilcoxon, Individual comparisons by ranking methods. Biometrics, 80–83 (1945)

Chapter 8
Knowledge-Based Optimization Algorithm

8.1 Introduction

Extracting knowledge (EK) is represented as multidisciplinary procedures for recognizing novel, meaningful, potentially useful, and robust knowledge in data [1]. EK is a discipline closely linked to data mining and machine learning approaches where data is examined to uncover hidden relations among the data [2]. A general obstacle in EK is that the complexity of its data withdraws the visualization of the existent relations in the information. Under such cases, the aim is to decrease the number of input vectors into a subset of archetype vectors by employing an EK strategy.

A Self-Organizing Map (SOM) [3] is an unsupervised neural network which has been extensively employed in several domains such as EK [4] and machine learning [5]. SOM performs vector quantization to decrease the number of data vectors into a representative small set of feature vectors established in a fixed lattice. SOM presents many advantages over other EK proposals. SOMs map complex high-dimensional data into a fixed low-dimensional lattice, which allows detecting unexpected patterns within the input information [3]. SOM keeps the data structure so that adjacent components of the original data are located nearby to neural units in the lattice structure [3]. Hence, SOM links multiple input data with a single layer of neural units, which produces naturally the visualization of complex data compositions [6]. These features allow to recognize and identify unexpected structures in data formations, determine relationship or variation between sets of vectors, and examine their natural patterns.

Recently, many optimization approaches based on evolutionary theories have been presented with impressive performance indexes. These methods analyze theories deduced from abstractions of natural or social aspects in real-life, which can be described as search procedures according to an optimization viewpoint [7]. Such algorithms imitate for instance the collective behavior of birds and fish in the Particle Swarm Optimization (PSO) algorithm [8], the social cooperations of bees in the Artificial Bee Colony (ABC) method [9], the process of artistic production in the Harmony Search (HS) [10], the echolocation manner of bats in the Bat

E. Cuevas et al., *Recent Metaheuristics Algorithms for Parameter Identification*, Studies in Computational Intelligence 854, https://doi.org/10.1007/978-3-030-28917-1_8

Algorithm (BAT) method [11], the transmission mechanism of fireflies in the Firefly (FF) method [12], the collective interaction of spiders in the Social Spider Optimization (SSO) [13], the collective animal relations in the Collective Animal Behavior (CAB) [14], the distinctive and natural evolution theory in population of individuals in the Differential Evolution (DE) [15] and Genetic Algorithms (GA) [16], Cuckoo Search (CS) [17], Differential Search (DS) [18], Moth-Flame Optimization (MFO) [19], Multi-Verse Optimizer (MVO) [20], Sine Cosine Algorithm (SCA) [21] and the Covariance Matrix Adaptation Evolutionary Strategy (CMA-ES) [22].

In the last decade, many researchers have focused their research work based on the inclusion of the knowledge extracted during the evolutionary process to improve the search strategies. The improvement of evolutionary methods can be split into two aspects: effectiveness, by controlling the search toward promising search zones from the entire search space, and efficiency, by reducing the computational complexity of such methods. The evolutionary strategies that employ some type of knowledge in their development can be classified into two big groups [23]. The first group compares those methods that combine knowledge before the opening of the evolutionary process. Such procedures are known as Evolutionary Methods with a Priori Information (EMAI). Under these techniques, the added information prevails fixed during the overall optimization task. Different EMAI strategies have been introduced. Kobeaga et al. in [24] submitted a GA that combines knowledge of reasonable solutions to set the codification stage of the individuals and the crossover operation. In [25–27], researchers have investigated different approaches that use expert experience to discover the locations of the initial population members employing to different engineering domains. On the other hand, the second group comprises evolutionary strategies that capture and carry knowledge in every iteration of the optimization task. These methods are called Evolutionary Techniques with Dynamic Information (ETDI). Approaches based on ETDI sources [28–31] have been suggested to explain the difficulty of parameter calibration in evolutionary techniques including EK to the determination of the parameter limits. In such methodologies, many techniques have been adopted to extract knowledge from the evolutionary process such as: factorial design of experiments [28], Bayesian Networks [30] and Fuzzy logic systems [31]. Some other ETDI approaches [32–35] have studied the use of information obtained during the optimization process to adjust the arrangement of evolutionary operators. They have studied methods such as statistical analysis [32, 34], genetic programming [33] and regression trees [35] for the description and retrieval of knowledge in the input data. All these methods combine a system for extracting intelligence in a canonical evolutionary algorithm. Under such circumstances, the extracted information is then employed to change indirectly the search mechanism. Hence, such algorithms do not examine the collected knowledge to control directly the search procedure. Moreover, most of those strategies apply simplistic machine learning routines considering only incomplete information of the evolutionary process such as the best element and some individuals of the population in the current generation.

Evolutionary optimization algorithms perform as stochastic search strategies for the exploration of a promissory solution in the entire search space. Since the location of the global optimum value is previously undiscovered, they try to examine a large number of locations before obtaining the optimal solution of an optimization problem. From this large collection of produced search positions, optimization strategies do not extract meaningful information from it. The use of this information could effectively control the searching procedure to encounter promissory zones utilizing a low amount of iterations. This chapter introduces an evolutionary optimization algorithm, called EA-SOM, in which information extracted during the optimization operation is applied to guide the search strategy. In the proposal, a SOM is applied as EK method to identify promising search zones through the reduction of dimensionality involved in the optimization task. The idea behind this reduction mechanism is to define good quality neighborhoods in compressed locations. Consequently, in each iteration, the proposed methodology considers a subset of the entire population to train the SOM. Once trained, the neural layer from the SOM which matches the location of the best solution is redefined. Then, by using local information of the neural layer, a new population of promissory solutions is generated. The performance of the proposed method is compared to some state-of-the-art optimization algorithms evaluated over a set of 23 well-known benchmark functions and three real-world engineering design problems. The numerical results suggest that the proposed method reaches the best evolutionary balance among exploration and exploitation stages considering accuracy, robustness, and population diversity.

This chapter is structured as follows: In Sect. 8.2, the main features of the SOMs are briefly explained. In Sect. 8.3, the proposed EA-SOM method is presented. Section 8.4 analyzes the computational procedure for the EA-SOM. In Sect. 8.5 the experimental study and the comparative analysis are reported. Finally, in Sect. 8.6, some conclusions are presented.

8.2 Self-organizing Map

In this section, it is presented a brief introduction to the fundamental thoughts of self-organization maps (SOMs). The report principally examines those elements especially crucial for the reduction of data and clustering. A Self-Organizing Map (SOM) [3] is considered as an unsupervised neural network broadly employed in disciplines such as EK [4] and machine learning [5]. Such neural networks are considered effective methods to examine complex multidimensional data.

One important benefit of SOMs against other EK techniques is that the acquired knowledge maintains the organization of data of the original input components. Unedr such circumstances, SOMs are broadly employed as clustering methods or data compression methods. SOM performs vector quantization to decrease the amount of data vectors into a low dimensional set of representative vectors ordered in a fixed lattice. SOMs map complex high-dimensional data into a low dimensional lattice, which enables getting hidden patterns inside the input data [3]. SOM keeps the organization of data so that related components in the original input data closely relates to neural units in the output layer of the network [3]. Hence, SOM connects many input data with a neural layer, which facilitates the visualization of complex

Fig. 8.1 Graphical representation of the lattice of a Self-Organization Map (SOM)

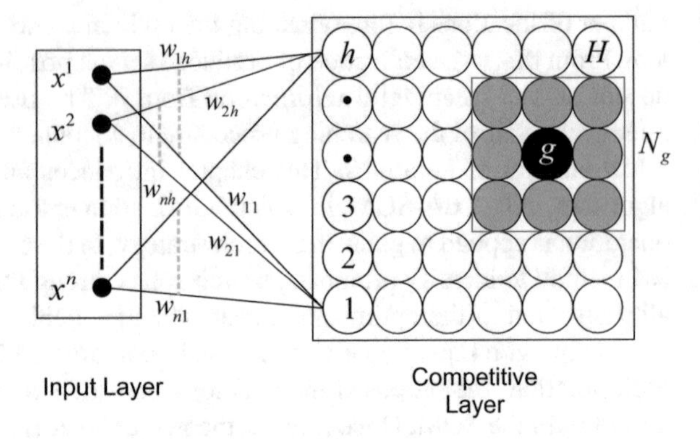

data formations [6]. These features allow to identify and recognize undiscovered topologies in data, discover similarity or dissimilarity between groups of vectors, and interpret their natural patterns.

SOMs are composed of two distinct layers: The first layer consists of an input layer meanwhile the second layer represents a competitive layer. Typically, the competitive layer comprises a collection of neurons organized as a two-dimensional lattice. SOM´s architecture is described in Fig. 8.1.

A SOM architecture comprises an output lattice structure, composed by competitive neurons of $H = h \cdot h$ units. Within this procedure, the SOM receives as input an n-dimensional vector $\mathbf{x}$. Where each element of the data vector $\mathbf{x}$ is connected to each neuron from the lattice.

Within the structure of the lattice, A weight w_{ij} is associated with a related link among the ith dimension of the input vector to the jth neuron. Under such requirements, each neuron involves a weight vector $\mathbf{w}_j \left(w_j^1, \ldots, w_j^n\right)$ with n different components.

The stage of training of a SOM comprises two main parts: the weight adjustment and the topology information of the neurons. The learning mechanism comprises two steps: similarity computation (I) and weight adaptation (II). In the beginning, the weights are randomly initialized. Later, the input data vector is injected to the neural structure. In the similarity computation (I) stage, the Euclidean-based distance among the weights and the data vector is computed. Thereafter, the neuron g that has accomplished the smallest distance value between the neurons is then picked and recognized as the winner neuron. In the second step, weight adaptation (II), the corresponding weights from the winner neuron g are modified. Additionally, a topological formation of N_g neurons in the closer area of the winner neuron g is calculated. Ultimately, the weights of the neurons within the structure N_g are also adjusted. Figure 8.1 graphically depicts such notions.

The entire training process for the SOMs can be compiled as follows [36]:

1. Set the initial iteration as $k = 0$. Then, initialize all the weights w_{ij} randomly. Initialize the learning rate α_0. And set the number of iterations for the learning routines.

2. Iterate Step 3 through Step 6 until the maximum number of iteration has been accomplished.
3. From the training set, select a random input vector $\mathbf{x}$.
4. Compute the winner neuron which weight vector corresponds to the nearest input vector. Such formulation is calculated as:

$$D_{\mathrm{Min}}(k) = \min_{j \in H}\{\|\mathbf{x}(k) - \mathbf{w}_j(k)\|\} \tag{8.1}$$

5. Modify the weights of the winner neuron and its nearby neighbors as:

$$\mathbf{w}_i(k+1) = \mathbf{w}_i(k) + \alpha(k) \cdot (\mathbf{x} - \mathbf{w}_i(k)), \quad \forall i \in N_g, g, \tag{8.2}$$

where $\alpha(k)$ denotes the portion of the learning rate which decreases according to the number of iterations involved in the training process. The mathematical formulation of such computation is as:

$$\alpha(k) = \alpha_0 \cdot e^{-k/K}, \tag{8.3}$$

$$d(k) = d_0 \cdot e^{-k/3K}, \tag{8.4}$$

where α_0 and d_0 are the initial values of the learning portion and the distance, respectively.
6. Update $k = k + 1$. If $k < K$ execute Step 3, otherwise the training process is finish.

To graphically visualize the data-reduction achieved in SOMs, Fig. 8.2 exposes this idea. First, the figure presents a SOM which is trained using 100 random points (Fig. 8.2a). In the training stage, the neural network considers the position of the sample as the input vector as well as its fitness value. Figures 8.2a and b describe five areas of the search space A, B, C, D and E. As an outcome of the training process, Fig. 8.2c depicts the process at which the distinct areas are assembled in the lattice. Comparing the contour plots related to Fig. 8.2b with the clusters of Fig. 8.2c, it can be demonstrated that SOMs preserve the topological structure of the search space.

8.3 The Proposed EA-SOM Method

Since the SOM's use historical information related from a process, it is reduced the extensive necessity of acquiring new data to infer its internal features and relations. Using extracted information to carry the search tactics in evolutionary computing methodologies represents an attractive idea that has the ability to improve the performance of optimization techniques. In this segment, the entire method of the EA-SOM algorithm is exhibited. The EA-SOM has been devised to determine the global solution for a nonlinear optimization problem with box constraints under the subsequent

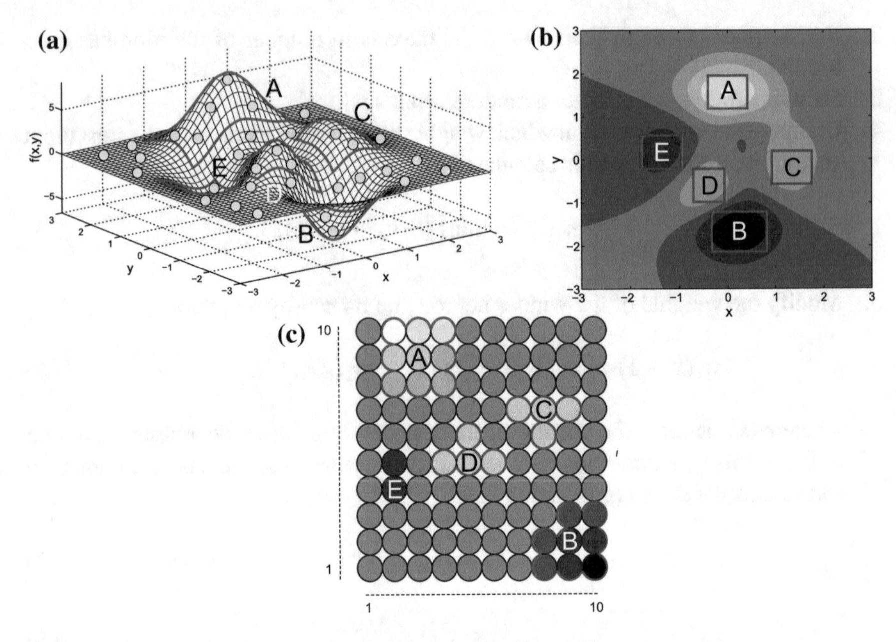

Fig. 8.2 This is how a data-reduction and clustering mechanism work in SOMs. **a** The training process, **b** Contour representation and **c** The data-reduction result considering a lattice of 10×10 neural units

mathematical formulation [37]:

$$\begin{array}{ll} \text{maximize } f(\mathbf{x}), & \mathbf{x} = (x^1, \ldots, x^n) \in \mathbb{R}^n \\ \text{subject to } \mathbf{x} \in \mathbf{X} & \end{array} \tag{8.5}$$

where $f : \mathbb{R}^n \rightarrow \mathbb{R}$ s a nonlinear relation which describes an established search space, limited by the lower (l_i) and upper (u_i) limits. EA-SOM is an iterative method that emerges from an initial iteration ($k = 0$) to a maximum number of generations ($k = Maxgen$). At the beginning ($k = 0$), the method provides a collection of solutions similar to four times the number of solutions exhibited. Such an initial collection is assumed to train the SOM at the beginning. In the method, the SOM is applied as EK procedure to recognize promissory search zones during the data reduction process. Hence, in each iteration, the proposed algorithm employs a subset of the historical positions to train the SOM. Then, the local information is used of this neural unit an entire population of candidate solutions is produced.

The method is split into six steps: (8.3.1) Initialization, (8.3.2) training, (8.3.3) EK, (8.3.4) solution production and (8.3.5) construction of the new training collection. The procedures 8.3.2–8.3.5 are repeatedly performed until the stop criterion has been reached.

8.3.1 Initialization

In the start point ($k = 0$), EA-SOM starts generating a collection $\mathbf{P}(k)$ of $(N \cdot 2)$ solutions ($\{\mathbf{p}_1(k), \mathbf{p}_2(k), \ldots, \mathbf{p}_{(N \cdot 2)}(k)\}$) among the pre-defined lower (l_i) and upper (u_i) limits of the search space. In $\mathbf{P}(k)$, solution $\mathbf{p}_i(k)$ ($i \in [1, \ldots, (N \cdot 2)]$) describes an n-dimensional solution $\{p_i^1(k), p_i^2(k), \ldots, p_i^n(k)\}$ that resembles to a potential solution for the optimization problem. Through the evolution procedure, the state of a given solution $\mathbf{p}_i(k)$ can be evaluated by examining the objective function $f(\mathbf{p}_i(k))$ Therefore, as the optimization procedure is performed, the best solution g prevails. All the initial elements of $\mathbf{P}(k)$ are deposited in a historical memory $\mathbf{H}(k)$ that involves all solutions created through the evolution procedure.

8.3.2 Training

In the proposal, the SOM is trained to recognize and identify promissory search zones during the data-reduction procedure. applying the learning approach defined in Sect. 8.2, the SOM is trained regarding a collection $\mathbf{T}(k)$ of $(N \cdot 2)$ data. In the training procedure, the parameters α_0, d_0 are respectively fixed to 1 and 10. The SOM network is trained just considering 100 epochs.

At the initial point, the training set $\mathbf{T}(0)$ resembles to the initial population. Nonetheless, throughout each iteration ($k \neq 0$), the training collection $\mathbf{T}(k)$ is produced regarding the scheme detailed in the point 8.3.5.

The amount of training examples that are employed in the learning procedure has an essential involvement in the quality of the degraded search space. The adoption of all possible historical data would provide the best resemblance. Nevertheless, the application of this large amount of data tends to increase the computational effort. Under such circumstances, in this chapter, the amount of training examples has been established to $(N \cdot 2)$, to accomplish the best balance among accuracy and computational effort.

8.3.3 Extracting Information (EK)

SOM preserves the organization of data such that similar components in the input data maintain closer to neurons in the lattice [3]. Hence, SOM links many input vectors with a single neural layer, which allows the visualization of complex data structures [6].

Once trained, he promissory search zones mapped in the lattice ($n \times n$ lattice) are localized. Hence, the output layer of the SOM corresponds to a potential zone where the best solution $\mathbf{g}$ is recognized. This unit w resembles to the neural unit whose weights are closest to $\mathbf{g}$, as:

$$w = \min_{w \in n \times n} \{\|\mathbf{g} - \mathbf{w}_w\|\} \tag{8.6}$$

In order to obtain codified knowledge in the lattice, local information approximately the neuron $\mathbf{w}_w$ is identified. Hence, the two neighbor neurons to $\mathbf{w}_w$ with the closest weights to $\mathbf{g}$ are also detected. Both elements $\mathbf{w}_A$ and $\mathbf{w}_B$ escribe the competitive neural units with the minimal distances to $\mathbf{g}$ regardless the presence of element $\mathbf{w}_w$. Adjacent with the neurons $\mathbf{w}_w$, $\mathbf{w}_A$ and $\mathbf{w}_B$, their respective distances to g are also maintained. Those distances are calculated as:

$$D_w = \|\mathbf{g} - \mathbf{w}_w\|, D_A = \|\mathbf{g} - \mathbf{w}_A\|, D_B = \|\mathbf{g} - \mathbf{w}_B\| \tag{8.7}$$

Once recognized, the neural component w matches to the best solution $\mathbf{g}$, its location is recognized as a reference within the spatial mapping of the $n \times n$ output layer. The purpose is to compare the position of the original input with those mapped in the output layer. This connection is characterized by a position rank R determined as follows:

$$R = 1 - \frac{|m - w|}{n} \tag{8.8}$$

where m corresponds to the middle element of the lattice ($m = n \cdot n/2$). Hence, the rank R keeps a value near to one in the center of the output layer meanwhile in the border a value close to zero.

8.3.4 Solution Production

In the proposed approach, the EK feature is applied to lead the search mechanism. Later, with the discovery of the promissory search areas and its reference in the output lattice, a collection of N new solutions is produced. In SOMs, areas of the input data are mapping to just only one neuron in the output layer. Under such requirements, the new solutions are created contemplating that their locations hold a high probability to be within of an attraction radius nearby of the winner neuron $\mathbf{w}_w$. This influence range δ directly depends on the distribution of the neurons $\mathbf{w}_w$, $\mathbf{w}_A$ and $\mathbf{w}_B$ that resembles to the promissory zones in the data. Consequently, δ is computed as the averaged distance produced by the three elements as follows:

$$\delta = \frac{D_w + D_A + D_B}{3} \tag{8.9}$$

Figure 8.3 exhibits a graphical description of δ in the meaning of a two-dimensional function. After the computation of δ, each solution j from N is generated as follows:

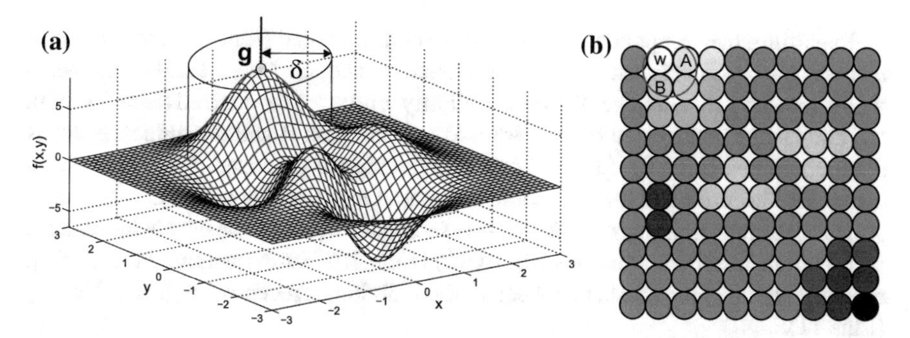

Fig. 8.3 Graphical representation of δ considering a two-dimensional function, **a** fitness function and **b** lattice in a 10×10 lattice

$$\mathbf{p}_j(k+1) = \mathbf{w}_w + \delta \cdot R \cdot (r \cdot 2 - 1), \tag{8.10}$$

where r corresponds to a random number within [0,1].

In this chapter, it is implemented the simplistic search strategy to demonstrate the potential of EK from historical information through the optimization procedure.

8.3.5 Construction of the New Training Set

With the formation of the new population, the subsequent action is to design the new training collection $\mathbf{T}(k + 1)$ for the coming generation. In each generation there are produced N new individuals, then, after a certain amount of iterations a big number solutions are stored in $\mathbf{H}(k)$. The training collection involves $(N \cdot 2)$ solutions. Hence, the first N solutions of $\mathbf{T}(k + 1)$ match to the N solutions currently generated. The remaining N are randomly chosen from the current historical information $\mathbf{H}(k)$.

Subsequent to the formation of $\mathbf{T}(k + 1)$, the recently created elements $\mathbf{P}(k + 1)$ are deposited in a historical memory so that: $\mathbf{H}(k + 1) = \mathbf{H}(k) \cup \mathbf{P}(k + 1)$.

8.4 Computational Procedure

The proposed algorithm has been conceived as an iterative procedure that uses many steps in its execution. Algorithm 8.1 reviews the entire process in pseudo-code. The proposal does not require any external parameter that demands to be calibrated before the execution. Hence, it only requires the number solutions N (input data size) and the number of generations *Maxgen* (Line 1). Then, the initialization procedure is accomplished by each solution in a random fashion among boundaries (Line 2). Such solutions represent the initial population $\mathbf{P}(0)$ for the first training stage as well as the first elements deposited in $\mathbf{H}(k)$ (Line 3).

Regarding the components of **T**(k), the SOM is trained (Line 5). Then, the best element **g** from **T**(k) is selected (Line 6). Once trained, the three neural components $\mathbf{w}_w$, $\mathbf{w}_A$ and $\mathbf{w}_B$ that involves the best similarity with **g** are recognized (Line 7). With the knowledge contributed by $\mathbf{w}_w$, $\mathbf{w}_A$ and $\mathbf{w}_B$, the rank R and the influence radius δ are computed (Line 8). Later, N new individuals are produced. Such individuals are created according to their positions (Line 9). Once the new population N is constructed, the new training set $\mathbf{T}(k+1)$ is assembled by joining the new N individuals and random historical solutions from $\mathbf{H}(k)$ (Line 10). Subsequently, the historical memory $\mathbf{H}(k+1)$ is refreshed, collecting the individuals recently produced $\mathbf{P}(k+1)$ (Line 11).

Algorithm 8.1 Pseudo-code for the proposed method

1. **Input:** N, *Maxgen*, k=0.
2. **P**(k) ← **Initialize**(($N \cdot 2$));
3. **H**(k) ← $\mathbf{P}^k$, **T**(k) ← **P**(k)
4. **while** k <=*Maxgen* **do**
5. **SOM** ← **TrainSOM**(**T**(k))
6. **g** ← **SelectBestParticle**(**T**(k));
7. $\mathbf{w}_w$, $\mathbf{w}_A$, $\mathbf{w}_B$ ← **DetectNeuralUnits**(**SOM**)
8. R, δ ← **CalculateInf**($\mathbf{w}_w$, $\mathbf{w}_A$, $\mathbf{w}_B$)
9. **P**(k+1) ← **GenerateNewSolutions**(N, $\mathbf{w}_w$);
10. **T**(k+1) ← **ConstructT**(**P**(k+1),**H**(k))
11. **H**(k+1) ← $\mathbf{H}(k) \cup \mathbf{P}(k+1)$
12. k ← k+1
13. **end while**
14. **Output: g**

8.5 Experimental Study

Optimization techniques based on evolutionary theories have been conceived as search strategies to solve complex optimization problems where classical mathematical methods are unsuitable or ineffective in the presence of many local minima. Regarding such methodologies, the performance of a new evolutionary strategy must be properly examined. Usually, these procedures are assessed by evaluating a collection of test functions with several complexities regarding as the stop criterion the number of iterations *Maxgen*. This segment manifests the numerical results of the proposed EA-SOM evaluated a test suit formed by 23 continuous functions regularly employed in numeric optimization [38, 39] as well as three real-world design engineering problems [40, 41] utilized to measure the performance of many optimization procedures. Table B.10 in Appendix B exposes the set of test functions employed in the experimental comparison. In the table, n denotes the n-dimensional search space in which the functions are judged, $f(x^*)$ determines the minimal value of the function evaluated at location x^* and $\boldsymbol{S}$ corresponds to the pre-defined boundaries.

The numerical results achieved by the proposed method are analyzed and statistically verified against the numerical results of 9 evolutionary algorithms, namely; Particle Swarm Optimization (PSO) [8], Differential Evolution (DE) [15], Artificial Bee Colony (ABC) [9], Differential Search (DS) [18], Cuckoo Search (CS) [17], Sine Cosine Algorithm (SCA) [21], Multi-Verse Optimizer (MVO) [20], Moth-Flame Optimization (MFO) [19] and the Covariance Matrix Adaptation Evolutionary Strategy (CMA-ES) [22]. In addition, the performance of the algorithms is analyzed evaluating three real-world engineering problems; Three-bar truss design, Tension/compression spring design and Welded beam design. Tables B.7, B.8, and B.9 in Appendix B illustrate the formulation for all the engineering problems. Section 8.5.1 analyzes the numerical results achieved by EA-SOM against its competitors regarding a collection of test functions. Section 8.5.2 describes the convergence results for each algorithm. Finally, in Sect. 8.5.3 the experimental results of the real-world engineering design problems are presented.

8.5.1 Performance Comparison

In this subsection, the performance of EA-SOM is analyzed against the performance of 9 state-of-art evolutionary methods considering a set of 23 test functions. The selected benchmark suite includes unimodal, multimodal and hybrid benchmark functions with several complexities. In order to do a straight comparison among the algorithms, each evaluation of a test function considers $Maxgen = 1000$ as stop criterion [13, 42–44]. The experiments are assessed considering $n = 30$ dimensional search space and performed 30 independent runs to eliminate the stochastic effect. In addition, statistical validation of the numerical results is conducted by the Wilcoxon rank sum test [37].

For the comparison study, the parameter configuration has been established related to their reported guidelines. Such arrangements present the best potential performance of each evolutionary strategy. Such configurations are as follows:

1. ABC: The algorithm has been implemented considering the population limit at 50 individuals.
2. DE: The considered variant is DE/rand/bin [15] where $cr = 0.5$ and $dw = 0.2$.
3. PSO: The weight factor linearly decreasing from $[0.9 - 0.2]$ and the constants set to $c_1 = 2$ and $c_2 = 2$ [8].
4. CS: The $pa = 0.25$ [17].
5. DS: The algorithm considers the $sp = 1$ and $cp = 0.5$ [18].
6. MFO: The source code has been taken by [19].
7. MVO: The source code has been taken from [20].

8. SCA: The implementation follows the source code [21].
9. CMA-ES: The source code has taken from [22].
10. EA-SOM: The population size has been configured by 50 individuals (N), and the lattice size as $n \cdot n$ neurons.

In the study, the population size for all the algorithms has been configured to 50 individuals. Table 8.1 shows the numerical results of EA-SOM and its competitors. In Table, the average fitness value is described as $\bar{f}$, the standard deviation as σ_f, the best fitness value as f_{Best} and the worst fitness corresponds to f_{Worst}. Additionally, the best entries are highlighted.

Regarding Table 8.1, the proposed approach offers better performance than the compared methods for the majority of test functions. These results demonstrate that the performance of EA-SOM is capable of finding optimal values in the presence of many local optima. On the other hand, due to the EK scheme, Table 8.1 hints that EA-SOM delivers a robust stability between exploration-exploitation stages creating EA-SOM to beat the rest of the remaining algorithms in ten test functions which present results with the lowest standard deviations. Hence, the generated solutions by EA-SOM are recognized to be consistent each run.

In functions $f_{16} - f_{18}$ and f_{23}, the suggested approach gives the worst performance; the DS algorithm strikes the proposed approach in function f_{16}, for functions f_{17} and f_{18} the CMA-ES method beats the proposal and for function f_{23}, ABC exhibits the best index. In case of functions f_6, f_8, f_9 and f_{12}, the proposed EA-SOM offers quite comparable performance than ABC, CS, DS, MFO, SCA and CMA-ES. For the rest of the test functions, EA-SOM beats the other algorithms providing more consistent solutions.

An examination of Table 8.1 reveals that the intended EA-SOM algorithm takes in most cases a better answer than the tested approaches. This fact implies that the proposed strategy works at least equal to other state-of-art- techniques.

In addition to the numerical results exposed by Table 8.1, a non-parametric statistical test is also performed in order to validate the numerical results of the aforementioned table. The popular Wilcoxon rank sum test [37] has been employed to the 5% of significance value considering the average fitness values for each test function. As each function is performed 30 runs to eliminate the stochastic influence, the best fitness value of each run has been recognized to lead the rank sum test. Table 8.2 presents the p-values taken by the rank sum for a pair-wise comparison between the algorithms. In Table 8.2, the comparison sets nine groups; EA-SOM versus ABC, EA-SOM versus DE, EA-SOM versus PSO, EA-SOM versus CS, EA-SOM versus DS, EA-SOM versus MFO, EA-SOM versus MVO, EA-SOM versus SCA and EA-SOM versus CMA-ES. In this validation, the stated null hypothesis H_0 designate that there is no significant difference among a pair of methods. On the other hand, the proposed alternative hypothesis H_1 recognizes the presence of a significant difference among a pair of algorithms. To aid the statistical outcomes of the non-parametric analysis, Table 8.2 utilizes the symbols ▲, ▼, and ►. ▲ Represents that the proposed algorithm performs significantly better than its competitors for a given function. ▼ Indicates that the proposed method performs worse than the tested algorithms, and

Table 8.1 Results of the minimization of Table B.10 with $n = 30$

		ABC	DE	PSO	CS	DS	MFO	MVO	SCA	CMA-ES	EA-SOM
f_1	$\bar{f}$	1.52E−11	1.23E+04	1.66E+03	9.27E−03	1.99E−05	1.20E+03	1.84E−01	6.52E−03	1.48E−18	**0.00E+00**
	σ_f	1.72E−11	2.38E+03	6.65E+02	2.85E−03	4.32E−05	3.28E+03	4.55E−02	1.83E−02	1.11E−18	**0.00E+00**
	f_{Best}	1.05E−12	7.28E+03	8.23E+02	4.78E−03	1.88E−06	1.94E−06	1.13E−01	9.45E−07	3.55E−19	**0.00E+00**
	f_{Worst}	7.08E−11	1.74E+04	3.32E+03	1.67E−02	2.40E−04	1.00E+04	2.92E−01	9.22E−02	6.64E−18	**0.00E+00**
f_2	$\bar{f}$	1.13E−06	5.20E+01	4.16E+01	1.67E+00	3.45E−04	2.85E+01	3.16E−01	6.33E−06	3.59E−09	**0.00E+00**
	σ_f	4.77E−07	6.63E+00	1.89E+01	1.22E+00	2.78E−04	1.97E+01	8.69E−02	1.29E−05	1.44E−09	**0.00E+00**
	f_{Best}	4.14E−07	3.99E+01	2.31E+01	6.05E−01	7.66E−05	5.58E−05	1.98E−01	2.43E−09	1.50E−09	**0.00E+00**
	f_{Worst}	2.22E−06	6.60E+01	1.03E+02	5.54E+00	1.27E−03	8.00E+01	5.45E−01	6.03E−05	6.84E−09	**0.00E+00**
f_3	$\bar{f}$	1.14E+04	5.79E+04	7.96E+03	4.37E+02	2.91E+03	1.56E+04	1.86E+01	3.48E+03	3.27E−04	**0.00E+00**
	σ_f	2.75E+03	9.38E+03	3.64E+03	9.09E+01	1.28E+03	1.05E+04	7.64E+00	3.42E+03	8.34E−04	**0.00E+00**
	f_{Best}	6.00E+03	4.51E+04	2.13E+03	2.59E+02	9.98E+02	2.31E+02	9.17E+00	4.36E+01	1.07E−06	**0.00E+00**
	f_{Worst}	1.56E+04	7.93E+04	2.03E+04	6.77E+02	4.97E+03	4.00E+04	3.64E+01	1.10E+04	4.12E−03	**0.00E+00**
f_4	$\bar{f}$	4.46E+01	6.86E+01	2.23E+01	3.39E+00	5.67E+00	5.65E+01	6.28E−01	1.15E+01	2.84E−07	**0.00E+00**
	σ_f	8.00E+00	3.73E+00	4.30E+00	6.72E−01	1.90E+00	9.87E+00	2.07E−01	6.61E+00	8.36E−08	**0.00E+00**
	f_{Best}	1.83E+01	5.95E+01	1.50E+01	2.21E+00	2.76E+00	3.25E+01	3.11E−01	1.20E+00	1.81E−07	**0.00E+00**
	f_{Worst}	5.60E+01	7.65E+01	3.50E+01	4.79E+00	9.96E+00	8.01E+01	1.14E+00	2.50E+01	6.10E−07	**0.00E+00**
f_5	$\bar{f}$	2.44E+01	2.88E+01	3.68E+02	2.47E+01	2.66E+01	1.57E+04	3.03E+01	2.79E+01	5.40E+01	**2.87E+01**
	σ_f	1.78E+00	4.66E−02	1.87E+02	5.80E−01	9.92E+00	3.39E+04	1.34E+01	4.46E−01	1.03E+02	**4.11E−03**
	f_{Best}	1.91E+01	2.75E+01	1.40E+02	2.31E+01	2.24E+01	9.12E+00	2.34E+01	2.70E+01	2.10E+01	**2.87E+01**
	f_{Worst}	2.72E+01	2.89E+01	1.05E+03	2.55E+01	7.89E+01	9.01E+04	8.14E+01	2.87E+01	5.35E+02	**2.87E+01**
f_6	$\bar{f}$	0.00E+00	1.19E+04	1.93E+03	0.00E+00	0.00E+00	1.60E+03	6.07E+00	6.67E−02	0.00E+00	**0.00E+00**

(continued)

Table 8.1 (continued)

		ABC	DE	PSO	CS	DS	MFO	MVO	SCA	CMA-ES	EA-SOM
	σ_f	0.00E+00	2.24E+03	6.78E+02	0.00E+00	0.00E+00	3.70E+03	3.12E+00	3.65E−01	0.00E+00	**0.00E+00**
	f_{Best}	**0.00E+00**	8.45E+03	7.65E+02	**0.00E+00**	**0.00E+00**	**0.00E+00**	1.00E+00	**0.00E+00**	**0.00E+00**	**0.00E+00**
	f_{Worst}	0.00E+00	1.70E+04	3.34E+03	0.00E+00	0.00E+00	1.00E+04	1.70E+01	2.00E+00	0.00E+00	**0.00E+00**
f_7	$\bar{f}$	2.21E−01	9.27E+00	5.38E+00	3.36E−02	3.77E−02	1.84E+00	1.38E−02	2.48E−02	1.29E−02	**7.57E−05**
	σ_f	5.28E−02	2.57E+00	2.34E+00	1.10E−02	1.42E−02	3.91E+00	6.42E−03	2.21E−02	3.63E−03	**6.38E−05**
	f_{Best}	9.31E−02	3.56E+00	2.16E+00	1.59E−02	1.41E−02	3.01E−02	6.10E−03	3.22E−03	6.53E−03	**3.50E−06**
	f_{Worst}	3.07E−01	1.57E+01	1.14E+01	6.45E−02	6.47E−02	1.89E+01	3.71E−02	1.12E−01	1.94E−02	**2.39E−04**
f_8	$\bar{f}$	2.07E+00	7.40E+00	8.09E+00	2.07E+00	2.07E+00	2.10E+00	2.07E+00	2.36E+00	1.48E+01	**2.07E+00**
	σ_f	2.65E−08	1.01E+00	1.93E+00	4.47E−06	2.60E−03	1.26E−02	6.13E−05	9.78E−02	1.63E+00	**1.42E−03**
	f_{Best}	**2.07E+00**	5.18E+00	4.63E+00	**2.07E+00**	**2.07E+00**	2.08E+00	**2.07E+00**	2.23E+00	1.13E+01	**2.07E+00**
	f_{Worst}	2.07E+00	9.68E+00	1.25E+01	2.07E+00	2.08E+00	2.13E+00	2.07E+00	2.67E+00	1.73E+01	**2.08E+00**
f_9	$\bar{f}$	7.13E−02	4.49E+00	1.52E+00	7.13E−02	7.20E−02	9.99E−02	7.15E−02	3.57E−01	7.13E−02	**7.52E−02**
	σ_f	1.06E−10	7.30E−01	6.29E−01	6.66E−07	1.13E−03	1.27E−02	4.37E−05	1.12E−01	3.35E−17	**1.29E−03**
	f_{Best}	**7.13E−02**	3.42E+00	6.05E−01	**7.13E−02**	**7.13E−02**	8.32E−02	7.15E−02	2.16E−01	**7.13E−02**	**7.13E−02**
	f_{Worst}	7.13E−02	6.50E+00	2.60E+00	7.13E−02	7.40E−02	1.33E−01	7.16E−02	6.32E−01	7.13E−02	**7.74E−02**
f_{10}	$\bar{f}$	1.67E−01	2.75E+02	1.74E+02	8.52E+01	8.37E+00	1.43E+02	1.10E+02	2.02E+01	1.31E+02	**0.00E+00**
	σ_f	3.79E−01	1.47E+01	3.12E+01	1.02E+01	2.59E+00	4.01E+01	3.58E+01	2.75E+01	6.61E+01	**0.00E+00**
	f_{Best}	8.20E−09	2.45E+02	9.56E+01	6.49E+01	2.56E+00	6.87E+01	5.78E+01	5.95E−06	5.97E+00	**0.00E+00**
	f_{Worst}	1.02E+00	3.06E+02	2.42E+02	1.08E+02	1.83E+01	2.42E+02	2.09E+02	8.81E+01	1.88E+02	**0.00E+00**
f_{11}	$\bar{f}$	1.55E−05	1.67E+01	1.13E+01	3.91E+00	1.48E−03	1.39E+01	9.64E−01	1.64E+01	4.41E−10	**8.88E−16**
	σ_f	8.37E−06	4.70E−01	1.66E+00	1.30E+00	1.31E−03	8.28E+00	7.69E−01	7.50E+00	1.10E−10	**0.00E+00**

(continued)

Table 8.1 (continued)

		ABC	DE	PSO	CS	DS	MFO	MVO	SCA	CMA-ES	EA-SOM
	f_{Best}	2.04E−06	1.59E+01	7.60E+00	2.20E+00	2.14E−04	5.12E−04	1.03E−01	4.16E−05	2.42E−10	**8.88E−16**
	f_{Worst}	3.54E−05	1.77E+01	1.47E+01	6.92E+00	4.89E−03	2.00E+01	2.61E+00	2.02E+01	7.13E−10	**8.88E−16**
f_{12}	$\bar{f}$	1.20E−06	1.12E+02	1.73E+01	1.02E−01	7.08E−03	1.45E+01	4.47E−01	1.75E−01	0.00E+00	**0.00E+00**
	σ_f	5.13E−06	2.24E+01	4.68E+00	2.77E−02	1.00E−02	3.35E+01	1.10E−01	2.00E−01	0.00E+00	**0.00E+00**
	f_{Best}	5.15E−11	7.27E+01	8.75E+00	4.79E−02	6.01E−07	5.09E−06	2.38E−01	2.26E−06	**0.00E+00**	**0.00E+00**
	f_{Worst}	2.80E−05	1.79E+02	2.77E+01	1.81E−01	3.44E−02	9.09E+01	6.10E−01	6.93E−01	0.00E+00	**0.00E+00**
f_{13}	$\bar{f}$	−3.00E+00	7.96E−01	1.71E+00	−2.72E+00	−3.00E+00	−2.59E+00	−2.16E+00	−3.00E+00	−3.00E+00	**−3.00E+00**
	σ_f	6.40E−14	3.31E−01	1.01E+00	1.65E−01	5.82E−08	4.96E−01	3.98E−01	2.65E−05	0.00E+00	0.00E+00
	f_{Best}	**−3.00E+00**	5.39E−02	−3.24E−02	−2.95E+00	**−3.00E+00**	**−3.00E+00**	−2.85E+00	**−3.00E+00**	**−3.00E+00**	**−3.00E+00**
	f_{Worst}	−3.00E+00	1.44E+00	3.92E+00	−2.42E+00	−3.00E+00	−6.00E−01	−1.08E+00	−3.00E+00	−3.00E+00	−3.00E+00
f_{14}	$\bar{f}$	2.75E+00	3.59E+01	3.34E+00	2.75E+00	2.75E+00	1.97E+01	2.75E+00	2.75E+00	2.75E+00	2.75E+00
	σ_f	1.34E−07	1.27E+01	8.86E−01	5.95E−06	9.84E−09	2.26E+01	1.05E−06	2.55E−04	0.00E+00	4.39E−05
	f_{Best}	**2.75E+00**	1.49E+01	2.80E+00	**2.75E+00**	**2.75E+00**	**2.75E+00**	**2.75E+00**	**2.75E+00**	**2.75E+00**	**2.75E+00**
	f_{Worst}	2.75E+00	6.25E+01	7.68E+00	2.75E+00	2.75E+00	9.97E+01	2.75E+00	2.75E+00	2.75E+00	2.75E+00
f_{15}	$\bar{f}$	4.59E−15	5.05E+00	2.21E+00	3.85E−06	6.86E−09	1.04E+00	7.72E−05	3.35E−07	7.35E−22	0.00E+00
	σ_f	3.41E−15	9.97E−01	7.48E−01	1.69E−06	1.22E−08	1.77E+00	2.54E−05	7.04E−07	6.16E−22	0.00E+00
	f_{Best}	9.96E−16	3.10E+00	1.13E+00	1.81E−06	2.97E−10	7.32E−10	4.76E−05	2.93E−10	1.80E−22	**0.00E+00**
	f_{Worst}	1.39E−14	7.40E+00	3.93E+00	1.03E−05	6.53E−08	4.00E+00	1.69E−04	3.03E−06	3.48E−21	0.00E+00
f_{16}	$\bar{f}$	−1.22E+04	−4.40E+03	−5.81E+03	−8.65E+03	−1.23E+04	−8.68E+03	−7.92E+03	−4.07E+03	−4.39E+03	−4.73E+03
	σ_f	1.19E+02	3.64E+02	9.44E+02	2.11E+02	1.91E+02	7.68E+02	8.60E+02	2.85E+02	2.66E+02	5.25E+02
	f_{Best}	−1.25E+04	−5.45E+03	−8.26E+03	−9.31E+03	**−1.26E+04**	−1.05E+04	−1.00E+04	−4.83E+03	−4.93E+03	−6.00E+03

(continued)

Table 8.1 (continued)

		ABC	DE	PSO	CS	DS	MFO	MVO	SCA	CMA-ES	EA-SOM
	f_{Worst}	−1.20E+04	−3.73E+03	−4.14E+03	−8.38E+03	−1.19E+04	−7.20E+03	−5.56E+03	−3.52E+03	−3.93E+03	−4.11E+03
f_{17}	$\bar{f}$	5.04E−13	1.64E+07	3.69E+02	1.15E+00	3.46E−03	2.93E−01	9.20E−01	1.20E+00	1.10E−18	8.04E−01
	σ_f	3.97E−13	1.08E+07	9.05E+02	4.23E−01	1.89E−02	5.15E−01	1.00E+00	1.94E+00	9.55E−19	2.14E−01
	f_{Best}	8.06E−14	1.36E+06	1.31E+01	5.40E−01	1.09E−08	4.87E−07	7.55E−04	3.03E−01	**1.52E−19**	4.46E−01
	f_{Worst}	2.12E−12	4.05E+07	3.42E+03	2.18E+00	1.04E−01	1.98E+00	4.59E+00	1.11E+01	5.05E−18	1.30E+00
f_{18}	$\bar{f}$	7.02E−11	4.89E+07	6.62E+04	1.59E−01	6.59E−06	2.03E−01	3.79E−02	2.73E+00	9.59E−18	4.46E−02
	σ_f	2.41E−10	1.87E+07	7.98E+04	6.15E−02	1.72E−05	6.10E−01	2.22E−02	5.80E−01	8.11E−18	7.77E−03
	f_{Best}	5.99E−13	7.68E+06	6.98E+01	5.53E−02	2.36E−08	2.64E−05	7.68E−03	2.10E+00	**1.21E−18**	2.62E−02
	f_{Worst}	1.33E−09	1.02E+08	2.83E+05	2.95E−01	9.49E−05	3.60E+00	1.21E−01	5.04E+00	3.12E−17	5.96E−02
f_{19}	$\bar{f}$	9.72E−07	1.35E+04	4.79E+03	3.64E+00	9.59E−04	2.53E+04	5.13E−01	1.77E−03	2.39E−09	0.00E+00
	σ_f	4.09E−07	3.10E+03	4.07E+03	3.60E+00	7.20E−04	2.25E+04	1.22E−01	2.11E−03	7.80E−10	0.00E+00
	f_{Best}	2.67E−07	5.22E+03	1.21E+03	6.68E−01	2.12E−04	4.48E−04	3.09E−01	1.09E−06	1.08E−09	**0.00E+00**
	f_{Worst}	1.70E−06	1.89E+04	2.23E+04	1.51E+01	2.98E−03	1.00E+05	7.33E−01	8.34E−03	4.05E−09	0.00E+00
f_{20}	$\bar{f}$	2.90E+01	6.30E+02	3.54E+02	1.09E+02	3.09E+01	1.29E+02	1.45E+02	2.90E+01	3.13E+01	2.90E+01
	σ_f	1.00E−05	5.96E+01	5.38E+01	8.76E+00	4.78E+00	8.78E+01	2.45E+01	4.52E−03	4.24E+00	4.80E−05
	f_{Best}	**2.90E+01**	5.15E+02	2.33E+02	8.77E+01	**2.90E+01**	5.30E+01	9.13E+01	**2.90E+01**	**2.90E+01**	**2.90E+01**
	f_{Worst}	2.90E+01	7.81E+02	4.66E+02	1.23E+02	4.73E+01	4.01E+02	1.98E+02	2.90E+01	4.53E+01	2.90E+01
f_{21}	$\bar{f}$	2.78E+02	4.96E+07	1.12E+05	6.88E+01	1.56E+02	8.20E+06	5.13E+01	3.13E+02	3.20E+01	3.20E+01
	σ_f	3.26E+01	1.72E+07	1.64E+05	8.21E+00	4.96E+01	5.80E+07	6.52E+00	6.28E+02	3.18E−10	3.90E−06
	f_{Best}	1.92E+02	2.10E+07	1.84E+03	5.93E+01	9.01E+01	4.79E+01	4.06E+01	**3.20E+01**	**3.20E+01**	**3.20E+01**
	f_{Worst}	3.38E+02	7.77E+07	6.82E+05	9.59E+01	2.63E+02	4.10E+08	6.80E+01	3.55E+03	3.20E+01	3.20E+01

(continued)

Table 8.1 (continued)

		ABC	DE	PSO	CS	DS	MFO	MVO	SCA	CMA-ES	EA-SOM
f_{22}	$\bar{f}$	2.90E+01	7.46E+02	7.10E+02	1.59E+02	2.99E+01	9.66E+02	1.66E+02	2.90E+01	3.15E+01	2.90E+01
	σ_f	4.15E−06	1.07E+02	5.79E+02	1.50E+01	3.56E+00	5.61E+02	3.47E+01	3.79E−03	5.11E+00	0.00E+00
	f_{Best}	**2.90E+01**	5.45E+02	3.00E+02	1.27E+02	**2.90E+01**	4.28E+01	8.95E+01	**2.90E+01**	**2.90E+01**	**2.90E+01**
	f_{Worst}	2.90E+01	9.86E+02	2.97E+03	1.84E+02	4.35E+01	2.53E+03	2.51E+02	2.90E+01	4.27E+01	2.90E+01
f_{23}	$\bar{f}$	−8.36E+01	9.86E+06	5.81E+04	−6.34E+01	−8.26E+01	1.84E+08	−7.68E+01	6.67E+00	−8.29E+01	−7.41E+01
	σ_f	1.17E−01	6.42E+06	2.83E+05	7.02E+00	3.69E−01	2.74E+08	1.65E+01	9.99E+01	3.11E−01	1.97E+00
	f_{Best}	**−8.38E+01**	1.87E+06	1.40E+03	−7.29E+01	−8.31E+01	−8.27E+01	−8.17E+01	−3.23E+01	−8.35E+01	−7.73E+01
	f_{Worst}	−8.34E+01	2.82E+07	1.56E+06	−4.76E+01	−8.19E+01	1.02E+09	−1.30E+01	5.25E+02	−8.23E+01	−7.05E+01

Table 8.2 $p-$values produced by Wilcoxon rank sum test over the averaged fitness value $\bar{f}$ for each function from Table 8.1

EA−SOM versus	ABC	DE	PSO	CS	DS	MFO	MVO	SCA	CMA-ES
f_1	5.66E−10▲	5.66E−10▲	5.66E−10▲	5.66E−10▲	5.66E−10▲	5.66E−10▲	5.66E−10▲	5.66E−10▲	5.66E−10▲
f_2	5.66E−10▲	5.66E−10▲	5.66E−10▲	5.66E−10▲	5.66E−10▲	5.66E−10▲	5.66E−10▲	5.66E−10▲	5.66E−10▲
f_3	5.66E−10▲	5.66E−10▲	5.66E−10▲	5.66E−10▲	5.66E−10▲	5.66E−10▲	5.66E−10▲	5.66E−10▲	5.66E−10▲
f_4	5.66E−10▲	5.66E−10▲	5.66E−10▲	5.66E−10▲	5.66E−10▲	5.66E−10▲	5.66E−10▲	5.66E−10▲	5.66E−10▲
f_5	6.64E−09▲	6.64E−09▲	6.64E−09▲	6.64E−09▲	1.19E−07▲	1.67E−06▲	1.58E−04▲	1.12E−08▲	5.64E−03▲
f_6	4.60E−01►	5.66E−10▲	5.66E−10▲	2.45E−01►	7.16E−01►	8.08E−02▲	5.16E−10▲	3.39E−01►	1.07E−01►
f_7	6.64E−09▲	6.64E−09▲	6.64E−09▲	6.64E−09▲	6.64E−09▲	6.64E−09▲	6.64E−09▲	6.64E−09▲	6.64E−09▲
f_8	6.64E−01►	6.64E−09▲	6.64E−09▲	2.32E−01►	3.17E−01►	6.63E−09▲	5.56E−09►	6.64E−09▲	6.64E−09▲
f_9	4.82E−01►	6.64E−09▲	6.64E−09▲	3.21E−01►	8.30E−01►	6.64E−09▲	5.94E−01►	6.64E−09▲	1.79E−01►
f_{10}	5.66E−10▲	5.66E−10▲	5.66E−10▲	5.66E−10▲	5.66E−10▲	5.66E−10▲	5.66E−10▲	5.66E−10▲	5.66E−10▲
f_{11}	5.66E−10▲	5.66E−E−10▲	5.66E−10▲	5.66E−10▲	5.66E−10▲	5.66E−10▲	5.66E−10▲	5.66E−10▲	5.66E−10▲
f_{12}	5.66E−10▲	5.66E−10▲	5.66E−10▲	5.66E−10▲	5.66E−10▲	5.66E−10▲	5.66E−10▲	5.66E−10▲	1.13E−01►
f_{13}	3.56E−01►	5.66E−10▲	5.66E−10▲	5.66E−10▲	4.60E−01►	6.72E−01►	5.66E−10▲	4.31E−01►	9.54E−01►
f_{14}	8.25E−01►	6.64E−09▲	6.64E−09▲	4.08E−01►	1.87E−01►	2.10E−01►	5.84E−01►	6.41E−01►	1.45E−01►
f_{15}	5.66E−10▲	5.66E−10▲	5.66E−10▲	5.66E−10▲	5.66E−10▲	5.66E−10▲	5.66E−10▲	5.66E−10▲	5.66E−10▲
f_{16}	6.64E−09▼	1.04E−02▲	2.23E−05▼	6.64E−09▼	6.64E−09▼	6.64E−09▼	8.62E−09▼	6.02E−06▲	1.53E−01▲
f_{17}	6.64E−09▼	6.64E−09▲	6.64E−09▲	2.26E−03▼	6.64E−09▼	8.41E−05▼	4.29E−01▼	3.92E−01▼	6.64E−09▼
f_{18}	6.64E−09▼	6.64E−09▲	6.64E−09▲	8.62E−09▼	6.64E−09▼	2.44E−04▼	9.94E−02▼	6.64E−09▲	6.64E−09▼
f_{19}	5.66E−10▲	5.66E−10▲	5.66E−10▲	5.66E−10▲	5.66E−10▲	5.66E−10▲	5.66E−10▲	5.66E−10▲	5.66E−10▲
f_{20}	3.56E−01►	6.64E−09▲	6.64E−09▲	6.64E−09▲	4.04E−01►	6.64E−09▲	6.64E−09▲	4.21E−01►	2.36E−01►

(continued)

Table 8.2 (continued)

EA–SOM versus	ABC	DE	PSO	CS	DS	MFO	MVO	SCA	CMA-ES
f_{21}	6.64E−09▲	6.64E−09▲	6.64E−09▲	6.64E−09▲	6.64E−09▲	6.64E−09▲	6.64E−09▲	2.50E−01►	1.81E−01►
f_{22}	5.32E−01►	5.66E−10▲	5.66E−10▲	5.66E−10▲	3.98E−01►	5.66E−10▲	5.66E−10▲	4.25E−01►	1.47E−01►
f_{23}	6.64E−09▼	6.64E−09▲	6.64E−09▲	5.08E−08▲	6.64E−09▼	4.55E−03▼	1.19E−07▼	6.64E−09▲	6.64E−09▼
▲	12	23	22	16	12	17	16	16	12
▼	4	0	1	3	4	4	4	1	3
►	7	0	0	4	7	2	3	6	8

▶ describes that the rank sum test is not capable of distinguishing among the results of the proposed technique with the results of a given algorithm. Regarding to the p-values from Table 8.2, it is confirmed that the proposed method performs better for the majority of the test functions. Nonetheless, there are some exceptions.

For function f_{16}, the groups: EA-SOM versus ABC, EA-SOM versus PSO, EA-SOM versus CS, EA-SOM versus DS, EA-SOM versus MFO and EA-SOM versus MVO show negative performance of the proposed technique, In function f_{17}, the groups formed by: EA-SOM versus ABC, EA-SOM versus CS, EA-SOM versus DS, EA-SOM versus MFO, EA-SOM versus MVO, EA-SOM versus SCA and EA-SOM versus CMA-ES display better index than EA-SOM. Ultimately, for function f_{18}, the performance of the EA-SOM approach gives the worst performance in the groups conformed by: EA-SOM versus ABC, EA-SOM versus CS, EA-SOM versus DS, EA-SOM versus MFO, EA-SOM versus MVO and EA-SOM versus CMA-ES. For functions f_6, f_8, f_9, and f_{12}, the ABC, CS, DS, MVO, SCA and CMA-ES presents similar performance than EA-SOM meaning that the Wilcoxon test is not capable to deliver a significance interval.

As a conclusion, the EK scheme of EA-SOM spreads the exploration and exploitation abilities of the tested evolutionary algorithms delivering the best performance regarding the total amount of experiments.

8.5.2 Convergence

The numerical results of the comparison in Sect. 8.5.1 are based on fitness values. Such results cannot fully describe the performance of evolutionary strategies. Therefore, a convergence analysis should be incorporated. In this subsection, the convergence examination over ABC, DE, PSO, CS, DS, MFO, MVO, SCA, CMA-ES and EA-SOM is performed. The central aim of this is to judge the velocity which certain method leads the optimum solution. This examination is required since accurate solutions do not assure acceptable performance.

In this analysis, the delivered performance of each method is evaluated within a 30-dimensional search space over each function using $Maxgen = 1000$. To construct the converge plots, each function is executed 30 times to eliminate the stochastic effect. The convergence data was chosen regarding the average fitness value $\bar{f}$. Figure 8.4, depicts the convergence plots for the suite of functions.

Figure 8.4 hints that the convergence velocity of EA-SOM by solving the suit of functions is the fastest than the convergence speed of the rest of the methods in most of the functions. As can be concluded, the performance of the proposal gives the best fitness values in a smaller number of generations. This exceptional behavior resembles the capacity of the method to create a stable model of the regions through the use of historical information.

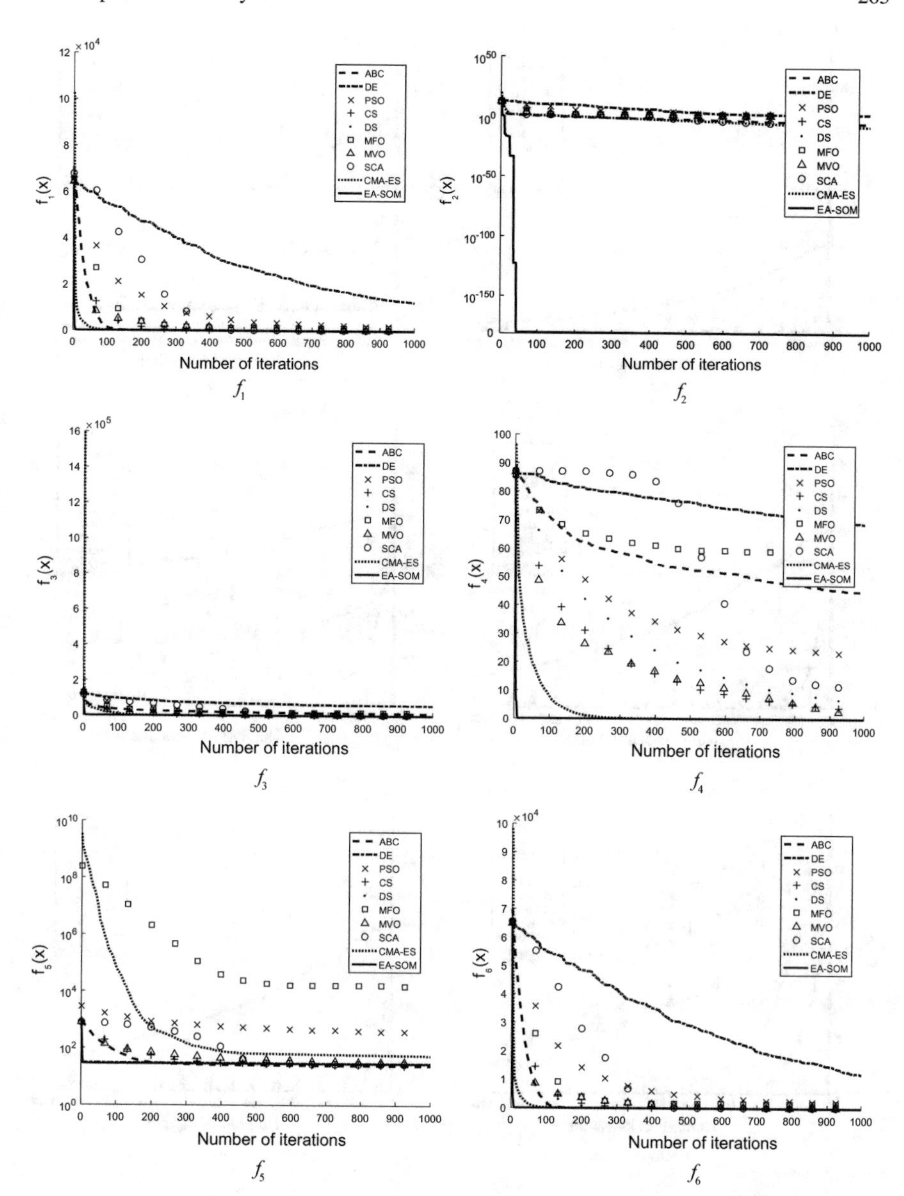

Fig. 8.4 Convergence test graph for functions of Table B.10

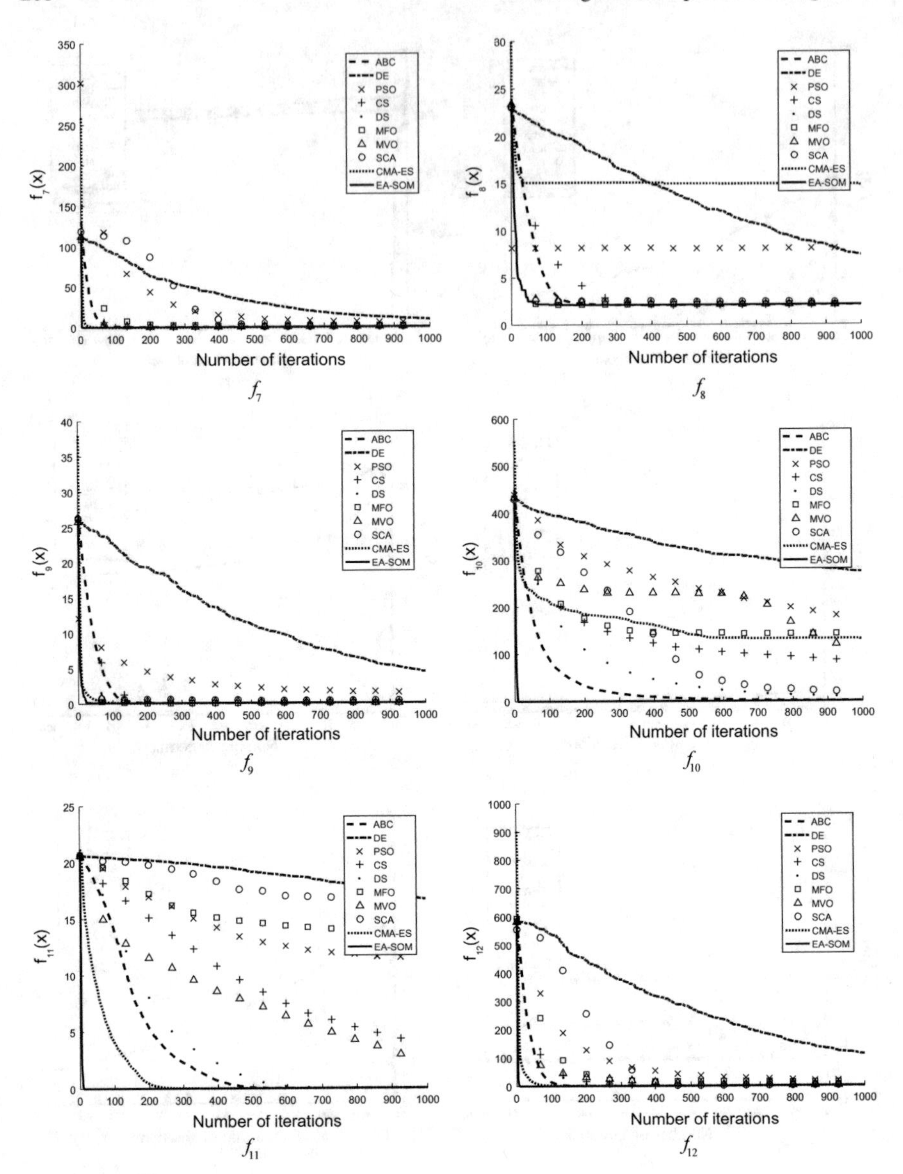

Fig. 8.4 (continued)

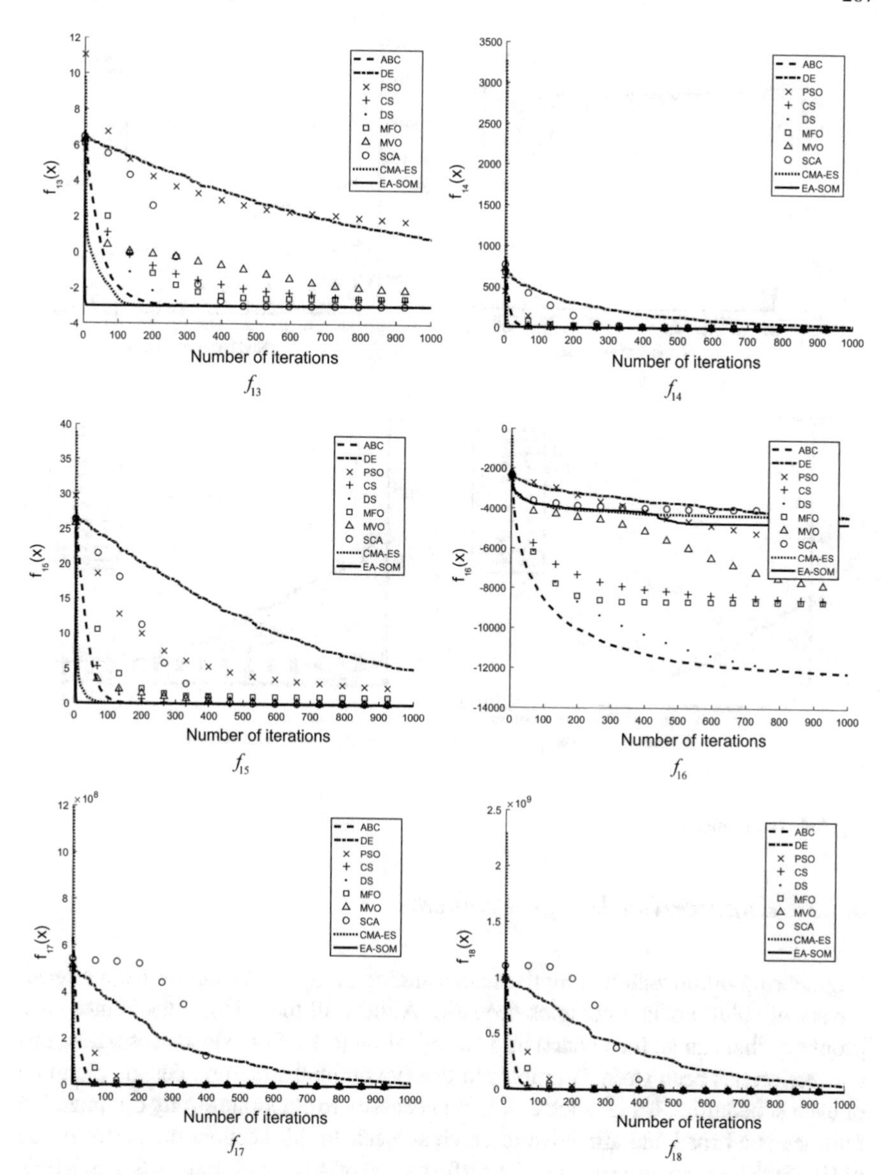

Fig. 8.4 (continued)

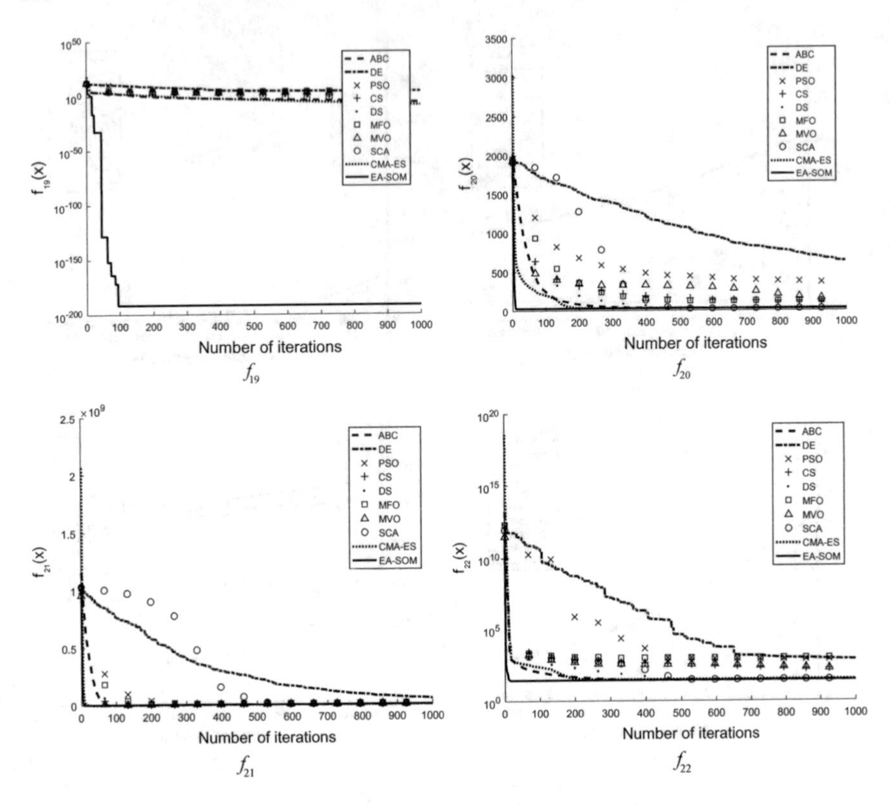

Fig. 8.4 (continued)

8.5.3 Engineering Design Problems

Engineering optimization comprises determining an optimal solution from a feasible set of solutions in a complex scenario. Almost all the science disciplines have problems that can be formulated into optimization tasks. Some instances where this conversion has been strongly employed involve optimal structural design, planning or even scheduling. Under such cases, the necessity to elaborate strong optimization routines is a broad and attractive research subject. In this section, the performance of EA-SOM is compared against the performance of ABC, DE, PSO, CS, DS, MFO, MVO, SCA and CMA-ES considering 3 commonly used optimization design problems [39, 41]. In Sect. 8.5.3.1 the performance effects of the three-bar truss design problem are manifested. Section 8.5.3.2 examines the tension/compression spring design problem. Ultimately, the experimental outcomes of the welded beam design problem are exhibited in Sect. 8.5.3.3.

8.5.3.1 Three-Bar Truss Design Problem

Three-bar truss design problem is recognized as an engineering problem to judge the robustness of an optimization methodology. The principal aim of this problem is to minimize the amount of a weighted three-bar truss subject to stress constraints on each of the truss segments by modifying cross-sectional areas x_1, x_2. The problem examines an objective function with three inequalities and two decision variables as specified in Table B.7. The numeric simulation of the evaluation for the constrained objective function is accomplished regarding the parameter configuration detailed in Sect. 8.5.1 using 1000 generations. The numerical results are displayed in Table 8.3 and Fig. 8.5 displays the three-bar truss.

As it can be determined, Table 8.3 records that the proposed algorithm takes extremely similar results than CS, DS and MVO though 30 executions. In the table, it is manifested the minimum fitness value, the parameters, and constraints for the optimization problem. The statistical results of the results are presented in Table 8.4. In the table, it is shown the worst, mean, standard deviation and the best fitness values. As it can be assumed in Table 8.4, EA-SOM achieves competitive results versus the remaining algorithms.

8.5.3.2 Tension/Compression Spring Design Problem

Spring design problem offers an interesting optimization problem to test the flexibility of an optimization technique. The main objective is to minimize the weight tension/compression spring. The problem is described in Table B.8. In this optimization problem, three decision variables: wire diameter $w(x_1)$, mean coil diameter $d(x_2)$ and the number of active coils $L(x_3)$ are involved in the process while it contains one linear constraint and three non-linear inequality constraints. The experimental outcomes are exhibited in Table 8.5 and Fig. 8.6 represents a schematic of the problem.

From Table 8.5, it can be assumed that EA-SOM works very similar than CS, DS, MFO, MVO, and SCA. Table 8.6 manifests the statistical results of the methods. From the results in Table 8.6, it can be confirmed that EA-SOM provides more uniform results versus its competitors.

8.5.3.3 Welded Beam Design Problem

Welded beam problem describes a complex optimization task whose aim is to obtain the minimal cost of a welded beam. The graphical description of the welded beam is presented in Fig. 8.7 and its mathematical definition is exposed in Table B.9. The design method requires four decision variables: the width $h(x_1)$, the length $l(x_2)$ of the welded area, the depth $t(x_3)$ and the thickness $b(x_4)$. The experimental results are manifested in Table 8.7.

Table 8.3 Results of the minimization of the three-bar truss problem

Parameter	ABC	DE	PSO	CS	DS	MFO	MVO	SCA	CMA-ES	EA-SOM
x_1	0.8643	0.8708	0.8573	0.8698	0.8700	0.8702	0.8697	0.8730	0.8721	0.8697
x_2	0.2278	0.2149	0.2710	0.2166	0.2162	0.2157	0.2167	0.2101	0.2122	0.2166
$g(x_1)$	−1.49E−04	−3.60E−04	−0.0275	−2.2204e−16	−6.2774e−10	0	−2.5611e−07	−5.7400e−05	−2.43E−04	−1.5336e−09
$g(x_2)$	−1.6858	−1.7029	−1.6396	−1.7005	−1.7010	−1.7018	−1.7005	−1.7091	−1.7064	−1.7005
$g(x_3)$	−0.3143	−0.2975	−0.3878	−0.2995	−0.2990	−0.2982	−0.2995	−0.2909	−0.2938	−0.2995
$f(x)$	279.7452	279.7472	280.5426	**279.7245**	**279.7245**	279.7246	**279.7245**	279.7312	279.7411	**279.7245**

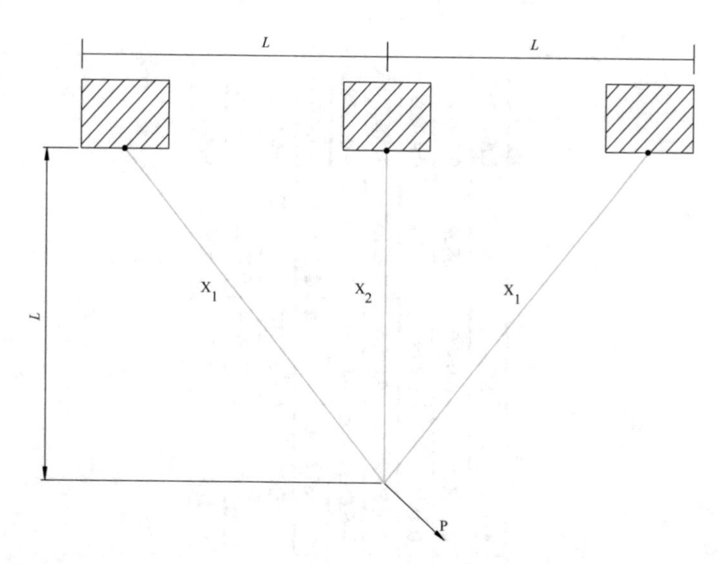

Fig. 8.5 Three-bar truss problem

Table 8.4 Statistical results for the three-bar truss problem

Algorithm	Worst	Mean	Std	Best
ABC	279.7849	279.7372	0.0125	279.7452
DE	280.1867	279.8096	0.0902	279.7472
PSO	282.8226	93.2079	29.9639	280.5426
CS	279.7245	279.7245	1.3843e−13	**279.7245**
DS	280.0971	279.7484	0.0677	**279.7245**
MFO	280.1945	279.7763	0.1028	279.7246
MVO	279.7270	279.7251	5.7728e−04	**279.7245**
SCA	282.8427	281.4059	1.5624	279.7312
CMA-ES	279.7884	191.7808	88.7932	279.7411
EA-SOM	279.7294	279.7245	0.0049	**279.7245**

From Table 8.7, it can be confirmed that EA-SOM method gives comparable results to CS and MFO across 30 runs. Table 8.8 reports the statistical results achieved by the rest of the evolutionary algorithms. From the outcomes, it can be concluded that EA-SOM provides more consistent outcomes versus the rest of its competitors.

Table 8.5 Results of the minimization of the tension/compression spring problem

Parameter	ABC	DE	PSO	CS	DS	MFO	MVO	SCA	CMA-ES	EA-SOM
w	0.0500	0.0500	0.0516	0.0500	0.0500	0.0500	0.0500	0.0500	1.6858	0.0500
d	0.4532	0.4278	0.3560	0.4800	0.4800	0.4800	0.4800	0.4796	0.9031	0.4800
L	4.8539	5.8778	11.3295	4.0563	4.0577	4.0563	4.0580	4.0806	6.4315	4.0580
$g(x_1)$	−0.0070	−0.0257	−0.000006	−1.5477e−13	−4.8364e−06	−0.0025	−6.6620e−05	−0.0035	1.0000	−4.1608e−04
$g(x_2)$	−0.1001	−0.1896	−0.000013	−2.0761e−14	−2.1545e−04	−1.6653e−15	−2.1897e−04	−0.0015	2.2629	−5.9926e−06
$g(x_3)$	−6.0440	−5.5282	−4.0523	−6.5135	−6.5126	−6.4950	−6.5121	−6.4810	−44.1382	−6.5103
$g(x_4)$	−0.6645	−0.6815	−0.7282	−0.6467	−0.6467	−0.6467	−0.6467	−0.6469	0.7259	−0.6467
$f(x)$	0.0078	0.0084	0.0127	**0.0073**	**0.0073**	**0.0073**	**0.0073**	**0.0073**	1.00E+06	**0.0073**

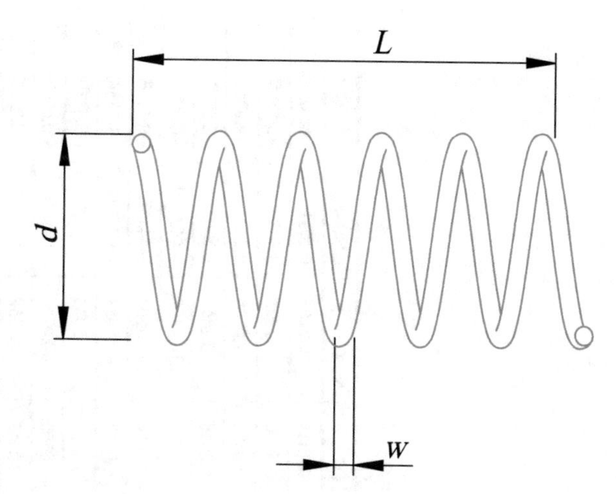

Fig. 8.6 Tension/compression spring design problem

Table 8.6 Statistical results for the tension/compression spring problem

Algorithm	Worst	Mean	Std	Best
ABC	0.0084	0.0076	0.0002	0.0078
DE	1.00E+06	2.00E+05	4.07E+05	0.0084
PSO	1.00E+06	1.67E+05	3.79E+05	0.0127
CS	0.0073	0.0073	2.3681e−12	**0.0073**
DS	0.0081	0.0074	2.2511e−04	**0.0073**
MFO	0.0073	0.0073	1.5149e−07	**0.0073**
MVO	1.00E+06	4.0000e+05	4.9827e+05	**0.0073**
SCA	0.0077	0.0074	9.4466e−05	**0.0073**
CMA-ES	1.00E+06	6.00E+05	4.98E+05	1.00E+06
EA-SOM	0.0073	0.0073	6.4271e−13	**0.0073**

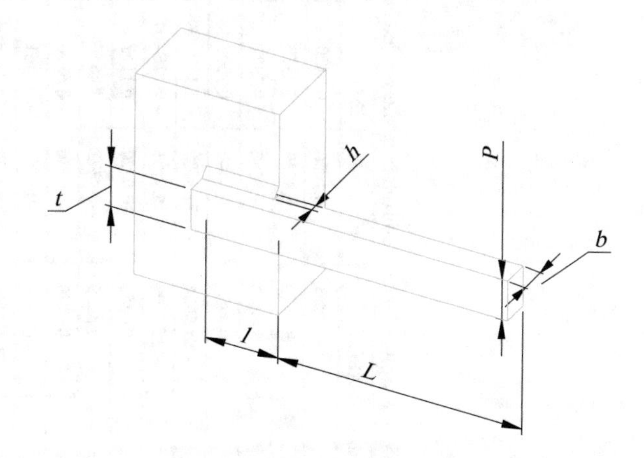

Fig. 8.7 Welded beam design problem

Table 8.7 Results of the minimization of the welded beam problem

Parameter	ABC	DE	PSO	CS	DS	MFO	MVO	SCA	CMA-ES	EA-SOM
h	0.1993	0.1626	0.8120	0.2057	0.2057	0.2057	0.2045	0.2004	0.2031	0.2057
l	3.8115	6.8239	6.4319	3.4705	3.4706	3.4705	3.5040	3.7893	3.5430	3.4705
t	8.6442	8.7650	9.2837	9.0366	9.0385	9.0366	9.0375	9.0748	9.0335	9.0366
b	0.2259	0.2346	0.7008	0.2057	0.2057	0.2057	0.2058	0.2062	0.2062	0.2057
$g(x_1)$	−81.6604	−2977.1639	−1.16E+04	−0.0062	−0.0322	−0.0062	−23.2984	−616.4397	−44.5785	−0.0062
$g(x_2)$	−135.7300	−2035.9101	−2.17E+04	−0.1334	−14.1393	−0.1334	−9.1686	−315.5598	−44.6635	−0.1188
$g(x_3)$	−0.0265	−0.0720	0.1112	−1.2836e−06	−5.0225e−05	−1.2836e−06	−0.0012	−0.0058	−0.0030	−1.1836e−06
$g(x_4)$	−3.3229	−2.9372	1.4640	−3.4330	−3.4326	−3.4406	−3.4297	−3.3945	−3.4237	−3.4330
$g(x_5)$	−0.0743	−0.0376	−0.6870	−0.0807	−0.0807	−0.0807	−0.0795	−0.0754	−0.0781	−0.0807
$g(x_6)$	−0.2350	−0.2361	−0.2461	−0.2355	−0.2356	−0.2355	−0.2355	−0.2358	−0.2355	−0.2355
$g(x_7)$	−1706.7356	−2718.3738	−2.35E+05	−0.0157	−1.9242	−0.0157	−2.5351	−55.4759	−38.0282	−0.0069
$f(x)$	1.8403	2.2593	2.0118	**1.7249**	1.7252	**1.7249**	1.7278	1.7694	1.7334	**1.7249**

Table 8.8 Statistical results for the welded beam problem

Algorithm	Worst	Mean	Std	Best
ABC	2.0848	1.9269	0.0849	1.8403
DE	3.6161	2.6481	0.3445	2.2593
PSO	3.3931	2.1544	0.3594	2.0118
CS	1.7250	1.7249	2.2364e−05	**1.7249**
DS	2.2552	1.8165	0.1205	1.7252
MFO	2.3171	1.7854	0.1282	**1.7249**
MVO	1.8169	1.7412	0.0182	1.7278
SCA	1.9711	1.8519	0.0493	1.7694
CMA-ES	1.11E+172	3.69E+170	NA	1.7334
EA-SOM	1.7249	1.7249	5.6400e−06	**1.7249**

NA means not available

8.6 Conclusions

In this chapter, a novel evolutionary optimization methodology called EA-SOM has been presented. In EA-SOM, an Extracted Knowledge (EK) mechanism is employed through its process to efficiently lead the search tactics to solve optimization problems. In the proposal, a Self-Organization Map (SOM) is employed as EK procedure to recognize and identify the promissory search zones by its data-reduction scheme. Hence, each iteration, the proposed methodology employs a set of solutions to train the SOM. Once trained, the neurons from the output layer that relates to the best solution are exploited. Then, by applying local information of this neural network, a whole population of potential solutions is provided. With the use of the EK mechanism, the novel proposal increases the convergence speed of multimodal functions by handling a reduced number of function calls.

The performance of the proposal has been analyzed considering several state-of-the-art evolutionary techniques evaluating a suite of 23 benchmark functions as well as three engineering design problems. The results confirm that the proposed method achieves the best balance among evolutionary stages regarding the accuracy and computational effort over its competitors. This outstanding performance demonstrates the potential of using historical information to build a strong representation for recognizing the promissory zones in the search space.

References

1. J. Han, M. Kamber, *Data mining: concepts and techniques* (Morgan Kaufmann, San Francisco, CA, USA, 2001)
2. G. Manco, P. Rullo, L. Gallucci, M. Paturzo, Rialto: a knowledge discovery suite for data analysis. Expert Syst. Appl. **59**, 145–164 (2016)

3. T. Kohonen, Self-organized formation of topologically correct feature maps. Biol. Cybernetics **43**, 59–69 (1982)
4. Q. Zhou, Y. Wang, P. Jiang, X. Shao, S.-K. Choi, H. Jiexiang, Longchao Cao, X. Meng, An active learning radial basis function modeling method based on self-organization maps for simulation-based design problems. Knowl.-Based Syst. **131**, 10–27 (2017)
5. S. Delgado, C. Higuera, J. Calle-Espinosa, F. Morán, F. Montero, A SOM prototype-based cluster analysis methodology. Expert Syst. Appl. **88**, 14–28 (2017)
6. A.A. Akinduko, E.M. Mirkes, A.N. Gorban, SOM: stochastic initialization versus principal components. Inf. Sci. **364–365**, 213–221 (2016)
7. S.J. Nanda, G. Panda, A survey on nature inspired metaheuristic algorithms for partitional clustering. Swarm Evol. Comput. **16**, 1–18 (2014)
8. J. Kennedy, R. Eberhart, Particle swarm optimization, in *Proceedings of the 1995 IEEE International Conference on Neural Networks*, vol. 4, pp. 1942–1948, December 1995
9. D. Karaboga, An Idea Based on Honey Bee Swarm for Numerical Optimization. TechnicalReport-TR06. Engineering Faculty, Computer Engineering Department, Erciyes University, 2005
10. Z.W. Geem, J.H. Kim, G.V. Loganathan, A new heuristic optimization algorithm: harmony search. Simulations **76**, 60–68 (2001)
11. X.S. Yang, A new metaheuristic bat-inspired algorithm, in *Nature Inspired Cooperative Strategies for Optimization (NISCO 2010), Studies in Computational Intelligence*, vol. 284, ed. by C. Cruz, J. González, G.T.N. Krasnogor, D.A. Pelta (Springer, Berlin, 2010), pp. 65–74
12. X.S. Yang, Firefly algorithms for multimodal optimization, in: *Stochastic Algorithms: Foundations and Applications*, SAGA 2009, Lecture Notes in Computer Sciences, vol. 5792, pp. 169–178, 2009
13. E. Cuevas, M. Cienfuegos, D. Zaldívar, M. Pérez-Cisneros, A swarm optimization algorithm inspired in the behavior of the social-spider. Expert Syst. Appl. **40**(16), 6374–6384 (2013)
14. E. Cuevas, M. González, D. Zaldivar, M. Pérez-Cisneros, G. García, An algorithm for global optimization inspired by collective animal behaviour. Discrete Dyn. Nat. Soc. (2012). Art. no. 638275
15. R. Storn, K. Price, Differential Evolution–A Simple and Efficient Adaptive Scheme for Global Optimisation Over Continuous Spaces. TechnicalReportTR-95–012, ICSI, Berkeley, CA, 1995
16. D.E. Goldberg, *Genetic Algorithm in Search Optimization and Machine Learning* (Addison-Wesley, 1989)
17. X.-S. Yang, S. Deb, Cuckoo search via L´evy flights, in *Proceedings World Congress on Nature and Biologically Inspired Computing (NABIC '09)* (2009), pp. 210–214
18. P. Civicioglu, Transforming geocentric cartesian coordinates to geodetic coordinates by using differential search algorithm. Comput. Geosci. **46**, 229–247 (2012)
19. S. Mirjalili, Moth-flame optimization algorithm: a novel nature-inspired heuristic paradigm. Knowl.-Based Syst. **89**, 228–249 (2015)
20. S. Mirjalili, S.M. Mirjalili, A. Hatamlou, Multi-verse optimizer: a nature-inspired algorithm for global optimization. Neural Comput. Appl. **27**(2), 495–513 (2016)
21. S. Mirjalili, SCA: a sine cosine algorithm for solving optimization problems. Knowl.-Based Syst. **96**, 120–133 (2016)
22. N. Hansen, S. Kern, Evaluating the CMA evolution strategy on multimodal test functions, in *Proceedings of 8th International Conference on Parallel Problem Solving from Nature—PPSN VIII*, vol. 3242/2004, no. 0, (2004) pp. 282–291
23. R. Giraldez, J.S. Aguilar-Ruiz, J.C. Riquelme, Knowledge-based fast evaluation for evolutionary learning. IEEE Trans. Syst. Man Cybern. Part C **35**(2), 254–261 (2005)
24. G. Kobeaga, M. Merino, J.A. Lozano, An efficient evolutionary algorithm for the orienteering problem. Comput. Oper. Res. **90**, 42–59 (2018)
25. R. Thomsen, G.B. Fogel, T. Krink, A clustal alignment improver using evolutionary algorithms, in *Proceedings of the 4th Congress Evolutionary Computation (CEC'2002)*, vol. 1, (2002) pp. 121–126

26. L. Wang, S. Wang, A knowledge-based multi-agent evolutionary algorithm for semiconductor final testing scheduling problem. Knowl.-Based Syst. **84**, 1–9 (2015)
27. M. Deveci, N.Ç. Demirel, Evolutionary algorithms for solving the airline crew pairing problem. Comput. Ind. Eng. **115**, 389–406 (2018)
28. M. Mobin, S.M. Mousavi, M. Komaki, M. Tavana, A hybrid desirability function approach for tuning parameters in evolutionary optimization algorithms. Measurement **114**, 417–427 (2018)
29. T. Agasiev, A. Karpenko, The program system for automated parameter tuning of optimization algorithms. Procedia Comput. Sci. **103**, 347–354 (2017)
30. E. Yeguas, M.V. Luzón, R. Pavón, R. Laza, G. Arroyo, F. Díaz, Automatic parameter tuning for evolutionary algorithms using a bayesian case-based reasoning system. Appl. Soft Comput. **18**, 185–195 (2014)
31. M.S. Nobile, P. Cazzaniga, D. Besozzi, R. Colombo, G. Mauri, G. Pasi, Fuzzy self-tuning PSO: a settings-free algorithm for global optimization. Swarm Evol. Comput. (2018). In press
32. S. Elsayed, R. Sarker, C.C. Coello, T. Ray, Adaptation of operators and continuous control parameters in differential evolution for constrained optimization. Soft Comput. 1–22 (2018). In press
33. L. Hong, J.H. Drake, J.R. Woodward, E. Özcan, A hyper-heuristic approach to automated generation of mutation operators for evolutionary programming. Appl. Soft Comput. **62**, 162–175 (2018)
34. Z. Hu, J. Yang, H. Sun, L. Wei, Z. Zhao, An improved multi-objective evolutionary algorithm based on environmental and history information. Neurocomputing **222**, 170–182 (2017)
35. C. Coello, R. Landa, Adding knowledge and efficient data structures to evolutionary programming: a cultural algorithm for constrained optimization, in *Proceeding GECCO'02 Proceedings of the 4th Annual Conference on Genetic and Evolutionary Computation* (2002), pp. 201–209
36. S. Haykin, *Neural Networks: A Comprehensive Foundation* (Prentice Hall, New York, 1999)
37. F. Wilcoxon, Individual comparisons by ranking methods. Biometrics, 80–83 (1945)
38. G. Anescu, Further scalable test functions for multidimensional continuous optimization (2017, Nov)
39. M.D. Li, H. Zhao, X.W. Weng, T. Han, A novel nature-inspired algorithm for optimization: virus colony search. Adv. Eng. Softw. **92**, 65–88 (2016)
40. X.-S. Yang, *Engineering Optimization : An Introduction with Metaheuristic Applications* (Wiley, London, 2010). Wiley InterScience (Online service)
41. A. Askarzadeh, A novel metaheuristic method for solving constrained engineering optimization problems: crow search algorithm. Comput. Struct. **169**, 1–12 (2016)
42. E. Rashedi, H. Nezamabadi-pour, S. Saryazdi, GSA: a gravitational search algorithm. Inf. Sci. (Ny) **179**(13), 2232–2248 (2009)
43. S. Yu, S. Zhu, Y. Ma, D. Mao, A variable step size firefly algorithm for numerical optimization. Appl. Math. Comput. **263**, 214–220 (2015)
44. X.-S. Yang, M. Karamanoglu, X. He, Flower pollination algorithm: a novel approach for multiobjective optimization. Eng. Optim. **46**(9), 1222–1237 (2014)

Appendix A
Systems Data

Tables A.1, A.2 and A.3.

Table A.1 10-bus test system data

Line no.	From bus i	To bus $i+1$	R (Ω)	X (Ω)	P_L (kW)	Q_L (kVAR)
1	1	2	1.35309	1.3235	1840	460
2	2	3	1.17024	1.1446	980	340
3	3	4	0.84111	0.8227	1790	446
4	4	5	1.52348	1.0276	1598	1840
5	2	9	2.01317	1.3579	1610	600
6	9	10	1.68671	1.1377	780	110
7	2	6	2.55727	1.7249	1150	60
8	6	7	1.0882	0.7340	980	130
9	6	8	1.25143	0.8441	1640	200

E. Cuevas et al., *Recent Metaheuristics Algorithms for Parameter Identification*, Studies in Computational Intelligence 854, https://doi.org/10.1007/978-3-030-28917-1

Table A.2 33-bus test system data

Line no.	From bus i	To bus $i+1$	R (Ω)	X (Ω)	P_L (kW)	Q_L (kVAR)
1	1	2	0.0922	0.0477	100	60
2	2	3	0.4930	0.2511	90	40
3	3	4	0.3660	0.1864	120	80
4	4	5	0.3811	0.1941	60	30
5	5	6	0.8190	0.7070	60	20
6	6	7	0.1872	0.6188	200	100
7	7	8	1.7114	1.2351	200	100
8	8	9	1.0300	0.7400	60	20
9	9	10	1.0400	0.7400	60	20
10	10	11	0.1966	0.0650	45	30
11	11	12	0.3744	0.1238	60	35
12	12	13	1.4680	1.1550	60	35
13	13	14	0.5416	0.7129	120	80
14	14	15	0.5910	0.5260	60	10
15	15	16	0.7463	0.5450	60	20
16	16	17	1.2890	1.7210	60	20
17	17	18	0.7320	0.5740	90	40
18	2	19	0.1640	0.1565	90	40
19	19	20	1.5042	1.3554	90	40
20	20	21	0.4095	0.4784	90	40
21	21	22	0.7089	0.9373	90	40
22	3	23	0.4512	0.3083	90	50
23	23	24	0.8980	0.7091	420	200
24	24	25	0.8960	0.7011	420	200
25	6	26	0.2030	0.1034	60	25
26	26	27	0.2842	0.1447	60	25
27	27	28	1.0590	0.9337	60	20
28	28	29	0.8042	0.7006	120	70
29	29	30	0.5075	0.2585	200	600
30	30	31	0.9744	0.9630	150	70
31	31	32	0.3105	0.3619	210	100
32	32	33	0.3410	0.5302	60	40

Table A.3 69-bus test system data

Line no.	From bus i	To bus $i+1$	R (Ω)	X (Ω)	P_L (kW)	Q_L (kVAR)
1	1	2	0.00050	0.0012	0.00	0.00
2	2	3	0.00050	0.0012	0.00	0.00
3	3	4	0.00150	0.0036	0.00	0.00
4	4	5	0.02510	0.0294	0.00	0.00
5	5	6	0.36600	0.1864	2.60	2.20
6	6	7	0.38100	0.1941	40.40	30.00
7	7	8	0.09220	0.0470	75.00	54.00
8	8	9	0.04930	0.0251	30.00	22.00
9	9	10	0.81900	0.2707	28.00	19.00
10	10	11	0.18720	0.0619	145.00	104.00
11	11	12	0.71140	0.2351	145.00	104.00
12	12	13	1.03000	0.3400	8.00	5.00
13	13	14	1.04400	0.3400	8.00	5.00
14	14	15	1.05800	0.3496	0.00	0.00
15	15	16	0.19660	0.0650	45.00	30.00
16	16	17	0.37440	0.1238	60.00	35.00
17	17	18	0.00470	0.0016	60.00	35.00
18	18	19	0.32760	0.1083	0.00	0.00
19	19	20	0.21060	0.0690	1.00	0.60
20	20	21	0.34160	0.1129	114.00	81.00
21	21	22	0.01400	0.0046	5.00	3.50
22	22	23	0.15910	0.0526	0.00	0.00
23	23	24	0.34630	0.1145	28.00	20.00
24	24	25	0.74880	0.2475	0.00	0.00
25	25	26	0.30890	0.1021	14.00	10.00
26	26	27	0.17320	0.0572	14.00	10.00
27	3	28	0.00440	0.0108	26.00	18.60
28	28	29	0.06400	0.1565	26.00	18.60
29	29	30	0.39780	0.1315	0.00	0.00
30	30	31	0.07020	0.0232	0.00	0.00
31	31	32	0.35100	0.1160	0.00	0.00
32	32	33	0.83900	0.2816	14.00	10.00
33	33	34	1.70800	0.5646	19.50	14.00
34	34	35	1.47400	0.4873	6.00	4.00
35	3	36	0.00440	0.0108	26.00	18.55
36	36	37	0.06400	0.1565	26.00	18.55

(continued)

Table A.3 (continued)

Line no.	From bus i	To bus $i+1$	R (Ω)	X (Ω)	P_L (kW)	Q_L (kVAR)
37	37	38	0.10530	0.1230	0.00	0.00
38	38	39	0.03040	0.0355	24.00	17.00
39	39	40	0.00180	0.0021	24.00	17.00
40	40	41	0.72830	0.8509	1.20	1.00
41	41	42	0.31000	0.3623	0.00	0.00
42	42	43	0.04100	0.0478	6.00	4.30
43	43	44	0.00920	0.0116	0.00	0.00
44	44	45	0.10890	0.1373	39.22	26.30
45	45	46	0.00090	0.0012	39.22	26.30
46	4	47	0.00340	0.0084	0.00	0.00
47	47	48	0.08510	0.2083	79.00	56.40
48	48	49	0.28980	0.7091	384.70	274.50
49	49	50	0.08220	0.2011	384.70	274.50
50	8	51	0.09280	0.0473	40.50	28.30
51	51	52	0.33190	0.1140	3.60	2.70
52	9	53	0.17400	0.0886	4.35	3.50
53	53	54	0.20300	0.1034	26.40	19.00
54	54	55	0.28420	0.1447	24.00	17.20
55	55	56	0.28130	0.1433	0.00	0.00
56	56	57	1.59000	0.5337	0.00	0.00
57	57	58	0.78370	0.2630	0.00	0.00
58	58	59	0.30420	0.1006	100.00	72.00
59	59	60	0.38610	0.1172	0.00	0.00
60	60	61	0.50750	0.2585	1244.00	888.00
61	61	62	0.09740	0.0496	32.00	23.00
62	62	63	0.14500	0.0738	0.00	0.00
63	63	64	0.71050	0.3619	227.00	162.00
64	64	65	1.04100	0.5302	59.00	42.00
65	11	66	0.20120	0.0611	18.00	13.00
66	66	67	0.00470	0.0014	18.00	13.00
67	12	68	0.73940	0.2444	28.00	20.00
68	68	69	0.00470	0.0016	28.00	20.00

Appendix B
Optimization Problems

Tables B.1, B.2, B.3, B.4, B.5, B.6, B.7, B.8, B.9 and B.10.

Table B.1 Unimodal test functions

Function	S	Dim	Minimum
$f_1(\mathbf{x}) = \sum_{i=1}^{n} x_i^2$	$[-100, 100]^n$	$n = 50$ $n = 100$	$\mathbf{x}^* = (0, \ldots, 0)$; $f(\mathbf{x}^*) = 0$
$f_2(\mathbf{x}) = \sum_{i=1}^{n} \lvert x_i \rvert + \prod_{i=1}^{n} \lvert x_i \rvert$	$[-10, 10]^n$	$n = 50$ $n = 100$	$\mathbf{x}^* = (0, \ldots, 0)$; $f(\mathbf{x}^*) = 0$
$f_3(\mathbf{x}) = \sum_{i=1}^{n} \left(\sum_{j=1}^{i} x_j \right)^2$	$[-100, 100]^n$	$n = 50$ $n = 100$	$\mathbf{x}^* = (0, \ldots, 0)$; $f(\mathbf{x}^*) = 0$
$f_4(\mathbf{x}) = \max_i \{\lvert x_i \rvert, 1 \le i \le n\}$	$[-100, 100]^n$	$n = 50$ $n = 100$	$\mathbf{x}^* = (0, \ldots, 0)$; $f(\mathbf{x}^*) = 0$
$f_5(\mathbf{x}) = \sum_{i=1}^{n-1} \left[100(x_{i+1} - x_i^2)^2 + (x_i - 1)^2 \right]$	$[-30, 30]^n$	$n = 50$ $n = 100$	$\mathbf{x}^* = (1, \ldots, 1)$; $f(\mathbf{x}^*) = 0$
$f_6(\mathbf{x}) = \sum_{i=1}^{n} (\lfloor x_i + 0.5 \rfloor)^2$	$[-100, 100]^n$	$n = 50$ $n = 100$	$\mathbf{x}^* = (0, \ldots, 0)$; $f(\mathbf{x}^*) = 0$
$f_7(\mathbf{x}) = \sum_{i=1}^{n} i x_i^4 + random(0, 1)$	$[-1.28, 1.28]^n$	$n = 50$ $n = 100$	$\mathbf{x}^* = (0, \ldots, 0)$; $f(\mathbf{x}^*) = 0$

E. Cuevas et al., *Recent Metaheuristics Algorithms for Parameter Identification*, Studies in Computational Intelligence 854, https://doi.org/10.1007/978-3-030-28917-1

Table B.2 Multimodal test functions

Function	S	Dim	Minimum
$f_8(\mathbf{x}) = \sum_{i=1}^{n} -x_i \sin\left(\sqrt{\lvert x_i \rvert}\right)$	$[-500, 500]^n$	$n = 50$ $n = 100$	$\mathbf{x}^* = (420, \ldots, 420)$; $f(\mathbf{x}^*) = -418.9829 \times n$
$f_9(\mathbf{x}) = \sum_{i=1}^{n} \left[x_i^2 - 10\cos(2\pi x_i) + 10\right]$	$[-5.12, 5.12]^n$	$n = 50$ $n = 100$	$\mathbf{x}^* = (0, \ldots, 0)$; $f(\mathbf{x}^*) = 0$
$f_{10}(\mathbf{x}) = -20\exp\left(-0.2\sqrt{\frac{1}{n}\sum_{i=1}^{n} x_i^2}\right) - \exp\left(\frac{1}{n}\sum_{i=1}^{n}\cos(2\pi x_i)\right) + 20 + \exp$	$[-32, 32]^n$	$n = 50$ $n = 100$	$\mathbf{x}^* = (0, \ldots, 0)$; $f(\mathbf{x}^*) = 0$
$f_{11}(\mathbf{x}) = \frac{1}{4000}\sum_{i=1}^{n} x_i^2 - \prod_{i=1}^{n}\cos\left(\frac{x_i}{\sqrt{i}}\right) + 1$	$[-600, 600]^n$	$n = 50$ $n = 100$	$\mathbf{x}^* = (0, \ldots, 0)$; $f(\mathbf{x}^*) = 0$
$f_{12}(\mathbf{x}) = \frac{\pi}{n}\left\{ \begin{array}{l} 10\sin(\pi y_1) + \\ \sum_{i=1}^{n-1}(y_i - 1)^2\left[1 + 10\sin^2(\pi y_{i+1})\right] + (y_n - 1)^2 \end{array} \right\}$ $+ \sum_{i=1}^{n} u(x_i, 10, 100, 4)$ $y_i = 1 + \frac{(x_i+1)}{4}$ $u(x_i, a, k, m) = \begin{cases} k(x_i - a)^m & x_i > a \\ 0 & -a \le x_i \le a \\ k(-x_i - a)^m & x_i < a \end{cases}$	$[-50, 50]^n$	$n = 50$ $n = 100$	$\mathbf{x}^* = (0, \ldots, 0)$; $f(\mathbf{x}^*) = 0$

(continued)

Table B.2 (continued)

Function	S	Dim	Minimum
$f_{13}(\mathbf{x}) = 0.1\left\{\begin{array}{l}\sin^2(3\pi x_1)\\ +\sum_{i=1}^{n}(x_i-1)^2\left[1+\sin^2(3\pi x_i+1)\right]\\ +(x_n-1)^2\left[1+\sin^2(2\pi x_n)\right]\end{array}\right\} + \sum_{i=1}^{n} u(x_i, 5, 100, 4);$ $u(x_i, a, k, m) = \begin{cases} k(x_i-a)^m & x_i > a \\ 0 & -a < x_i < a \\ k(-x_i-a)^m & x_i < -a \end{cases}$	$[-10, 10]^n$	$n = 50$ $n = 100$	$\mathbf{x}^* = (1, \ldots, 1);$ $f(\mathbf{x}^*) = 0$
$f_{14}(\mathbf{x}) = \frac{1}{2}\sum_{i=1}^{n}\left(x_i^4 - 16x_i^2 + 5x_i\right)$	$[-5, 5]^n$	$n = 50$ $n = 100$	$\mathbf{x}^* = (-2.90, \ldots, -2.90);$ $f(\mathbf{x}^*) = -39.16599 \times n$

Table B.3 Hybrid test functions

Function	S	Dim	Minimum
$f_{15}(\mathbf{x}) = f_1(\mathbf{x}) + f_2(\mathbf{x}) + f_9(\mathbf{x})$	$[-100, 100]^n$	$n = 50$ $n = 100$	$\mathbf{x}^* = (0, \ldots, 0)$; $f(\mathbf{x}^*) = 0$
$f_{16}(\mathbf{x}) = f_9(\mathbf{x}) + f_5(\mathbf{x}) + f_{11}(\mathbf{x})$	$[-100, 100]^n$	$n = 50$ $n = 100$	$\mathbf{x}^* = (0, \ldots, 0)$; $f(\mathbf{x}^*) = n - 1$
$f_{17}(\mathbf{x}) = f_3(\mathbf{x}) + f_5(\mathbf{x}) + f_{10}(\mathbf{x}) + f_{13}(\mathbf{x})$	$[-100, 100]^n$	$n = 50$ $n = 100$	$\mathbf{x}^* = (0, \ldots, 0)$; $f(\mathbf{x}^*) = (1.1 \times n) - 1$
$f_{18}(\mathbf{x}) = f_2(\mathbf{x}) + f_5(\mathbf{x}) + f_9(\mathbf{x}) + f_{10}(\mathbf{x}) + f_{11}(\mathbf{x})$	$[-100, 100]^n$	$n = 50$ $n = 100$	$\mathbf{x}^* = (0, \ldots, 0)$; $f(\mathbf{x}^*) = n - 1$
$f_{19}(\mathbf{x}) = f_1(\mathbf{x}) + f_2(\mathbf{x}) + f_8(\mathbf{x}) + f_{10}(\mathbf{x}) + f_{12}(\mathbf{x})$	$[-100, 100]^n$	$n = 50$ $n = 100$	$\mathbf{x}^* = (1, \ldots, 1)$; $f(\mathbf{x}^*) = 0$

Table B.4 Unimodal benchmark functions

Function	S	Dimension	Minimum
$f_1(\mathbf{x}) = \sum_{i=1}^{n} x_i^2$	$[-100, 100]^n$	$n = 30$	$\mathbf{x}^* = (0, \ldots, 0)$; $f(\mathbf{x}^*) = 0$
$f_2(\mathbf{x}) = \sum_{i=1}^{n} \lvert x_i \rvert + \prod_{i=1}^{n} \lvert x_i \rvert$	$[-10, 10]^n$	$n = 30$	$\mathbf{x}^* = (0, \ldots, 0)$; $f(\mathbf{x}^*) = 0$
$f_3(\mathbf{x}) = \sum_{i=1}^{n} \left(\sum_{j=1}^{i} x_i^2 \right)^2$	$[-100, 100]^n$	$n = 30$	$\mathbf{x}^* = (0, \ldots, 0)$; $f(\mathbf{x}^*) = 0$
$f_4(\mathbf{x}) = \max\{\lvert x_i \rvert, 1 \leq i \leq n\}$	$[-100, 100]^n$	$n = 30$	$\mathbf{x}^* = (0, \ldots, 0)$; $f(\mathbf{x}^*) = 0$
$f_5(\mathbf{x}) = \sum_{i=1}^{n} \left[100\left(x_{i+1} - x_i^2\right)^2 + (x_i - 1)^2 \right]$	$[-30, 30]^n$	$n = 30$	$\mathbf{x}^* = (1, \ldots, 1)$; $f(\mathbf{x}^*) = n$
$f_6(\mathbf{x}) = \sum_{i=1}^{n} ([x_i + 0.5])^2$	$[-100, 100]^n$	$n = 30$	$\mathbf{x}^* = (0, \ldots, 0)$; $f(\mathbf{x}^*) = 0$
$f_7(\mathbf{x}) = \sum_{i=1}^{n} i x_i^4 + random(0, 1)$	$[-1.28, 1.28]^n$	$n = 30$	$\mathbf{x}^* = (0, \ldots, 0)$; $f(\mathbf{x}^*) = 0$
$f_8(\mathbf{x}) = \sum_{i=1}^{n} x_i + \sum_{i=1}^{n} \frac{x_i}{(-x_i + \sum_{i1=1}^{n} x_{i1})^2}$	$\left[10e^{-6}, 2\right]^n$	$n = 30$	$\mathbf{x}^* = \left(\frac{1}{n-1}, \ldots, \frac{1}{n-1} \right)$; $f(\mathbf{x}^*) = \frac{2n}{n-1}$

Table B.5 Multimodal benchmark functions

Function	S	Dimension	Minimum
$f_9(\mathbf{x}) = \sum_{i=1}^{n} x_i^2 + \sum_{i=1}^{n} \frac{x_i^2}{(-x_i + \sum_{i1=1}^{n} x_{i1})^4}$	$[10e^{-6}, 2]^{\mathrm{n}}$	$n = 30$	$\mathbf{x}^* = \left(\frac{1}{n-1}, \ldots, \frac{1}{n-1}\right)$; $f(\mathbf{x}^*) = \frac{2n}{(n-1)^2}$
$f_{10}(\mathbf{x}) = \sum_{i=1}^{n} \left[x_i^2 - 10\cos(2\pi x_i) + 10\right]$	$[-5.12, 5.12]^{\mathrm{n}}$	$n = 30$	$\mathbf{x}^* = (0, \ldots, 0)$; $f(\mathbf{x}^*) = 0$
$f_{11}(\mathbf{x}) = -20\exp\left(-0.2\sqrt{\frac{1}{n}\sum_{i=1}^{n} x_i^2}\right) - \exp\left(\frac{1}{n}\sum_{i=1}^{n}\cos(2\pi x_i)\right) + 20 + e$	$[-32, 32]^{\mathrm{n}}$	$n = 30$	$\mathbf{x}^* = (0, \ldots, 0)$; $f(\mathbf{x}^*) = 0$
$f_{12}(\mathbf{x}) = -0.1\sum_{i=1}^{n}\cos(5\pi x_i) - \sum_{i=1}^{n} x_i^2$	$[-1, 1]^{\mathrm{n}}$	$n = 30$	$\mathbf{x}^* = (0, \ldots, 0)$; $f(\mathbf{x}^*) = -0.1 \times n$
$f_{13}(\mathbf{x}) = -0.1\sum_{i=1}^{n}\cos(5\pi x_i) - \sum_{i=1}^{n} x_i^2$	$[-1, 1]^{\mathrm{n}}$	$n = 30$	$\mathbf{x}^* = (0, \ldots, 0)$; $f(\mathbf{x}^*) = -0.1 \times n$
$f_{14}(\mathbf{x}) = \sum_{i=1}^{n}\left[x_i^2\left(2x_i^2 + x_{i+1} + 2\right) - x_i x_{i-1}(3x_i + 3x_{i-1} - x_{i+1})\right]^2 + \frac{1}{n^2}\sum_{i=1}^{n}(x_i - 1)^2 + e; x_{n+1} = x_1, x_0 = x_n$	$[-1, 2]^{\mathrm{n}}$	$n = 30$	$\mathbf{x}^* = (1, \ldots, 1)$; $f(\mathbf{x}^*) = 1$

(continued)

Table B.5 (continued)

Function	S	Dimension	Minimum
$f_{15}(\mathbf{x}) = -(n+1)e^{-10\sqrt{n}\left[\sum_{i=1}^{n}(x_i-1)^2\right]^{1/2}} + \sum_{i=1}^{n} x_i^2$	$[-2, 2]^n$	$n = 30$	$\mathbf{x}^* = (1, \ldots, 1)$; $f(\mathbf{x}^*) = -1$
$f_{16}(\mathbf{x}) = \sum_{i=1}^{n} -x_i \sin\left(\sqrt{\lvert x_i \rvert}\right)$	$[-500, 500]^n$	$n = 30$	$\mathbf{x}^* = (420, \ldots, 420)$; $f(\mathbf{x}^*) = -418.9829 \times n$
$f_{17}(\mathbf{x}) = \frac{\pi}{n}\left\{10\sin(\pi y_1) + \sum_{i=1}^{n-1}(y_i - 1)^2\left[1 + 10\sin^2(\pi y_{i+1})\right] + (y_n - 1)^2\right\} + \cdots \sum_{i=1}^{n} u(x_i, 10, 100, 4)$; $y_i = 1 + \frac{x_i + 1}{4}$; $u(x_i, a, k, m) = \begin{cases} k(x_i - a)^m & x_i > a \\ 0 & -a < x_i < a \\ k(-x_i - a)^m & x_i < -a \end{cases}$	$[-50, 50]^n$	$n = 30$	$\mathbf{x}^* = (0, \ldots, 0)$; $f(\mathbf{x}^*) = 0$
$f_{18}(\mathbf{x}) = 0.1\left\{\sin^2(3\pi x_1) + \sum_{i=1}^{n}(x_i - 1)^2\left[1 + \sin^2(3\pi x_1 + 1)\right] + (x_n - 1)^2\left[1 + \sin^2(2\pi x_n)\right]\right\} + \cdots \sum_{i=1}^{n} u(x_i, 5, 100, 4)$	$[-50, 50]^n$	$n = 30$	$\mathbf{x}^* = (1, \ldots, 1)$; $f(\mathbf{x}^*) = 0$

Table B.6 Hybrid benchmark functions

Function	S	Dimension	Minimum
$f_{19}(\mathbf{x}) = f_1(\mathbf{x}) + f_2(\mathbf{x}) + f_{10}(\mathbf{x})$	$[-100, 100]^n$	$n = 30$	$\mathbf{x}^* = (0, \ldots, 0)$; $f(\mathbf{x}^*) = 0$
$f_{20}(\mathbf{x}) = f_5(\mathbf{x}) + f_{10}(\mathbf{x}) + f_{12}(\mathbf{x})$	$[-100, 100]^n$	$n = 30$	$\mathbf{x}^* = (0, \ldots, 0)$; $f(\mathbf{x}^*) = n - 1$
$f_{21}(\mathbf{x}) = f_3(\mathbf{x}) + f_5(\mathbf{x}) + f_{11}(\mathbf{x}) + f_{18}(\mathbf{x})$	$[-100, 100]^n$	$n = 30$	$\mathbf{x}^* = (0, \ldots, 0)$; $f(\mathbf{x}^*) = (1.1 \times n) - 1$
$f_{22}(\mathbf{x}) = f_2(\mathbf{x}) + f_5(\mathbf{x}) + f_{10}(\mathbf{x}) + f_{11}(\mathbf{x}) + f_{12}(\mathbf{x})$	$[-100, 100]^n$	$n = 30$	$\mathbf{x}^* = (0, \ldots, 0)$; $f(\mathbf{x}^*) = n - 1$
$f_{23}(\mathbf{x}) = f_1(\mathbf{x}) + f_2(\mathbf{x}) + f_{11}(\mathbf{x}) + f_{16}(\mathbf{x}) + f_{17}(\mathbf{x})$	$[-100, 100]^n$	$n = 30$	$\mathbf{x}^* = (1, \ldots, 1)$; $f(\mathbf{x}^*) = -3 \times n$

Table B.7 Three-bar truss design problem description

Function	S	Dimension	Constraints
$f(\mathbf{x}) = \left(2\sqrt{2}x_1 + x_2\right) \times L$ $L = 100\,\text{cm}, \quad P = 2\,\text{kN/cm}^2, \quad \sigma = 2\,\text{kN/cm}^2$	$0 \leq x_i \leq 1, i = 1, 2$	$n = 2$	$g_1(\mathbf{x}) = \frac{\sqrt{2}x_1 + x_2}{\sqrt{2}x_1^2 + 2x_1x_2} P - \sigma \leq 0$ $g_2(\mathbf{x}) = \frac{x_2}{\sqrt{2}x_1^2 + 2x_1x_2} P - \sigma \leq 0$ $g_3(\mathbf{x}) = \frac{1}{\sqrt{2}x_2 + x_1} P - \sigma \leq 0$

Table B.8 Tension/compression spring design problem description

Function	S	Dimension	Constraints
$f(\mathbf{x}) = (x_3 + 2)x_2x_1^2$	$0.05 \leq x_1 \leq 2$ $0.25 \leq x_2 \leq 1.3$ $2 \leq x_3 \leq 15$	$n = 3$	$g_1(\mathbf{x}) = 1 - \frac{x_2^3 x_3}{71{,}785x_1^4} \leq 0$ $g_2(\mathbf{x}) = \frac{4x_2^2 - x_1x_2}{12{,}566(x_2x_1^3 - x_1^4)} + \frac{1}{5108x_1^2} - 1 \leq 0$ $g_3(\mathbf{x}) = 1 - \frac{140.45x_1}{x_2^2 x_3} \leq 0$ $g_4(\mathbf{x}) = \frac{x_1 + x_2}{1.5} - 1 \leq 0$

Table B.9 Welded beam design problem description

Function	S	Dimension	Constraints
$f(\mathbf{x}) = 1.10471x_1^2x_2 + 0.04811x_3x_4(14 + x_2)$	$0.1 \leq x_i \leq 2, i = 1, 4$ $0.1 \leq x_i \leq 10, i = 2, 3$	$n = 4$	$g_1(\mathbf{x}) = \tau(x) - \tau_{\max} \leq 0$ $g_2(\mathbf{x}) = \sigma(x) - \sigma_{\max} \leq 0$ $g_3(\mathbf{x}) = x_1 - x_4 \leq 0$ $g_4(\mathbf{x}) = 0.10471x_1^2 + 0.04811x_3x_4(14 + x_2) - 5 \leq 0$ $g_5(\mathbf{x}) = 0.125 - x_1 \leq 0$ $g_6(\mathbf{x}) = \delta(x) - \delta_{\max} \leq 0$ $g_7(\mathbf{x}) = P - P_c(x) \leq 0;$ $\tau(x) = \sqrt{(\tau')^2 + 2\tau'\tau''\frac{x_2}{2R} + (\tau'')^2},$ $\tau' = \frac{P}{\sqrt{2}x_1x_2}, \tau'' = \frac{MR}{J}, M = P\left(L + \frac{x_2}{2}\right),$ $R = \sqrt{\frac{x_2^2}{2} + \left(\frac{x_1 + x_3}{2}\right)^2}, \delta(x) = \frac{4PL^3}{Ex_3^3x_4},$ $J = 2\left[\sqrt{2}x_1x_2\left\{\frac{x_2^2}{12} + \left(\frac{x_1 + x_3}{2}\right)^2\right\}\right], \sigma(x) = \frac{6PL}{x_4x_3^2},$ $P_c(x) = \frac{4.013E\sqrt{\frac{x_3^2x_4^6}{36}}}{L^2}\left(1 - \frac{x_3}{2L}\sqrt{\frac{E}{4G}}\right)$

Table B.10 Benchmark functions

Function	S	Dimension	Minimum
$f_1(\mathbf{x}) = \sum_{i=1}^{n} x_i^2$	$[-100, 100]^n$	$n = 30$	$\mathbf{x}^* = (0, \ldots, 0)$; $f(\mathbf{x}^*) = 0$
$f_2(\mathbf{x}) = \sum_{i=1}^{n} \|x_i\| + \prod_{i=1}^{n} \|x_i\|$	$[-10, 10]^n$	$n = 30$	$\mathbf{x}^* = (0, \ldots, 0)$; $f(\mathbf{x}^*) = 0$
$f_3(\mathbf{x}) = \sum_{i=1}^{n} \left(\sum_{j=1}^{i} x_i^2 \right)^2$	$[-100, 100]^n$	$n = 30$	$\mathbf{x}^* = (0, \ldots, 0)$; $f(\mathbf{x}^*) = 0$
$f_4(\mathbf{x}) = \max\{\|x_i\|, 1 \leq i \leq n\}$	$[-100, 100]^n$	$n = 30$	$\mathbf{x}^* = (0, \ldots, 0)$; $f(\mathbf{x}^*) = 0$
$f_5(\mathbf{x}) = \sum_{i=1}^{n} \left[100(x_{i+1} - x_i^2)^2 + (x_i - 1)^2 \right]$	$[-30, 30]^n$	$n = 30$	$\mathbf{x}^* = (1, \ldots, 1)$; $f(\mathbf{x}^*) = n$
$f_6(\mathbf{x}) = \sum_{i=1}^{n} ([x_i + 0.5])^2$	$[-100, 100]^n$	$n = 30$	$\mathbf{x}^* = (0, \ldots, 0)$; $f(\mathbf{x}^*) = 0$
$f_7(\mathbf{x}) = \sum_{i=1}^{n} i x_i^4 + random(0, 1)$	$[-1.28, 1.28]^n$	$n = 30$	$\mathbf{x}^* = (0, \ldots, 0)$; $f(\mathbf{x}^*) = 0$
$f_8(\mathbf{x}) = \sum_{i=1}^{n} x_i + \sum_{i=1}^{n} \frac{x_i}{(-x_i + \sum_{i1=1}^{n} x_{i1})^2}$	$[10e^{-6}, 2]^n$	$n = 30$	$\mathbf{x}^* = \left(\frac{1}{n-1}, \ldots, \frac{1}{n-1} \right)$; $f(\mathbf{x}^*) = \frac{2n}{n-1}$

(continued)

Table B.10 (continued)

Function	S	Dimension	Minimum
$f_9(\mathbf{x}) = \sum_{i=1}^{n} x_i^2 + \sum_{i=1}^{n} \frac{x_i^2}{\left(-x_i + \sum_{i1=1}^{n} x_{i1}\right)^4}$	$\left[10e^{-6}, 2\right]^n$	$n = 30$	$\mathbf{x}^* = \left(\frac{1}{n-1}, \ldots, \frac{1}{n-1}\right)$; $f(\mathbf{x}^*) = \frac{2n}{(n-1)^2}$
$f_{10}(\mathbf{x}) = \sum_{i=1}^{n} \left[x_i^2 - 10\cos(2\pi x_i) + 10\right]$	$[-5.12, 5.12]^n$	$n = 30$	$\mathbf{x}^* = (0, \ldots, 0)$; $f(\mathbf{x}^*) = 0$
$f_{11}(\mathbf{x}) = -20\exp\left(-0.2\sqrt{\frac{1}{n}\sum_{i=1}^{n} x_i^2}\right) - \exp\left(\frac{1}{n}\sum_{i=1}^{n}\cos(2\pi x_i)\right) + 20 + e$	$[-32, 32]^n$	$n = 30$	$\mathbf{x}^* = (0, \ldots, 0)$; $f(\mathbf{x}^*) = 0$
$f_{12}(\mathbf{x}) = -0.1\sum_{i=1}^{n}\cos(5\pi x_i) - \sum_{i=1}^{n} x_i^2$	$[-1, 1]^n$	$n = 30$	$\mathbf{x}^* = (0, \ldots, 0)$; $f(\mathbf{x}^*) = -0.1 \times n$
$f_{13}(\mathbf{x}) = -0.1\sum_{i=1}^{n}\cos(5\pi x_i) - \sum_{i=1}^{n} x_i^2$	$[-1, 1]^n$	$n = 30$	$\mathbf{x}^* = (0, \ldots, 0)$; $f(\mathbf{x}^*) = -0.1 \times n$
$f_{14}(\mathbf{x}) = \sum_{i=1}^{n}\left[\begin{array}{l} x_i^2\left(2x_i^2 + x_{i+1} + 2\right) - x_i x_{i-1} \\ (3x_i + 3x_{i-1} - x_{i+1}) \end{array}\right]^2 + \frac{1}{n^2}\sum_{i=1}^{n}(x_i - 1)^2 + e;\ x_{n+1} = x_1,\ x_0 = x_n$	$[-1, 2]^n$	$n = 30$	$\mathbf{x}^* = (1, \ldots, 1)$; $f(\mathbf{x}^*) = 1$

(continued)

Table B.10 (continued)

Function	S	Dimension	Minimum
$f_{15}(\mathbf{x}) = -(n+1)e^{-10\sqrt{n}\left[\sum_{i=1}^{n}(x_i-1)^2\right]^{1/2}} + \sum_{i=1}^{n} x_i^2$	$[-2, 2]^n$	$n = 30$	$\mathbf{x}^* = (1, \ldots, 1)$; $f(\mathbf{x}^*) = -1$
$f_{16}(\mathbf{x}) = \sum_{i=1}^{n} -x_i \sin\left(\sqrt{\lvert x_i \rvert}\right)$	$[-500, 500]^n$	$n = 30$	$\mathbf{x}^* = (420, \ldots, 420)$; $f(\mathbf{x}^*) = -418.9829 \times n$
$f_{17}(\mathbf{x}) = \frac{\pi}{n}\left\{10\sin(\pi y_1) + \sum_{i=1}^{n-1} \frac{(y_i-1)^2}{\left[1+10\sin^2(\pi y_{i+1})\right] + (y_n-1)^2}\right\}$ $+ \cdots \sum_{i=1}^{n} u(x_i, 10, 100, 4)$ $y_i = 1 + \frac{x_i+1}{4}$ $u(x_i, a, k, m) = \begin{cases} k(x_i-a)^m & x_i > a \\ 0 & -a < x_i < a \\ k(-x_i-a)^m & x_i < -a \end{cases}$	$[-50, 50]^n$	$n = 30$	$\mathbf{x}^* = (0, \ldots, 0)$; $f(\mathbf{x}^*) = 0$
$f_{18}(\mathbf{x}) = 0.1\left\{\begin{array}{l}\sin^2(3\pi x_1) + \\ \sum_{i=1}^{n} \frac{(x_i-1)^2\left[1+\sin^2(3\pi x_1+1)\right]+}{(x_n-1)^2\left[1+\sin^2(2\pi x_n)\right]}\end{array}\right\}$ $\sum_{i=1}^{n} u(x_i, 5, 100, 4)$	$[-50, 50]^n$	$n = 30$	$\mathbf{x}^* = (1, \ldots, 1)$; $f(\mathbf{x}^*) = 0$

(continued)

Table B.10 (continued)

Function	S	Dimension	Minimum
$f_{19}(\mathbf{x}) = f_1(\mathbf{x}) + f_2(\mathbf{x}) + f_{10}(\mathbf{x})$	$[-100, 100]^n$	$n = 30$	$\mathbf{x}^* = (0, \ldots, 0)$; $f(\mathbf{x}^*) = 0$
$f_{20}(\mathbf{x}) = f_5(\mathbf{x}) + f_{10}(\mathbf{x}) + f_{12}(\mathbf{x})$	$[-100, 100]^n$	$n = 30$	$\mathbf{x}^* = (0, \ldots, 0)$; $f(\mathbf{x}^*) = n - 1$
$f_{21}(\mathbf{x}) = f_3(\mathbf{x}) + f_5(\mathbf{x}) + f_{11}(\mathbf{x}) + f_{18}(\mathbf{x})$	$[-100, 100]^n$	$n = 30$	$\mathbf{x}^* = (0, \ldots, 0)$; $f(\mathbf{x}^*) = (1.1 \times n) - 1$
$f_{22}(\mathbf{x}) = f_2(\mathbf{x}) + f_5(\mathbf{x}) + f_{10}(\mathbf{x}) + f_{11}(\mathbf{x}) + f_{12}(\mathbf{x})$	$[-100, 100]^n$	$n = 30$	$\mathbf{x}^* = (0, \ldots, 0)$; $f(\mathbf{x}^*) = n - 1$
$f_{23}(\mathbf{x}) = f_1(\mathbf{x}) + f_2(\mathbf{x}) + f_{11}(\mathbf{x}) + f_{16}(\mathbf{x}) + f_{17}(\mathbf{x})$	$[-100, 100]^n$	$n = 30$	$\mathbf{x}^* = (1, \ldots, 1)$; $f(\mathbf{x}^*) = -3 \times n$

Printed in the United States
By Bookmasters

Printed in the United States
By Bookmasters